A Guide to
Golf Course Irrigation
System Design and Drainage

By

E.S. Pira
Professor Emeritus
Food Engineering Department
University of Massachusetts
Amherst, Massachusetts 01003

Ann Arbor Press, Inc.
Chelsea, Michigan

Library of Congress Cataloging-in-Publication Data

Catalog record is available from the Library of Congress

ISBN 1-57504-030-1

ANN ARBOR PRESS, INC.
121 South Main Street, Chelsea, Michigan 48118

Printed in the United States of America
1 2 3 4 5 6 7 8 9 0

A Guide to
Golf Course Irrigation
System Design and Drainage

About the Author

Edward S. Pira, Professor Emeritus, University of Massachusetts, received his BSAE degree from the University of Connecticut, in 1949, and his MS degree in 1959. He joined the Agricultural Engineering Department, Stockbridge School, at the University of Massachusetts in January, 1953, and was assigned to teach undergraduate courses in turfgrass irrigation and drainage. Mr. Pira's winter school, a ten-week short course, became very popular, and attracted students from the United States, Canada, Europe, Australia, South Africa, and South America. Graduates from his turfgrass programs serve as golf course superintendents on some of the most prestigious golf courses worldwide.

To assist in teaching the turfgrass irrigation and drainage program, many guest lecturers from the leading irrigation companies were utilized. Companies represented include Rain Bird, Toro, Buckner, and Hunter. Golf course architects, system designers, installers, consultants and soil conservation personnel also lectured on a regular basis presenting "state-of-the art" equipment installation methods and operating procedures.

Mr. Pira spent his sabbatical leave (1986) visiting numerous golf courses, irrigation companies and consultants to learn about new systems and operating procedures used for golf course irrigation. His leave was highlighted by a trip to St. Andrews, Scotland. He was a successful researcher, wrote many articles, and presented numerous papers pertaining to subsurface and drip irrigation. He holds two patents related to drip irrigation component design and installation.

During his tenure he received the Stockbridge Outstanding Teacher Award numerous times, and special recognition from the Irrigation Association, and the Golf Course Superintendents Association of America. He retired September 1, 1992.

Acknowledgments

In appreciation for sharing their experience and "know-how" related to current irrigation system components, controllers, installation methods, scheduling and operating procedures—and also for providing literature and photographs—I wish to recognize the following contributors:

Mark Loper and Staff—Turf Products Corp., Enfield, Connecticut

Paul Roche—S.V. Moffett Co., Inc., "Turf and Irrigation Equipment," Henrietta, New York

William Zuraw—Crumpin Fox Golf Course Superintendent, Bernardston, Massachusetts

The many superintendents at various golf courses that I visited, and the irrigation equipment companies that I correspond with.

To Eva (my girl Friday), who ran many errands and assisted in typing and editing the contents.

A special acknowledgment to:

Brian E. Vinchesi, Design Engineer, Irrigation Consulting & Engineering, Inc., Pepperell, Massachusetts in recognition for the suggestions, design information, and sketch and figure alterations which were considered for making revisions and updating *A Guide to Golf Course Irrigation System Design and Drainage.*

Introduction

If a customer were buying a suit of clothes it is unlikely that he would randomly select one off a rack or one that was tailor-made to fit someone else. Chances of a suit fitting correctly that was selected in this manner would be slim indeed.

Certainly, a much wiser way of buying the suit would be to have a competent tailor make it specifically for that individual. Before the tailor makes the suit he must gather certain information, such as:

- the customer's physical measurements
- the season it is to be worn
- the wear and tear to be imposed
- the price of the material and labor

From this data the tailor can proceed to make the suit.

In principle, the purchase of an irrigation system for a golf course is no different than buying a suit. Each golf course is unique and must be considered individually. The best procedure is to obtain the services of a competent engineer or irrigation company to tailor fit the system to meet the specific watering needs of your golf course.

It is the author's hope that this book will help golf course superintendents and students to attain the requisite knowledge to oversee the installation and maintenance of irrigation systems that fit the needs of their particular course.

Purpose

The design and installation of a fully automatic irrigation system for an 18-hole golf course can be priced up to $1,000,000 and higher. An investment of this magnitude dictates that the job be handled by qualified engineers and irrigation companies.

Generally, in practice, the decisions related to the selection and purchase of this costly and sophisticated watering system rest in the hands of the golf course committee. In many cases the committee may consist of members that possess little, if any, background in the principles of hydraulics, the watering needs of turfgrass, or irrigation system design. Too often a system is purchased on the basis of "lowest bid" or which company had the best salesman. In either case, the system may prove less than adequate, or may lack the necessary versatility required to properly irrigate as large and diverse an area as a golf course.

Since the golf course superintendent is the expert in growing turfgrass and will have the responsibility for operating as well as maintenance of the watering system, he should be involved at the outset, that is, from the initial planning stages to the follow-through on every phase of installation. The superintendent should insist on having a system with the appropriate capacity and versatility to meet the specific watering demands of his golf course. It is essential to know and understand the function of each component, the location of every pipe, valve, and fitting as installed, and also to know exactly what the system is designed to do. The irrigation schedule must be programmed in accordance with the specifications built into the system.

The primary purpose of this book is to serve as a course outline or lesson plan that will assist students preparing for careers as golf course superintendents and to give them an understanding of the principles of irrigation system design and operation.

Objectives

- To acquaint students with irrigation terms and their relationships.
- To acquaint students with the components of a system and its functions.
- To acquaint the students with types of systems, materials, and equipment, and factors to consider in their selection.
- To assist students in understanding the principles of irrigation system design and operation.
- To determine the irrigation schedule or program.
- To enable students to speak knowledgeably to irrigation system designers, hydraulic engineers, salesmen, and golf course committees.

Much of the information that is presented in this book has been selected from various sources that include commercial literature and irrigation design manuals on hand at the time of publication.

In an effort to avoid cluttering pages with numerous footnotes and references, and to promote ease of reading, a list of acknowledgments is provided for each division. However, oftentimes the tables, illustrations, and data cited were similar and appeared in more than one source; this caused difficulty in assigning credit.

Important Notice

The tables, charts, illustrations, and examples of commercial equipment were selected as "typical" or used for specific clarification purposes. In no way should their use be interpreted as an endorsement or recommendation of any product, manufacturer, or type of equipment.

Contents

Chapter 1

Preliminary Design Data and General Information

INTRODUCTION

The relationships that exist between turfgrass, soil, and good watering practices have been well established. In recognition of this relationship golf clubs, as in the past, continue to invest many thousands of dollars for moderately priced systems to a million or more for the highly sophisticated computerized control system, which may include weather stations and other accessories.

It is little wonder, then, that the golf course committee and/or members expect and even demand top-notch turfgrass and playing conditions. Oftentimes, the superintendent's job may hinge on his or her ability to grow high-quality turfgrass.

For these reasons it behooves the superintendent to become an expert in plant, soil, and water relationships. Furthermore, it is his responsibility to become *totally* involved from the inception, in the design, the type and degree of sophistication, and the installation, scheduling, operating and maintenance of the irrigation system.

PRE-DESIGN DATA AND GENERAL INFORMATION

This chapter contains a discussion and explanation of the following topics:

- Contour map and related information.
- As-built map for locating objects or components.

- Pressure types, including measurements, conversions and related problems.
- Pipeline pressure (friction) losses, including friction loss tables and related problems.
- Land measurement methods, formulae and problems.

Important: It is *essential* for the beginner to have a good understanding of the above, since most of them are involved in practically every aspect of irrigation system design as presented throughout this book.

MAP OF DESIGN AREA— DRAWN TO SCALE

An accurate and comprehensive map of your golf course *must* be available. The map should show the following:

- Layout of the fairways, greens, tees, and bunkers.
- Elevation differences (preferably a contour map as shown in Figure 1.1).
- Obstructions: gullies, trees, buildings, ledges, rock, etc.
- Prevailing wind direction and average speed.
- North direction.
- Actual land surface measurements.
- Water supply location—availability.
- Soil types and characteristics (review pages 225–227).

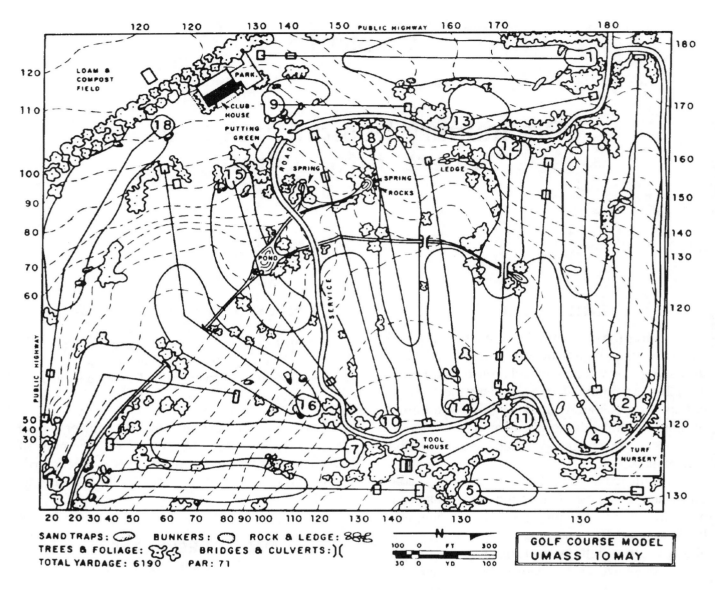

Figure 1.1

In event that a map of the golf course is not available, it is advisable to survey extremely large and irregular areas by means of an aerial photograph.

The scale usually desired is $1'' = 100'$ for an 18-hole course and $1'' = 50'$ for a 9-hole course.

The accuracy of an aerial photograph can be verified by a comparison of the *actual* measurements against the *scaled* distances between the same two points.

After such a survey has been obtained, any missing details can be filled in. Elevation differences can be obtained by using a surveyor's level. However,

for contour mapping of large complicated areas such as golf courses, the services of a registered surveyor should be obtained.

MAP OF DESIGN AREA—AS-BUILT

An accurate, as-built map of the irrigation system should be made by the installer and/or superintendent. This map should show the precise location of each component, and may have changed from the original plan due to obstacles or other conditions found in the field.

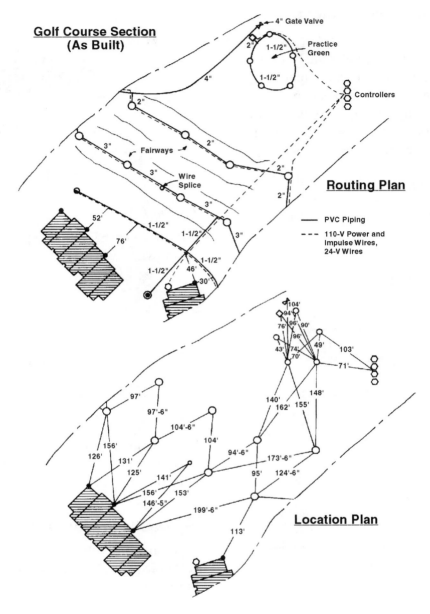

Figure 1.2

Triangulation Measurement Method

Triangulation is a simple, yet accurate, method for locating *buried* valves, wire splices, etc. Furthermore, triangulation can be used for placing objects on scaled plot plans.

This method consists of making two direct measurements from fixed reference points to the object. The best points are corners of buildings and/or other permanently established markers.

Figure 1.2 shows various objects for a small section of a golf course irrigation system and the triangulation method for locating the objects.

PRINCIPLES OF PRESSURE

It is essential to understand the basic principles of pressure before attempting to select sprinklers, pipes, pump, power units, and irrigation system design.

Pressure Types and Measurement

Physical Pressure

The expression "physical pressure" shall be defined as the weight in pounds or other equivalent weight units of a body, matter or other things in nature applied to an area, usually square inches or other equivalent area units. Expressed as pounds per square inch (psi) or other equivalent units.

Physical pressure:

$$\frac{\text{wt. (lb)}}{\text{area (in.}^2)} = \text{psi}$$

Example 1

Assume a golfer's weight = 200 lb

Figure 1.3

shoe area   heel = 8 in.^{2}
                   ball = 12 in.^{2}

Calculate the following:

Figure 1.4

$$\frac{200}{8} = 25 \text{ psi} \qquad \frac{200}{20} = 10 \text{ psi} \qquad \frac{200}{40} = 5 \text{ psi}$$

These illustrations show that golfers exert considerable pressure that can cause severe soil compaction problems, particularly on greens, where the greatest traffic is concentrated. The pressure would be extremely high if only the spike area of the golfers' shoes was calculated.

Example 2

Assume truck weight = 3000 lb

ground contact 30 in.^{2}/tire

Figure 1.5

$$\frac{3000}{(4 \times 30)} = 25 \text{ psi}$$

Example 3

1 ft^3 of water (62.4 lb) is supported on 4 1-in.^{2} blocks as shown.

Figure 1.6

$$\frac{62.4}{4} = 15.6 \text{ psi / block}$$

Note: Compaction caused by trucks, mowers, and other equipment can be reduced by using oversized tires or rollers.

Hydraulic Pressure

Hydraulic pressure as it relates to an irrigation system consists of "static pressure" and "dynamic pressure."

Static Water Pressure

Static water pressure is an indication of "potential" pressure available. It is easily defined as pressure with the water stopped, all outlets shut off. The principles and problems related to static pressure are presented in the next section.

Dynamic Water Pressure

Dynamic pressure is also known as "working pressure" and is defined as the pressure with water flowing. Basically, the difference between the two pressures is the consideration of friction (pressure) losses that occur when water flows in a pipeline. The principles and problems related to dynamic pressure are presented following the discussion and analysis of the Friction Loss Tables. See Appendix 4.

Static Water Pressure

Static water pressure is completely independent of the base area. It is strictly a function of the difference of elevation or height of the water column, commonly expressed as "feet of water" or "feet of head."

Important: The pressure measurement devices (gauges) must be in contact or inserted into the liquid. Static pressure is a measurement of the elevation difference of the liquid column only.

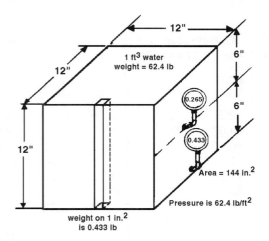

Figure 1.7. 1 ft³ water weighs 62.4 lb. Base area = 144 in.² Therefore, a pressure of 0.433 lb will be exerted on each square inch. 62.4 lb/144 in.² = 0.433 psi.

Note: Figure 1.7 shows that the pressure gauge inserted into the water 6 in. below the surface reads 0.2165 or 0.22 psi, whereas, at 12 in. down the gauge reading is 0.43 psi.

The static pressure exerted by the column of water is also independent of the size, shape, and length of the vessel or pipeline. Repeating, it is a function of the water column height. (See Figure 1.8.)

Relationship of Pressure (psi) and Head (ft)

A column of water 1 ft high (head = 1 ft) will exert a pressure of 0.433 psi.

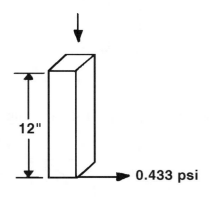

Figure 1.9

Applying a pressure of 0.433 psi will lift a column of water 1 ft high.

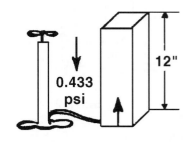

Figure 1.10

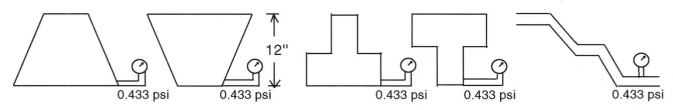

Figure 1.8

It takes a column of water 2.31 ft or head of 2.31 ft high to exert a pressure of 1 psi.

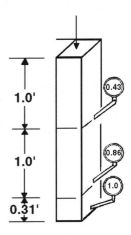

Figure 1.11

A pressure of 1 psi will lift a column of water 2.31 feet high.

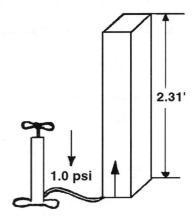

Figure 1.12

From our discussion of feet, psi and pressure there are two important conversion factors that have been determined.

0.433 psi equals 1 foot

and

2.31 ft equals 1 psi

These are important conversion factors to go back and forth between feet and psi measurements and to take into consideration elevation effects. Tables 1.1 and 1.2 outline the conversions.

Table 1.1. Conversion of Feet Head of Water into Pressure (psi)

Feet Head	lb/in.^{2}	Feet Head	lb/in.^{2}	Feet Head	lb/in.^{2}
1	0.43	60	25.99	200	86.62
2	0.87	70	30.32	225	97.45
3	1.30	80	34.65	250	108.27
4	1.73	90	38.98	275	119.1
5	2.17	100	43.31	300	129.93
6	2.60	110	47.64	325	140.75
7	3.03	120	51.97	350	151.58
8	3.40	130	56.3	400	173.24
9	3.90	140	60.63	500	216.55
10	4.33	150	64.96	600	259.85
20	8.66	160	69.29	700	303.16
30	12.99	170	73.63	800	346.47
40	17.32	180	77.96	900	389.78
50	21.65	190	83.29	1000	433.09

Table 1.2. Conversion of Pressure per Square Inch to Feet Head of Water

lb/in.^{2}	Feet Head	lb/in.^{2}	Feet Head	lb/in.^{2}	Feet Head
1	2.31	40	92.36	170	392.52
2	4.62	50	115.45	180	415.61
3	6.93	60	138.54	190	438.90
4	9.24	70	161.63	200	461.78
5	11.54	80	184.72	225	519.51
6	13.85	90	207.81	250	577.24
7	16.16	100	230.90	275	643.03
8	18.47	110	253.98	300	692.69
9	18.47	120	277.07	325	750.41
10	20.78	125	288.62	350	808.13
15	34.63	130	300.16	375	865.89
20	46.18	140	323.25	400	922.58
25	57.72	150	346.34	500	1154.48
30	69.27	160	369.43	1000	2308.00

Notes: 1. Multiply in pounds per square inch by 2.31 or divide by 0.433 to obtain equivalent height in feet. 2. Multiply height in feet by 0.433 or divide by 2.31 to obtain pressure in pounds per square inch.

STUDENT PROBLEMS

Solve the following problems using the Conversion Tables and/or Conversion Factors in Appendix 4.

Problem 1a. Static Pressure (Sample Problem)

This sample problem shows the pressure gauge readings that would be found in the various points along the pipeline.

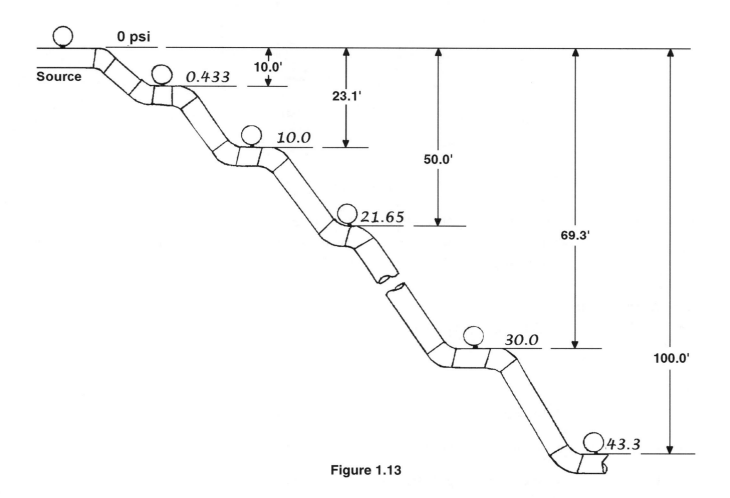

Figure 1.13

Problem 1b. Static Pressure

Fill in the blanks showing the height water will be lifted by the pressure given.

1 psi = ___ ft   50 psi = ___ ft

10 psi = ___ ft   100 psi = ___ ft

20 psi = ___ ft

Problem 2. Static Pressure

Fill in the pressure coming into each green shown. (Do not consider friction or other losses in pipes.)

Note: Only the *vertical elevation difference* between the source and outlet is considered.

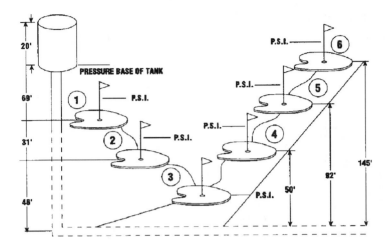

Figure 1.14

Problem 3. Static Pressure (see golf course map, Figure 1.15.)

The elevation of any spot on the golf course can be found by tracing the contour line out to the scale. Contour lines are continuous points of equal elevation shown on the golf course map.

Example Problem

From the golf course map determine the pressure needed to lift water from pond level (elev. 40 ft) to Green No. 1 (elev. 175 ft).

Procedure:

175 ft – 40 ft = 135 ft lift = 58.5 psi.

Important: The base pressure in the irrigation system is determined as follows:

Downhill Flow – Base Pressure = Operating Pressure + Static Pressure – Friction Losses

Uphill Flow – Base Pressure = Operating Pressure – Static Pressure – Friction Losses

Note: Base pressure means the pressure at the base of a sprinkler, outlet and/or a given point in the irrigation system.

Problem 4. Base Pressure (omit friction losses)

Assume a pump was located 5 ft above the water level of the pond (elev. = 40 ft) and was operating at 80 psi. Determine the *base pressure* at: Green 1. _23.7_ psi, Green 15. _?_ psi, and Green 6. _?_ psi.

Procedure:

Base Pressure = Operating Pressure (80 psi) − Static Lift (psi) = 80 psi − (175 ft − 45 ft) × 0.433 = 23.7 psi

Assume that the direction of flow in Fairway 8 is from green to tee. If the pressure at Green 8 was 80 psi, determine the *base pressure* at the stream _101.7_ psi, and at the tee _?_ psi.

Procedure:

Base Pressure = 80 psi + (120 ft − 70 ft) × 0.433 = 101.7 psi

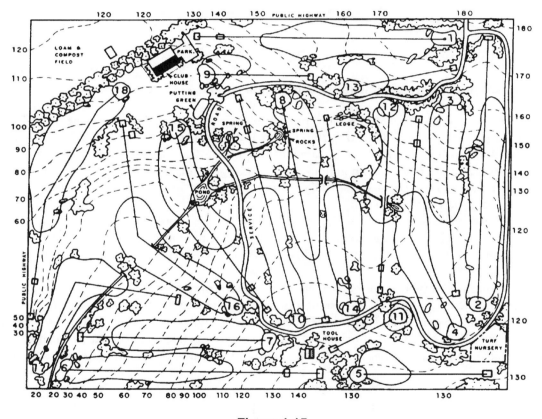

Figure 1.15

Important: Regardless of whether water is flowing or stopped the *static pressure differential* that exists remains constant.

Furthermore, as indicated above, a pump at an elevation of 45 ft would require a pressure of 56.3 psi just to lift the water to Green 1. This pressure must be added to the designed operating pressure.

The inference is that the *static pressure,* especially on hilly golf courses, plays a significant role in designing and operating the irrigation system.

Problem 5. Static Pressure

Assume that the same pipeline is used to supply Green 1 and Green 2. If the pressure at Green 1 is 50 psi (flow is from 1 to 2), the pressure at Green 2 = 90 psi (round off). If the pressure at Green 2 is 50 psi (flow is from 2 to 1), Green 1 = 10 psi.

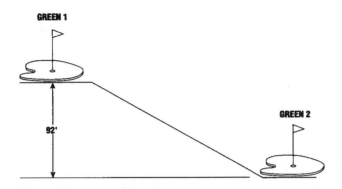

Figure 1.16

The above illustration involves some rather significant irrigation system design implications:

- *Pipe Selection*—If a pipe (Class 50) pressure-rated at 50 psi was installed based on the operating pressure given, it could possibly burst as the flow from Green 1 at 50 psi approached Green 2 at 90 psi.
- *Pressure Regulation*—It is essential to maintain uniform pressure throughout the entire irrigation system to obtain the proper area of coverage and gallons per minute (gpm) discharge from the sprinklers. To obtain uniform pressure it will be necessary to install *pressure regulators.* Built-in pressure regulation is available on some valve-in-head models.
- *Direction of Water Flow*—Where possible, the irrigation system should be designed to make water flow *downhill.* The pressure increase can be controlled by various means. On the other hand, water flowing uphill will lose pressure. To overcome this pressure loss it will be necessary to install a booster pump. See page 371.
- *Sprinkler Pressure*—The typical sprinkler table for a specific sprinkler model lists the operating pressure range. Exceeding the upper limit of 100 psi, in this case, will tend to vaporize the discharge water from the sprinkler. Below the lower limit of 60 psi the discharge water will not break up into proper size droplets. Either case should be avoided. Good irrigation system design practice dictates that the sprinkler model be selected from its mid-pressure range. This will allow for minor variations of pressure that may occur in the irrigation system.

Base pressure directly influences the distance that a sprinkler can throw water and the amount that it will use. The lower the pressure, the less the distance (area) the sprinkler can cover. The base pressure at the sprinkler that is needed or available is a prerequisite for selecting a sprinkler model for the irrigation system.

91DR Performance

psi	Radius (ft)	gpm	Nozzle
60	66	28.0	18x15
	72	36.5	22x15
	80	46.0	26x25
	86	55.4	30x15
70	68	30.6	18x15
	76	39.5	22x15
	84	49.8	26x15
	90	59.9	30x15
80	70	33.0	18x15
	79	42.3	22x15
	88	53.3	26x15
	93	64.1	30x15
90	72	35.2	18x15
	81	44.9	22x15
	90	56.5	26x15
	96	68.0	30x15
100	73	37.3	18x15
	82	47.5	22x15
	91	59.6	26x15
	98	71.8	30x15

HYDRAULIC TABLES

Introduction

Friction loss charts and tables have been in existence for a number of decades. Over the years they have been used extensively by competent hydraulic engineers, irrigation system designers, irrigation equipment dealers, and others as a tool for selecting and installing ultra millions of feet of pipe. This should serve as ample evidence that the values given in the tables are reliable. Therefore, for pipe sizing and calculating pipeline friction loss it is important

to adhere to the table values. *Note:* The friction loss tables are based on psi per 100 ft. The values are additive in proportion to the length. The velocity is a function of flow (gpm) per pipe size. Length has no bearing on velocity; therefore, the values are not additive. To deliver 100 gpm, regardless of length, in a 3 in. pipe vs. a 4 in. pipe the water must flow at a higher velocity (fps).

Friction Loss Values and Conversions

It is important that each table be carefully examined to determine exactly what information it is conveying. Some tables express the values in feet of loss per 100 or 1000 ft of length, and others as psi per 100 or 1000 ft of length.

As explained earlier (page 6), psi × 2.31 = ft, and ft × 0.433 = psi.

Example Problem

Assume the pressure at the beginning of the pipeline was 100 psi and the friction loss was 10 psi per 1000 ft. Express this in terms of feet.

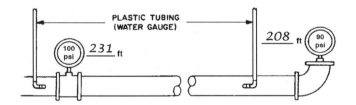

Figure 1.17

Friction Loss Tables and Conversions

Included in this section are partial friction loss tables for Class 160 PVC plastic pipe and for Schedule 40 standard steel pipe. For a detailed discussion of pipe types and friction losses refer to Chapter 5 "Pipe Selection." For complete friction loss tables for various types, classes and schedules PVC plastic, polyethylene, steel and unlined iron pipe refer to Appendix 4.

STUDENT PROBLEMS

Problem 1.

Fill in the missing blank spaces for each type of pipe listed. Use Friction Loss Tables in Appendix 4.

Type of Pipe	Size (in.)	gpm	Distance (ft)	Loss psi/100 ft	Total Loss (psi)	Total Loss (ft)	Velocity (ft/sec)
	1.25	25	100				
PVC	3	100	1000	0.79	7.9	18.2	4.0
Class	2	50	350				
160	6	400	50	1.0	0.5	1.3	6.4
	4	200	2500				

Fill in the blank spaces for each pipe listed. Use Appendix 4 for complete friction loss tables.

Type of Pipe	Size (in.)	gpm	Distance (ft)	Loss psi/100 ft	Total Loss (psi)	Total Loss (ft)	Velocity (ft/sec)
	1.25	25	100	7.3	7.3	16.8	5.4
Steel	3	100	1000				
Schedule	2	50	350				
40	6	400	50				
	4	200	2500	1.85	46.3	107.0	5.0

Problem 2. Conversion Factors

In the event that the friction loss table for a specific type of pipe was not available, an acceptable friction loss value can be determined from friction loss tables which contain conversion factors. Refer to the friction loss table for unlined iron pipe (C=100). See page 15. *Note:* The friction loss values found in this table are based on iron pipe (C=100) in psi per 100 ft and 1000 ft. To convert the friction loss for an equivalent pipe size to plastic pipe (C=150) or other types of pipe listed, use the "multiplication factor" shown.

Example

Determine the friction loss in psi and feet for a 4-in. plastic (PVC) Class 200 and C-Value = 150 pipe carrying 200 gpm a distance of 1000 ft.

Solution

Friction loss for a 4-in. iron pipe, distance 1000 ft = 19.1 psi

$$19.1 \times 0.45 = \underline{8.6 \text{ psi}} \times 2.31 = \underline{19.9 \text{ ft}} \text{ for PVC pipe}$$

FRICTION LOSS CHARACTERISTICS
PVC CLASS 160 IPS PLASTIC PIPE
(1120, 1220) SDR 26 C = 150
PSI LOSS PER 100 FEET OF PIPE (PSI/100 FT)

Sizes 1" thru 5". Flow GPM 1 thru 1250.

SIZE	1.00		1.25		1.50		2.00		2.50		3.00		3.50		4.00		5.00		SIZE
OD	1.315		1.660		1.900		2.375		2.875		3.500		4.000		4.500		5.563		OD
ID	1.195		1.532		1.754		2.193		2.655		3.230		3.692		4.154		5.133		ID
WALL THK	0.060		0.064		0.073		0.091		0.110		0.135		0.154		0.173		0.214		WALL THK
Flow G.P.M.	Velocity F.P.S.	P.S.I. Loss	Velocity F.P.S.	P.S.I. Loss	Velocity F.P.S.	P.S.I. Loss	Velocity F.P.S.	P.S.I. Loss	Velocity F.P.S.	P.S.I. Loss	Velocity F.P.S.	P.S.I. Loss	Velocity F.P.S.	P.S.I. Loss	Velocity F.P.S.	P.S.I. Loss	Velocity F.P.S.	P.S.I. Loss	Flow G.P.M.
1	0.28	0.02	0.17	0.01	0.13	0.00													1
2	0.57	0.06	0.34	0.02	0.26	0.01	0.16	0.00											2
3	0.85	0.14	0.52	0.04	0.39	0.02	0.25	0.01											3
4	1.14	0.23	0.69	0.07	0.53	0.04	0.33	0.01	0.23	0.00									4
5	1.42	0.35	0.86	0.11	0.66	0.05	0.42	0.02	0.28	0.01									5
6	1.71	0.49	1.04	0.15	0.79	0.08	0.50	0.03	0.34	0.01	0.23	0.00							6
7	1.99	0.66	1.21	0.20	0.92	0.10	0.59	0.03	0.40	0.01	0.27	0.01							7
8	2.28	0.84	1.39	0.25	1.06	0.13	0.67	0.04	0.46	0.02	0.31	0.01							8
9	2.57	1.05	1.56	0.31	1.19	0.16	0.76	0.05	0.52	0.02	0.35	0.01	0.26	0.00					9
10	2.85	1.27	1.73	0.38	1.32	0.20	0.84	0.07	0.57	0.03	0.39	0.01	0.29	0.01					10
11	3.14	1.52	1.91	0.45	1.45	0.23	0.93	0.08	0.63	0.03	0.43	0.01	0.32	0.01					11
12	3.42	1.78	2.08	0.53	1.59	0.28	1.01	0.09	0.69	0.04	0.46	0.01	0.35	0.01	0.28	0.00			12
14	3.99	2.37	2.43	0.71	1.85	0.37	1.18	0.12	0.81	0.05	0.54	0.02	0.41	0.01	0.33	0.01			14
16	4.57	3.04	2.78	0.91	2.12	0.47	1.35	0.16	0.92	0.06	0.62	0.02	0.47	0.01	0.37	0.01			16
18	5.14	3.78	3.12	1.13	2.38	0.58	1.52	0.20	1.04	0.08	0.70	0.03	0.53	0.02	0.42	0.01			18
20	5.71	4.59	3.47	1.37	2.65	0.71	1.69	0.24	1.15	0.09	0.78	0.04	0.59	0.02	0.47	0.01			20
22	6.28	5.46	3.82	1.64	2.91	0.85	1.86	0.29	1.27	0.11	0.86	0.04	0.65	0.02	0.52	0.01	0.34	0.00	22
24	6.85	6.44	4.17	1.92	3.18	1.00	2.03	0.34	1.38	0.13	0.93	0.05	0.71	0.03	0.56	0.02	0.37	0.01	24
26	7.42	7.47	4.51	2.23	3.44	1.15	2.20	0.39	1.50	0.15	1.01	0.06	0.77	0.03	0.61	0.02	0.40	0.01	26
28	7.99	8.57	4.86	2.56	3.71	1.32	2.37	0.45	1.62	0.18	1.09	0.07	0.83	0.04	0.66	0.02	0.43	0.01	28
30	8.57	9.74	5.21	2.91	3.97	1.50	2.54	0.51	1.73	0.20	1.17	0.08	0.89	0.04	0.70	0.02	0.46	0.01	30
35	9.99	12.95	6.08	3.87	4.64	2.00	2.96	0.68	2.02	0.27	1.36	0.10	1.04	0.05	0.82	0.03	0.54	0.01	35
40	11.42	16.59	6.95	4.95	5.30	2.56	3.39	0.86	2.31	0.34	1.56	0.13	1.19	0.07	0.94	0.04	0.61	0.01	40
45	12.85	20.63	7.82	6.16	5.96	3.19	3.81	1.08	2.60	0.42	1.75	0.16	1.34	0.09	1.06	0.05	0.69	0.02	45
50	14.28	25.07	8.69	7.49	6.63	3.88	4.24	1.31	2.89	0.52	1.95	0.20	1.49	0.10	1.18	0.06	0.77	0.02	50
55	15.71	29.91	9.56	8.93	7.29	4.62	4.66	1.56	3.18	0.62	2.15	0.24	1.64	0.12	1.30	0.07	0.85	0.02	55
60	17.14	35.14	10.43	10.49	7.95	5.43	5.09	1.83	3.47	0.72	2.34	0.28	1.79	0.15	1.41	0.08	0.92	0.03	60
65	18.57	40.76	11.29	12.17	8.62	6.30	5.51	2.12	3.76	0.84	2.54	0.32	1.94	0.17	1.53	0.09	1.00	0.03	65
70	19.99	46.76	12.16	13.96	9.28	7.23	5.93	2.44	4.06	0.96	2.73	0.37	2.09	0.19	1.65	0.11	1.08	0.04	70
75			13.03	15.86	9.94	8.21	6.36	2.77	4.34	1.09	2.93	0.42	2.24	0.22	1.77	0.12	1.16	0.04	75
80			13.90	17.88	10.60	9.26	6.78	3.12	4.63	1.23	3.12	0.47	2.39	0.25	1.89	0.14	1.23	0.05	80
85			14.77	20.00	11.27	10.35	7.21	3.49	4.91	1.38	3.32	0.53	2.54	0.28	2.00	0.16	1.31	0.06	85
90			15.64	22.23	11.93	11.51	7.63	3.88	5.20	1.53	3.51	0.59	2.69	0.31	2.12	0.17	1.39	0.06	90
95			16.51	24.58	12.59	12.72	8.05	4.29	5.49	1.69	3.71	0.65	2.84	0.34	2.24	0.19	1.47	0.07	95
100			17.38	27.03	13.26	13.99	8.48	4.72	5.78	1.86	3.91	0.72	2.99	0.37	2.36	0.21	1.54	0.08	100
110			19.12	32.24	14.58	16.69	9.33	5.63	6.36	2.22	4.30	0.86	3.29	0.45	2.60	0.25	1.70	0.09	110
120					15.91	19.61	10.18	6.61	6.94	2.61	4.69	1.01	3.59	0.52	2.83	0.30	1.85	0.11	120
130					17.24	22.74	11.02	7.67	7.52	3.03	5.08	1.17	3.89	0.61	3.07	0.34	2.01	0.12	130
140					18.56	26.09	11.87	8.80	8.10	3.47	5.47	1.34	4.19	0.70	3.31	0.39	2.16	0.14	140
150					19.89	29.64	12.72	10.00	8.68	3.94	5.86	1.52	4.48	0.79	3.54	0.45	2.32	0.16	150
160							13.57	11.27	9.26	4.45	6.25	1.71	4.78	0.89	3.78	0.50	2.47	0.18	160
170							14.42	12.61	9.83	4.97	6.64	1.92	5.08	1.00	4.01	0.56	2.63	0.20	170
180							15.27	14.02	10.41	5.53	7.03	2.13	5.38	1.11	4.25	0.63	2.78	0.22	180
190							16.11	15.49	10.99	6.11	7.43	2.35	5.68	1.23	4.49	0.69	2.94	0.25	190
200							16.96	17.03	11.57	6.72	7.82	2.59	5.98	1.35	4.72	0.76	3.09	0.27	200
225							19.08	21.19	13.02	8.36	8.79	3.22	6.73	1.68	5.31	0.95	3.48	0.34	225
250									14.47	10.16	9.77	3.91	7.48	2.04	5.91	1.15	3.87	0.41	250
275									15.91	12.12	10.75	4.67	8.23	2.44	6.50	1.37	4.25	0.49	275
300									17.36	14.24	11.73	5.49	8.97	2.86	7.09	1.61	4.64	0.58	300
325									18.81	16.51	12.70	6.36	9.72	3.32	7.68	1.87	5.03	0.67	325
350											13.68	7.30	10.47	3.81	8.27	2.15	5.41	0.77	350
375											14.66	8.29	11.22	4.33	8.86	2.44	5.80	0.87	375
400											15.64	9.35	11.97	4.88	9.45	2.75	6.19	0.98	400
425											16.62	10.46	12.72	5.46	10.04	3.07	6.58	1.10	425
450											17.59	11.62	13.46	6.07	10.63	3.42	6.96	1.22	450
475											18.57	12.85	14.21	6.70	11.23	3.78	7.35	1.35	475
500											19.55	14.13	14.96	7.37	11.82	4.15	7.74	1.48	500
550													16.46	8.80	13.00	4.96	8.51	1.77	550
600													17.95	10.33	14.18	5.82	9.29	2.08	600
650													19.45	11.99	15.36	6.75	10.06	2.41	650
700															16.55	7.75	10.83	2.77	700
750															17.73	8.80	11.61	3.14	750
800															18.91	9.92	12.38	3.54	800
850																	13.16	3.96	850
900																	13.93	4.41	900
950																	14.71	4.87	950
1000																	15.48	5.36	1000
1050																	16.25	5.86	1050
1100																	17.03	6.39	1100
1150																	17.80	6.94	1150
1200																	18.58	7.51	1200
1250																	19.35	8.10	1250

Note: Shaded areas of chart indicate velocities over 5' per second. Use with Caution.

(Continued)

FRICTION LOSS CHARACTERISTICS
SCHEDULE 40 STANDARD STEEL PIPE C = 100
PSI LOSS PER 100 FEET OF PIPE (PSI/100 FT)

Sizes ½" thru 3¼"
Flow GPM 1 thru 600

	0.50	0.75	1.00	1.25	1.50	2.00	2.50	3.00	3.50	
SIZE	0.50	0.75	1.00	1.25	1.50	2.00	2.50	3.00	3.50	SIZE
OD	0.840	1.050	1.315	1.660	1.900	2.375	2.875	3.500	4.000	OD
ID	0.622	0.824	1.049	1.380	1.610	2.067	2.469	3.068	3.548	ID
WALL THK	0.109	0.113	0.133	0.140	0.145	0.154	0.203	0.216	0.226	WALL THK

Flow GPM	0.50 Vel FPS	0.50 PSI Loss	0.75 Vel FPS	0.75 PSI Loss	1.00 Vel FPS	1.00 PSI Loss	1.25 Vel FPS	1.25 PSI Loss	1.50 Vel FPS	1.50 PSI Loss	2.00 Vel FPS	2.00 PSI Loss	2.50 Vel FPS	2.50 PSI Loss	3.00 Vel FPS	3.00 PSI Loss	3.50 Vel FPS	3.50 PSI Loss	Flow GPM
1	1.05	0.91	0.60	0.23	0.37	0.07	0.21	0.02	0.15	0.01	0.09	0.00							1
2	2.10	3.28	1.20	0.84	0.74	0.26	0.42	0.07	0.31	0.03	0.19	0.01	0.13	0.00					2
3	3.16	6.95	1.80	1.77	1.11	0.55	0.64	0.14	0.47	0.07	0.28	0.02	0.20	0.01	0.13	0.00			3
4	4.21	11.85	2.40	3.02	1.48	0.93	0.85	0.25	0.62	0.12	0.38	0.03	0.26	0.01	0.17	0.01			4
5	5.27	17.91	3.00	4.56	1.85	1.41	1.07	0.37	0.78	0.18	0.47	0.05	0.33	0.02	0.21	0.01	0.16	0.00	5
6	6.32	25.10	3.60	6.39	2.22	1.97	1.28	0.52	0.94	0.25	0.57	0.07	0.40	0.03	0.26	0.01	0.19	0.01	6
7	7.38	33.40	4.20	8.50	2.59	2.63	1.49	0.69	1.10	0.33	0.66	0.10	0.46	0.04	0.30	0.01	0.22	0.01	7
8	8.43	42.77	4.80	10.89	2.96	3.36	1.71	0.89	1.25	0.42	0.76	0.12	0.53	0.05	0.34	0.02	0.25	0.01	8
9	9.49	53.19	5.40	13.54	3.33	4.18	1.92	1.10	1.41	0.52	0.85	0.15	0.60	0.06	0.39	0.02	0.29	0.01	9
10	10.54	64.65	6.00	16.46	3.70	5.08	2.14	1.34	1.57	0.63	0.95	0.19	0.66	0.08	0.43	0.03	0.32	0.01	10
11	11.60	77.13	6.60	19.63	4.07	6.07	2.35	1.60	1.73	0.75	1.05	0.22	0.73	0.09	0.47	0.03	0.35	0.02	11
12	12.65	90.62	7.21	23.07	4.44	7.13	2.57	1.88	1.88	0.89	1.14	0.26	0.80	0.11	0.52	0.04	0.38	0.02	12
14	14.76	20.56	8.41	30.69	5.19	9.48	2.99	2.50	2.20	1.18	1.33	0.35	0.93	0.15	0.60	0.05	0.45	0.03	14
16	16.87	54.39	9.61	39.30	5.93	12.14	3.42	3.20	2.51	1.51	1.52	0.45	1.07	0.19	0.69	0.07	0.51	0.03	16
18	18.98	92.02	10.81	48.88	6.67	15.10	3.85	3.98	2.83	1.88	1.71	0.56	1.20	0.23	0.78	0.08	0.58	0.04	18
20			12.01	59.41	7.41	18.35	4.28	4.83	3.14	2.28	1.90	0.68	1.33	0.29	0.86	0.10	0.64	0.05	20
22			13.21	70.88	8.15	21.90	4.71	5.77	3.46	2.72	2.10	0.81	1.47	0.34	0.95	0.12	0.71	0.06	22
24			14.42	83.27	8.89	25.72	5.14	6.77	3.77	3.20	2.29	0.95	1.60	0.40	1.04	0.14	0.77	0.07	24
26			15.62	96.57	9.64	29.83	5.57	7.86	4.09	3.71	2.48	1.10	1.74	0.46	1.12	0.16	0.84	0.08	26
28			16.82	10.78	10.38	34.22	5.99	9.01	4.40	4.26	2.67	1.26	1.87	0.53	1.21	0.18	0.90	0.09	28
30			18.02	25.88	11.12	38.89	6.42	10.24	4.72	4.84	2.86	1.43	2.00	0.60	1.30	0.21	0.97	0.10	30
35					12.97	51.74	7.49	13.62	5.50	6.44	3.34	1.91	2.34	0.80	1.51	0.28	1.13	0.14	35
40					14.83	66.25	8.56	17.45	6.29	8.24	3.81	2.44	2.67	1.03	1.73	0.36	1.29	0.18	40
45					16.68	82.40	9.64	21.70	7.08	10.25	4.29	3.04	3.01	1.28	1.95	0.44	1.45	0.22	45
50					18.53	00.16	10.71	26.37	7.87	12.46	4.77	3.69	3.34	1.56	2.16	0.54	1.62	0.27	50
55							11.78	31.47	8.65	14.86	5.25	4.41	3.68	1.86	2.38	0.65	1.78	0.32	55
60							12.85	36.97	9.44	17.46	5.72	5.18	4.01	2.18	2.60	0.76	1.94	0.37	60
65							13.92	42.88	10.23	20.25	6.20	6.00	4.35	2.53	2.81	0.88	2.10	0.43	65
70							14.99	49.18	11.01	23.23	6.68	6.89	4.68	2.90	3.03	1.01	2.26	0.50	70
75							16.06	55.89	11.80	26.40	7.16	7.83	5.01	3.30	3.25	1.15	2.43	0.56	75
80							17.13	62.98	12.59	29.75	7.63	8.82	5.35	3.72	3.46	1.29	2.59	0.64	80
85							18.21	70.47	13.37	33.29	8.11	9.87	5.68	4.16	3.68	1.44	2.75	0.71	85
90							19.28	78.33	14.16	37.00	8.59	10.97	6.02	4.62	3.90	1.61	2.91	0.79	90
95									14.95	40.90	9.07	12.13	6.35	5.11	4.11	1.78	3.07	0.88	95
100									15.74	44.97	9.54	13.33	6.69	5.62	4.33	1.95	3.24	0.96	100
110									17.31	53.66	10.50	15.91	7.36	6.70	4.76	2.33	3.56	1.15	110
120									18.88	63.04	11.45	18.69	8.03	7.87	5.20	2.74	3.88	1.35	120
130											12.41	21.68	8.70	9.13	5.63	3.17	4.21	1.56	130
140											13.36	24.87	9.37	10.47	6.06	3.64	4.53	1.79	140
150											14.32	28.26	10.03	11.90	6.50	4.14	4.86	2.04	150
160											15.27	31.84	10.70	13.41	6.93	4.66	5.18	2.30	160
170											16.23	35.63	11.37	15.01	7.36	5.22	5.50	2.57	170
180											17.18	39.61	12.04	16.68	7.80	5.80	5.83	2.86	180
190											18.14	43.78	12.71	18.44	8.23	6.41	6.15	3.16	190
200											19.09	48.14	13.38	20.28	8.66	7.05	6.48	3.47	200
225													15.08	25.22	9.75	8.76	7.29	4.32	225
250													16.73	30.65	10.83	10.65	8.10	5.25	250
275													18.40	36.57	11.92	12.71	8.91	6.27	275
300															13.00	14.93	9.72	7.36	300
325															14.08	17.32	10.53	8.54	325
350															15.17	19.87	11.34	9.79	350
375															16.25	22.57	12.15	11.13	375
400															17.33	25.44	12.96	12.54	400
425															18.42	28.46	13.77	14.03	425
450															19.50	31.64	14.58	15.60	450
475																	15.39	17.24	475
500																	16.20	18.96	500
550																	17.82	22.62	550
600																	19.44	26.57	600

(Continued)

Note: Shaded areas of chart indicate velocities over 5' per second. **Use with Caution.**

LOSS OF PRESSURE DUE TO FRICTION IN UNLINED IRON PIPE
Expressed in Pounds Per Square Inch

Per 100 Feet of Length

G.P.M.	1/2"	3/4"	1"	1¼"	1½"	2"	2½"	3"	4"
1	.9								
2	3.2	.8							
3	6.8	1.8	.6						
4	11.7	3.0	.9	.3					
5	17.8	4.5	1.4	.4	.1				
6	24.9	6.4	1.9	.5	.2				
7		9.1	2.5	.7	.3				
8		10.8	3.4	.8	.4	.1			
10		16.5	5.1	1.3	.6	.2			
12		23.0	7.1	1.8	.8	.3			
14		29.0	9.5	2.4	1.1	.4			
16			12.1	3.1	1.4	.5			
18			15.2	3.9	1.8	.6	.1		
20			18.2	4.8	2.2	.7	.2		
25			27.8	7.2	3.4	1.1	.3	.1	
30				10.2	4.8	1.6	.4	.2	
35				13.5	6.3	2.2	.5	.3	
40				17.3	8.1	2.8	.7	.4	
50				26.0	12.3	4.3	1.4	.6	.1
60					17.2	6.0	1.9	.8	.2
70					23.0	8.0	2.6	1.1	.2
80					29.5	10.3	3.4	1.4	.3
90						12.7	4.2	1.7	.4
100						15.5	5.2	2.1	.5
120						21.7	7.3	3.0	.7
140							9.6	3.9	.9
160							12.6	5.1	1.2
180							15.6	6.4	1.5
200							18.7	7.7	1.9
220							22.6	9.2	2.2
240								10.9	2.6
260								12.5	3.1
280								14.6	3.5
300								16.5	4.0
320								18.4	4.6
340								20.7	5.1
360								23.0	5.7

Per 1000 Feet of Length

G.P.M.	2"	2½"	3"	4"	6"	8"	10"
25	11.7	3.9	1.7	.4			
50	43.0	14.3	6.0	1.3	.2		
100		52.0	21.7	5.2	.7		
150			46.0	11.3	1.6		
200			77.0	19.1	2.6		
250				28.8	4.0	1.0	.3
300				40.7	5.6	1.5	.5
350				53.7	7.5	1.8	.6
400					9.6	2.3	.8
450					11.9	2.9	1.0
500					14.4	3.6	1.2
550					17.3	4.2	1.4
600					20.3	5.0	1.7
650					23.6	5.8	1.9
700					27.1	6.7	2.2
750					30.6	7.6	2.6
800					34.5	8.5	2.9
850					38.8	9.5	3.2
900					43.2	10.6	3.5
950						11.7	3.9
1000						12.9	4.3
1100						15.3	5.2
1200						18.0	6.1
1300						20.8	7.1
1400						23.8	8.1
1500						27.3	9.2
1600							10.3

Figures above shaded area are at velocities of less than 5 feet per second; in shaded area between 5 and 10 feet per second; below shaded area more than 10 feet per second.

Velocities above 5 feet per second should be used with caution. Velocities above 10 feet per second should be avoided.

These friction loss tables are based on the Williams and Hazen Formula, using the coefficient of c equal to 100 as standard. To find friction loss values for other than c = 100 multiply the result in these tables by the following factors:

Type of Pipe	C Factor	Multiplication Factor
Plastic	150	0.45
Asbestos Cement	140	0.53
Smooth and Straight Brass or Aluminum	120	0.714
New Smooth and Straight Iron	110	0.838
Ordinary Iron	100	1.000
Old Iron	90	1.224
Rough Iron	80	1.513

HYDRAULIC PRESSURE

Hydraulic pressure involves the science that deals with laws governing water or other liquids in motion and or operated under pressure.

Since sprinkler performance is directly related to the pressure at its base, it is *essential* to design the irrigation system on the working pressure (water flowing), never on *static pressure* (water not flowing).

Working Pressure Calculation Procedure

The working pressure at an outlet or specific point from the source or the starting point can be calculated by the following procedure:

- Uphill Flow:
   Subtract 0.433 psi for each foot of vertical rise or elevation difference (static pressure) above the given starting point. *Note:* Distance is not a factor.
   Working Pressure (WP) =
   operating pressure (OP) minus static pressure (SP) minus friction losses (FL) in the pipe, valves, and fittings
   or simply WP = OP – SP – FL

- Downhill Flow:
   Add 0.433 psi for each foot of vertical drop below the starting point. Again, distance is not a factor. *Note:* The working pressure can be calculated in terms of *psi* or *feet*.

$$WP = OP + SP - FL$$

Example

Use tables for steel (page 14), for PVC (page 13).
   Calculate the *working pressure* at points A, B, C, D, and E for the following pipes:

- 1.5-in. steel pipe (Sch. 40)
- 1.5-in. PVC (Class 160)
   Flow = 35 gpm

Procedure

Step 1. Solve for friction loss:

steel = 6.44 psi/100 ft
PVC = 2.00 psi/100 ft

Step 2. Solve for pressure at A, B, C, D, and E as shown in Table 1.4.

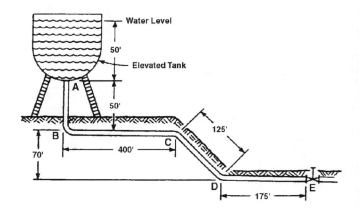

Figure 1.18

Table 1.4

Static Pressure	Steel Pipe	PVC Pipe
Point A = 21.6 psi	= 21.6 psi	= 21.6 psi
Point B = 43.3 psi	43.3–(6.44 x 0.5) = 40.1 psi	43.3–(2.0 x 0.5) = 42.3 psi
Point C = 43.3 psi	43.3–(6.44 x 4.50) = 14.3 psi	43.3–(2.0 x 4.50) = 38.8 psi
Point D = 73.6 psi	73.6–(6.44 x 5.75) = 36.6 psi	73.6–(2.0 x 5.75) = 62.1 psi
Point E = 73.6 psi	73.6–(6.44 x 7.50) = 25.2 psi	73.6–(2.0 x 7.50) = 58.6 psi

STUDENT PROBLEM

Running (Working) Pressure

Calculate the static (not running) pressure and running pressure at each green. Assume the 6 greens are piped as shown. Only one green is irrigated at a time. Sprinkler flow rate is 50 gpm. The incoming pressure (pressured system) at Green 1 is 100 psig. Compare the running pressure at each green for a 2-in. PVC Class 160 vs. 2.5 in. pipe.

Procedure (suggested):

1. Establish a "head" pressure line (100 psi) as shown.
2. Determine the number of feet below and above the head pressure line for each green, as shown.
3. Calculate static pressure (SP) ft × 43 = psi. Below line add to head pressure; if above, subtract from head pressure.
4. Determine the friction loss to each green. From PVC Class 160 table (page 13) find the friction loss value 1.3 psi/ 100 ft and multiply by the length, as shown.
5. Running pressure = static pressure minus friction loss. For example, Green 2.

   static pressure = 21.5 + 100 = 122 psi

   running pressure = 122 – 13 = 109 psi

6. Fill in the blank spaces: Calculate the static pressure at each green, and the running pressure for a 2-in. and 2.5-in. PVC pipe. Use format as found in example problem on previous page.

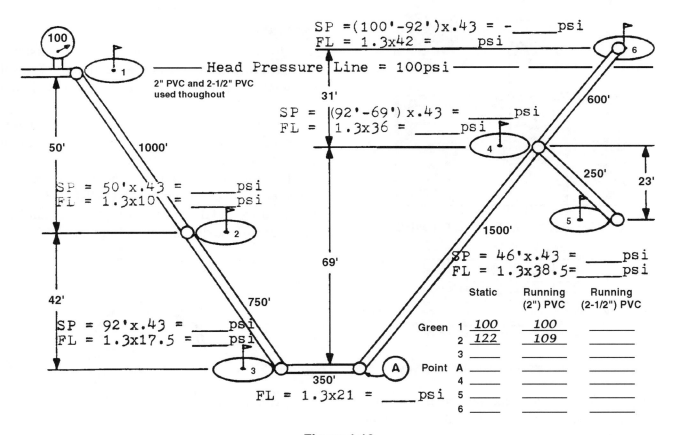

Figure 1.19

LAND MEASUREMENTS

Introduction

The importance of accurate measurements and calculations of land areas cannot be overemphasized. To properly design and install a golf course watering system it is mandatory to make exact measurements for placing sprinkler heads, pipelines and other components. In addition, accurate measurements are essential for ordering the proper quantities of pipe, control wiring, supplies and other materials.

It is important to note that land surveys (maps) are based on the *horizontal* distance and not the *actual* surface distance. For example, a measurement on a golf course map between the tee and green on a par 5 hole is the horizontal distance. However, if there are hills and/or valleys in between, the *actual* surface can be much greater.

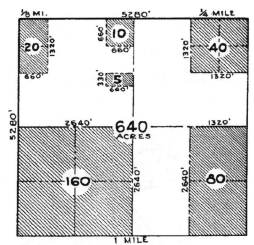

STANDARD SECTION OF LAND

TABLE OF MEASUREMENTS

MILE	CHAINS	RODS	FEET
1/8	10	40	660
1/4	20	80	1320
1/2	40	160	2640
1	80	320	5280

ONE CHAIN = 4 RODS, = 66 FEET   ONE ROD = 16½ FEET.
ONE ACRE = 43,560 SQ. FT. = 10 SQ. CHAINS.
208.7' X 208.7' = ONE ACRE.

Figure 1.20. U.S. government land survey.

The specifications for applying fertilizers, lime, pesticides, etc., are usually based on the actual land surface area.

Measuring Land Surface Distance
Tape

Tapes come in lengths of 50, 100, 300, and 1,000 ft and usually is made of steel. For ease in computing and recording distances, these tapes are graduated in decimal units.

A piece of land on a hillside really has more surface in a given area than a piece of level ground of the same calculated area. *All area measurements are made on the horizontal at all times.*

Breaking Chain

On slopes greater than 5% it is necessary to break chain (i.e., measure distances less than the full length of the tape).

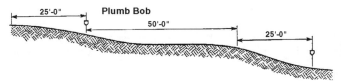

Figure 1.21

Odometer

An odometer is a device which measures distance by registering the number of revolutions of a wheel. The wheel circumference is 10 ft and it is divided into 10 equal parts.

On smooth ground the odometer gives precise results, but on rough ground or in rank vegetation measurements are less accurate.

It is important to note that actual pipe lengths are determined by the land surface distance and not by the horizontal distance. Hilly or rolling land has a greater surface distance than flat land.

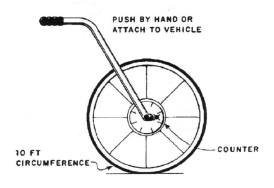

Figure 1.22

Units of Measure and Equivalents

Linear	ft	= 12 in.
	rod (rd)	= 16.5 ft, 5.5 yd
	chain (ch)	= 4 rd, or 66 ft
	mi	= 5,280 ft, 320 rd, 80 ch
Area	ft^2	= 144 in.^{2}
	rd^2	= 30.25 yd^2
	mi^2	= 1 sect or 640 ac
	ac	= 10 ch^2, 160 rd^2
		= 43,560 ft^2, 208.7 × 208.7 ft
Volume	1 gal	= 231 in.^{3}
and		= 0.1337 ft^3
Weight		= 8.33 lb
of Water	1 ft^3	= 1,728 in.^{2}
		= 7.48 gal
		= 62.4 lb
	1 ac-in.	= 27,154 gal
	1 ac-ft	= 43,560 ft^3
		= 325,850 gal
		= 12 ac-in.

Water in small volume (tanks, etc.) is usually measured in gal. Water in reservoirs is measured in ft^3 or in ac-ft. Water as applied to land is measured in ac-in. or ac-ft.

Rate	1 gpm	= 0.00223 ft^3/sec
of		= 1,440 gpd
Flow	l cfs	= 7.48 gps
		= 448.8 gpm (approx. 450)
		= 0.992 ac-in./hr (approx. 1)
		= 1,984 ac-ft/day (approx. 2)
Units of	1 ft head	
Pressure	in water	= 0.433 psi
	1 psi	= 2.31 ft in water
	atmos.	
	pressure	= 14.7 psi at sea level
	1 kW	= 1,000 W of electricity
	1 kWh	= 1.34 hp
Power	1 hp	= 33,000 ft-lb/min of work

Formulas   brake hp $= \dfrac{gpm \times head\ (ft)}{3960 \times pump\ eff.}$

precipitation rate (in./hr) $= \dfrac{gpm \times 96.3}{area\ in\ ft^2}$

cost per ac-ft $= \dfrac{1.032 \times total\ head \times power\ rate}{overall\ plant\ efficiency}$

area of a circle $= d^2 \times 0.7854$

vol. of cylinder in gal $= d^2 \times length \times 0.0034$

ACCURATE MEASUREMENTS OF LAND AREAS

Introduction

In order to accurately determine the land area it is essential to carefully study the area. Check the size, shape, elevation differences, obstructions, etc. Then, decide on a line of attack and proceed to take a sufficient number of field measurements that can be drawn to scale.

The area can be divided into simple geometric figures as shown. The area can be calculated by applying the appropriate equation(s).

Note: The area of odd-shaped fairways, greens, etc., can be solved by selecting the appropriate geometric figure found in the following figures.

Areas of Geometric Figures

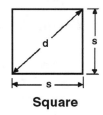

Square

$$A = s^2$$
$$A = 1/2d^2$$
$$s = 0.7071d = \sqrt{A}$$
$$d = 1.414s = 1.414\sqrt{A}$$

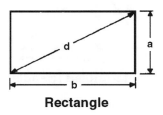

Rectangle

$$A = ab$$
$$A = a\sqrt{d^2 - a^2} = b\sqrt{d^2 - b^2}$$
$$d = \sqrt{a^2 + b^2}$$
$$a = \sqrt{d^2 - b^2} = A/b$$
$$b = \sqrt{d^2 - a^2} = a/a$$

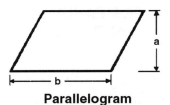

Parallelogram

$$A = ab$$
$$a = A / b$$
$$b = A / a$$

Note that dimension a is measured at right angles to line b.

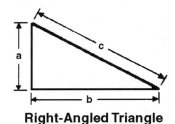

Right-Angled Triangle

$$A = \frac{ab}{2}$$
$$c = \sqrt{a^2 + b^2}$$
$$b = \sqrt{c^2 - a^2}$$
$$a = \sqrt{c^2 - b^2}$$

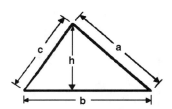

Acute-Angled Triangle

$$A = \frac{bh}{2} \text{ if h is known}$$
$$\text{if S} = \frac{1}{2}(a + b + c), \text{ then}$$
$$A = \sqrt{S(S-a)(S-b)(S-c)}$$

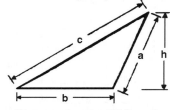

Obtuse-Angled Triangle

$$A = \frac{bh}{2} \text{ if h is known}$$
$$\text{if S} = \frac{1}{2}(a + b + c), \text{ then}$$
$$A = \sqrt{S(S-a)(S-b)(S-c)}$$

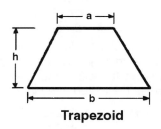

Trapezoid

$$A = \frac{(a+b)h}{2}$$

Note: In the United Kingdom, this figure is called a *trapezium* and the one below it is known as a *trapezoid,* the terms being reversed.

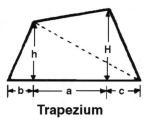

Trapezium

$$A = \frac{(H+h)a + bh + cH}{2}$$

A trapezium can also be divided into two triangles, as indicated by the dotted line. The area of each of these triangles is computed, and the results added to find the area of the trapezium.

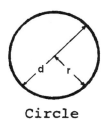

Circle

$$A = \pi r^2 = 3.1416r^2 = 0.7854d^2$$
$$C = 2\pi r = 6.2832r = 3.1416d$$
$$r = C/6.2832 = \left(\sqrt{A/3.1416} = 0.564\right)\left(\sqrt{A}\right)$$
$$d = C/3.1416 = \left(\sqrt{A/0.7854} = 1.128\right)\left(\sqrt{A}\right)$$

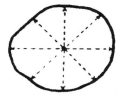

Not Perfect Circle

satisfactory approximation:

$$A = \pi r^2$$

The radius is measured at specific degrees around the green. Usually 18 to 36 measurements are made and the average is computed.

Trapezoidal Rule for Irregularly Shaped Areas

1. Parallel Sides

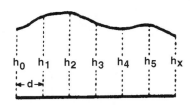

$$A = \frac{h_o}{2} + (\text{sum of h's})* + \left(\frac{h_x}{2}\right)d$$

* (sum of the numbered h's only)

2. Odd Length

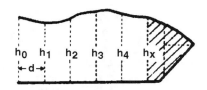

$$A = \frac{h_o}{2} + (\text{sum of h's}) + \left(\frac{h_x}{2}\right)d$$

+ odd area

3. All Sides Curved

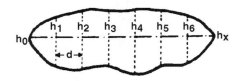

$$A = (\text{sum of h's})d$$

Note: $h_o = O$ and $h_x = O$

or

figure ends as odd areas

Long Narrow Curved Area

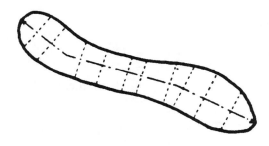

satisfactory approximation
$$A = \text{length} \times \text{width (average)}$$

Note: Width measured at several points but not at specific intervals.

Ratios for Unknown Length (Distance)

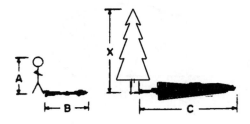

$$\frac{A \text{ (known)}}{B \text{ (known)}} = \frac{x \text{ (unknown)}}{C \text{ (known)}}$$

Note: Cross multiply, then divide.

Example Problems

Land Measurements (Mensuration)

Problem 1

Calculate the area (ft^2) of the apron, where 36 measurements, as shown, totaled 1548 ft, and apron diameter = 108.0 ft.

Procedure

$$\text{Area of a circle} = \pi r^2$$
$$\pi = 3.14$$
$$r = 1/2 \text{ diameter}$$
$$r^2 = r \times r$$

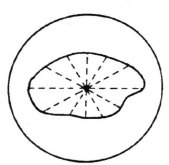

Hint: Apron area = total apron area minus green area.

Answer = 3350.45 ft^2

Problem 2

Calculate the area (ft^2) of the fairway.

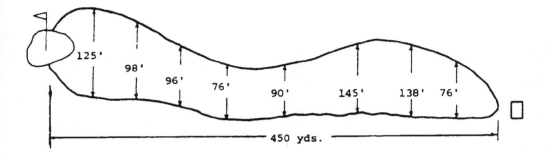

Hint: area = length (ft) × average width

Answer = 142,425 ft^2

Problem 3

Calculate the *acreage.*

Procedure

area of a right triangle = bh/2

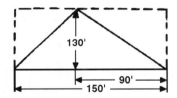

Note: When the height (h) is known, a rectangle can be made as shown (dashes).

$$1 \ ac = 43560 \ ft^2$$

Answer = 0.224 ac

Problem 4

Calculate the acreage area of an obtuse triangle.

Procedure

$$\text{triangle} = \sqrt{S(S-a)(S-b)(S-c)}$$

$$S = \text{a number} = \frac{1}{2}(a+b+c)$$

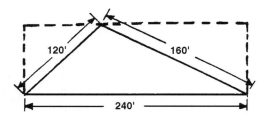

Note: Do not use the hypotenuse to form a rectangle.

Answer = 0.2 ac

Problem 5

Calculate the area in acres. Use the trapezoidal rule for parallel sides.

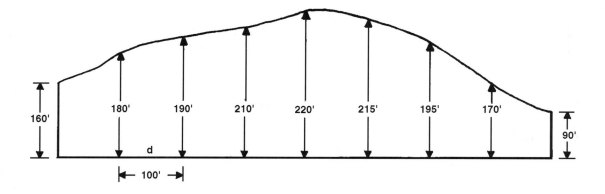

Answer = 3.46 ac

Problem 6

Calculate the area in acres.

Procedure

Divide the area into squares, rectangles, triangles, etc.
    Note: One possibility of dividing the area is shown by the dashed lines.

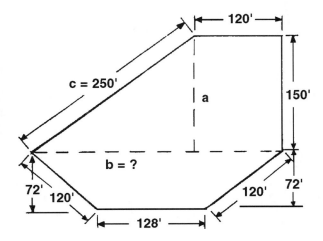

Hint: Solve for length of line b first.

Answer = 1.13 ac

Problem 7

Calculate the height of the tree.

Procedure

$$\text{ratio problem} = \frac{\text{height of man}}{\text{length of shadow}} = \frac{X}{\text{tree shadow}}$$

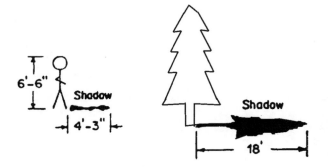

Hint: Convert measurements to equivalent units, either ft or in.

Answer = 27 ft, 6 in.

Problem 8

Calculate the pond length.

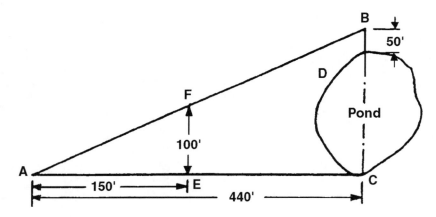

Hint: Set up a ratio as shown

$$L = -\frac{100 \text{ ft}}{150 \text{ ft}} = \frac{X}{440 \text{ ft}} - (50 \text{ ft})$$

Answer = 243.31 ft

Problem 9

Convert 10.854 ft to equivalent units of the architectural scale, i.e., _____ feet, _____ inches, _____ 16ths of an inch.

Note: Architects, builders, carpenters, etc., generally use the *architectural scale*—feet, inches and fractions of an inch.

Answer = 10 ft, 10.25 in.

Engineers, surveyors, etc., use the *engineer's scale* for land measurements—feet and 10ths of a foot (decimals).

It is quite possible that a conversion from one scale to the other scale may be necessary.

Problem 10

Convert 6 ft, 9.25 in. to feet and tenths of a foot.

Note: Conversions apply to the *fractions of a foot only.*

Answer = 6.77 ft

ACKNOWLEDGMENTS

1. "Massachusetts Irrigation Guide for Sprinkler Irrigation Design." Massachusetts Soil Conservation Service.
2. "Water—The Yearbook of Agriculture." USDA.
3. "Sprinkler Irrigation." 1st Edition. Shrunk, J.F. and McCulloch, A.W.
4. "Rainfall—Evapotranspiration Data." Toro Manufacturing Company.
5. "Ground Water Regulations." Massachusetts Department of Environmental Protection.
6. Federal Regulatory Program to Protect Waters and Wetlands of the United States.
7. "Friction Loss Characteristics Tables." Rainbird Sprinkler Manufacturing Company.

Chapter 2

Irrigation System Design

INTRODUCTION TO IRRIGATION SYSTEM COMPONENTS

General

The oversimplified description of an irrigation system should be interpreted in its proper perspective.

Obviously, to properly design and install a system that will "tailor-fit" the conditions of as large and complex an area as a golf course takes a good deal of skill, know-how, and use of sound engineering practices. For the selection of the type of system, the diversity wanted, and the choice of components best suited to fit the conditions involved requires considerable investigation, understanding, and compromise.

Figure 2.1 is provided to illustrate that an irrigation system can consist of a few basic components and/or automatic control devices.

Basically, the *manually* operated system consists of pipes, quick-coupling valves and fittings, sprinklers, pump, and power unit. In addition to these, the *automatically* operated system requires electronic and/or computerized controlling devices, remote controlled valves, and possibly a weather station.

STUDENT EXERCISE

Identify each of the items listed below by placing the number or marking the section on Figure 2.1. Examples can be found on the pages shown.

Identify	Page
1. screen and foot valve	204
2. tees	69
3. 45° angle elbow	71
4. 90° angle elbow (ell joint)	70
5. check valve	205
6. ball valve	203
7. drain valve	30
8. suction side of pump	173
9. static suction lift	173
10. gravel sump	30
11. swing joint	69
12. pop-up sprinkler	360
13. field controller	394
14. remote control valve	211
15. grounding rod	145
16. computer controller	409
17. quick-coupling valve	359
18. weather station	411

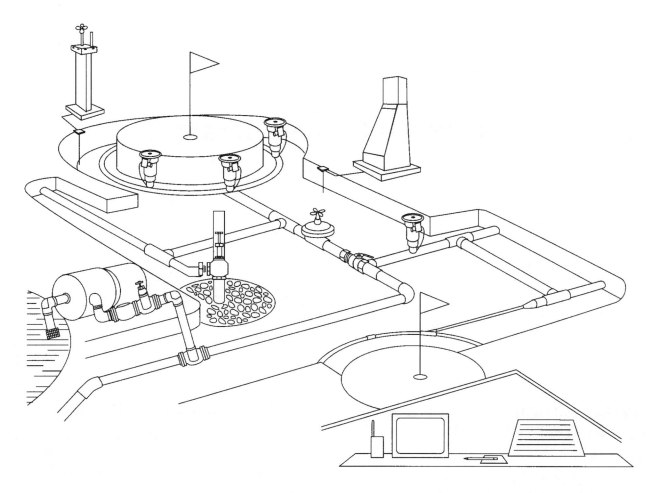

Figure 2.1. Irrigation system pipes, valves, fittings, sprinklers, drains, pumps and controllers (optional).

STUDENT PROBLEM—TERM PROJECT (REQUIRED)

General Information

At the offset, it should be understood that for any given golf course there may exist certain limiting conditions that could more or less dictate the type and/or capacity of the irrigation system to be designed, such as:

- quantity of water available
- water source—lake, pond, river, stream, canal, well, or city water
- location of the water source
- availability of 3-phase electric power
- existing pump and/or power unit
- amount of money available
- personal preferences

Assume for this undertaking that the water supply is adequate and limiting restrictions, if any, will be assigned.

Assignment

Design a single-row, fully automatic irrigation system for a "problem" golf course (to be provided). The final project must include:

- A completed map showing coverage of fairways, greens, tees, pipe layout, pipe sizes, irrigation schedule, and location of shut-off valves and drains.
- A complete report giving the step-by-step design procedure followed. The report must include all calculations and supplemental data explaining each step listed below.

Suggested Procedure

The following suggested procedure was developed to assist students with a logical step-by-step method for solving the assigned golf course irrigation system design problem (term project).

Finally, list the changes, adjustments, and alternatives recommended to improve the system design or its operation.

GOLF COURSE IRRIGATION SYSTEM DESIGN

Although there is no standard procedure when it comes to designing a system, it is common practice to complete the fairway irrigation system first, then to add on the greens and tees. The fairway system design should include the sprinkler models, the total gallons per minute needed, the pipe layout, and the pipe sizes.

Solving for Average Fairway Width

As a starting point the average fairway width of all the fairways to be irrigated is used to determine the following:

- The feasibility of a single-row vs. a multi-row system.
- To reduce the number of sprinkler models, which in turn results in stocking fewer spare parts.
- For selecting the proper size and sprinkler model.

Obviously, designing for the "average width" means that wide fairways may not get adequate coverage at the edges, whereas with narrow fairways the coverage may extend into the rough for some distance. However, following the initial design each fairway should be carefully studied and appropriate corrective measures should be taken, such as using smaller-diameter sprinklers for narrow fairways or multi-row patterns on wide fairways, if necessary.

Procedure

A contoured map of the golf course drawn to scale is required.

For fairways that are relatively uniform throughout, one or two measurements may be adequate.

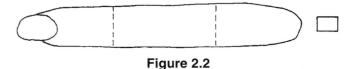

Figure 2.2

For irregularly shaped fairways, good judgment should dictate the number of measurements needed.

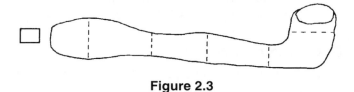

Figure 2.3

Note: Avoid diagonal measurements across the fairway, particularly where the dog-leg occurs.

Formula:

$$\text{average width} = \frac{\text{total average width of all fairways}}{\text{total number of fairways}}$$

Solving for Effective Coverage

The uniformity of water distribution is the weakest part of an irrigation system. Even without considering the effects from pressure variations, wind speed, and land slopes, the rotary heads vary from something near the sprinkler to zero at the outer end of the circle.

The precipitation rate expressed in inches per hour varies from the sprinkler outward, which means no sprinkler has a precipitation rate. It has been found that an average difference of 3 to 4 times the amount of water in one area compared to another may exist between sprinkler heads. A variation of 3 or 4 to 1 indicates that either puddling or wilting results.

In the New England area this variation may go for years without becoming noticeable because the evaporation rate is slower than the rate of lateral water movement of the soil. This allows the moisture levels in the soil to even out somewhat. In hot desert climates, the water is evaporated before it moves out, which results in brown turf.

- To make the best possible choice of sprinklers to select what will "tailor-fit" the job, it is essential to understand the principles of sprinkler water distribution patterns. (See Figure 2.4.)

In general, sprinklers have similar water distribution patterns. Usually the distribution is fairly uniform to about 65% of the diameter, and rapidly reduces to almost nothing at the outer fringes.

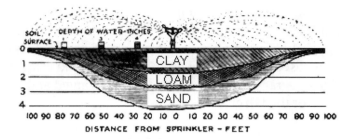

Figure 2.4. The sprinkler water distribution pattern. Water tends to go straight down. In sprinkler irrigation, less goes down at the edges because less water falls there.

Taking a bird's-eye-view looking at the top of the sprinkler head, it becomes apparent why the uniformity of distribution varies from the head out.

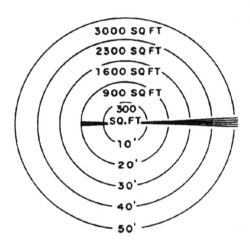

Figure 2.5

Note that the first 10-ft circle covers about 300 ft^2 and the last segment contains about 3000 ft^2. Although the sprinkler may place the water down evenly all the way out, the difference in area covered creates the sprinkler profiles from the head to zero at the end, as illustrated in Figure 2.6.

The "typical distribution curves" are based on a uniform rotation of the head. Should the head stop, slow up, or speed up for any reason, the typical profiles will change.

Based on the assumption that the water distribution pattern shown in Figures 2.7 and 2.8 is typical of all revolving sprinkler models, the following reasoning can be applied.

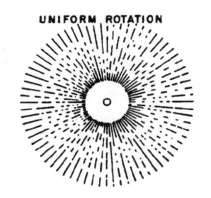

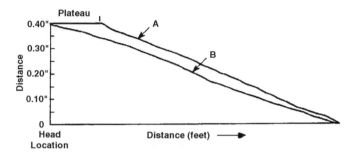

Figure 2.6

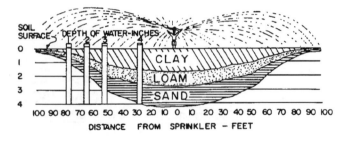

Figure 2.7.

Interpolation

Figure 2.9 indicates that when a sprinkler is operated long enough to add one inch of water to the pan located a distance of approximately 30 ft from the sprinkler (30% of the sprinkler diameter), the following amounts of water shown in Table 2.1 would be expected in the other pans.

Sprinkler Selection

A study of the previous illustrations shows that the typical water distribution patterns present some major considerations when it comes to choosing the size of the sprinkler for the job.

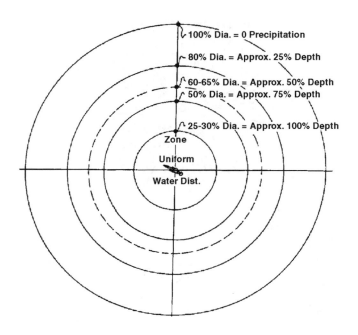

Figure 2.8

The following examples are included to help clarify the thinking and reasoning that should be applied before making a tentative sprinkler selection based on the average fairway width and effective coverage desired (depth of water wanted at the edge of the fairway).

Sprinkler Effective Coverage

The term *sprinkler effective coverage* shall be used to mean the relationship of the amount or depth of water obtained at a given distance from the sprinkler. It is expressed as a *percentage* of the wetted diameter. That is, at 0% to 30% = uniform amount or depth (assume that the sprinkler applied 1 in. of water here), at a distance of 50% = 3/4 of the amount as compared to that at the sprinkler (0.75 in.), at 60% = 1/2 the amount (0.5 in.), at 80% = 1/4 the amount (0.25 in.) and at 100% end of throw (0 in.) of water.

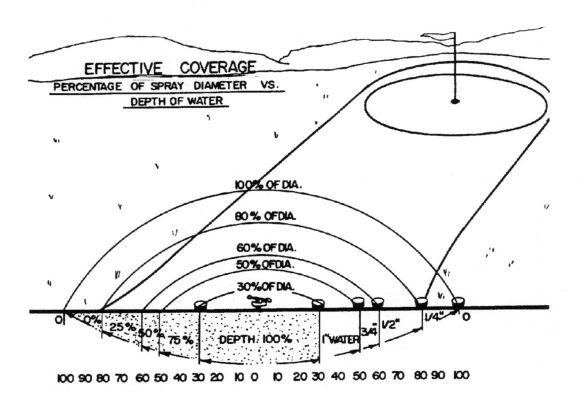

Figure 2.9

Table 2.1

Approximate Distance of Cans from Sprinkler, Radius (ft)	% Sprinkler Diameter	Approximate Depth of Water (in.)	Percentage of Depth
30	60/200 = 30%	1.0[a]	100%
50	100/200 = 50%	3/4	75%
60	120/200 = 60%	1/2	50%
80	160/200 = 80%	1/4	25%
100	200/200 = 100%	None	0%

[a] Water distribution fairly uniform throughout.

Desired Effective Coverage

The term *desired effective coverage* shall be used to mean the amount or depth of water (in.) desired at a given distance from the sprinkler, for example, at the edge of the fairway compared to the center. This decision is the responsibility of the irrigation system designer and/or the golf course superintendent.

Consider the following:

A 90-ft-diameter sprinkler was selected and used for a 90-ft-wide fairway.

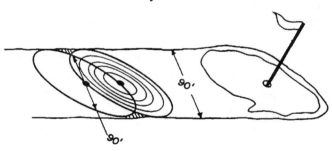

Figure 2.10

Under these conditions, no water should be expected at the edge of the fairway. In addition, there is incomplete coverage of the area.

Assume it is desired to have fairly uniform water distribution across the entire width of the fairway (90 ft).

As the water distribution remains fairly uniform to about 30% of the sprinkler diameter it would require a 300-ft-diameter sprinkler (90 ft = 30% of 300 ft). (See Figure 2.11.)

Obviously, either case would be undesirable, which suggests that a compromise must be made between the two extremes.

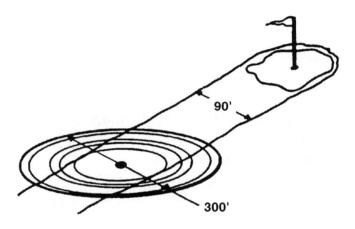

Figure 2.11

Another Consideration

Another factor that must be taken into account is the distance the water must exceed the edge of the fairway to provide the desired effective coverage.

An effective coverage of 70% of the sprinkler diameter is recommended as first choice. Table 2.2 shows that at 60%, the depth of water may be adequate but the distance of 30 ft beyond the edge of the fairway may be undesirable or hard to justify. However, at 80% the depth becomes low for sustaining turf the distance beyond the edge of the fairway, which appears to be marginal. Therefore, as a compromise, 70% was selected. The depth of water

Table 2.2

Fairway Width	% Sprinkler Diameter (Effective Coverage)	Total Sprinkler Diameter (ft)	Approximate % of Depth at Edge	Distance Water Exceeds Edge (ft)
90	100%	100	None	None
90	80%	90/0.80 = 112.5	1/4 or 25%	11
90	70%[a]	90/0.70 = 128.5	3/8 or 37%	19
90	60%	90/0.60 = 150	1/2 or 50%	30
90	50%	90/0.50 = 180	3/4 or 75%	45
90	30%	90/0.30 = 300	1.0 or 100%	105

[a]Recommended first choice.

Table 2.3

Base Pressure (psi)	Discharge Rate (gpm)	Diameter or Radius (ft)	Precipitation Rate (in./hr)	Maximal Spacing, One-Row (ft)	Nozzle Sizes (in.)

applied at the edges should be adequate where the least play occurs.

Sprinkler Models and Performance Characteristics

General Information

Sprinkler manufacturing companies will provide their catalogs, upon request, which list the models available and their performance at various pressures and with various nozzle sizes. The information in Table 2.3 is generally provided for a given model.

Pressure

The distance or diameter of coverage with a specific nozzle depends on the pressure at the nozzle itself. As water is discharged, the pressure head converts to velocity head, which sends water on its way with approximately the following speeds.

Pressure (psi)	Speed (mph)
40	52
50	58
60	65
70	70
80	74

Now, consider the possible effects that may result after a few years of operation as water (particularly if it contains sand or dirt) moves out of brass or plastic nozzles at speeds in excess of 50 mph.

Should the nozzle size increase, the flow will increase. If the pressure remains the same, a higher precipitation rate results. For nozzle discharge rates, see Table 2.4.

Generally, the increased flow causes higher friction losses in the supply pipelines, which reduces the base pressure at the sprinkler. This results in a reduction of area or diameter covered. It is the range of the nozzle that is affected, most giving a very poor water distribution profile. Checking and replacing worn nozzles is perhaps one of the least expensive costs in maintaining sprinkler performance.

Sprinklers—Part-Circle

Part-circle sprinklers are available, and can be used to advantage, particularly where water application is either restricted by borderlines and obstacles or where it is unwanted, as in bunkers, bodies of water, etc.

A good system design strives to provide water to the turf in a relatively uniform application.

Consider two heads identical in every way, except that one distributes water in a full 360-degree area around the head—a full-circle head—and the other only distributes water in a 180-degree area around the head—a half-circle head. (See Figure 2.12.)

If the nozzles of both heads provide the same gpm under pressure, twice as much water would be distributed in the half-circle area as in the full-circle area in the same span of time.

This would mean the application rate—or total water applied per unit of area, would be twice as great in the part-circle area as in the full-circle area. This defeats the primary purpose of the design: providing water to the turf in a relatively uniform application.

The only solution to this is to have a full-circle sprinkler which discharges twice as much water in a given time span as compared to a half-circle head. This is done in small turf heads, but when larger heads are considered, restrictions are imposed by physical laws, making possible applications impractical ones.

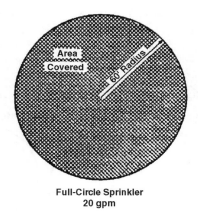

Full-Circle Sprinkler
20 gpm

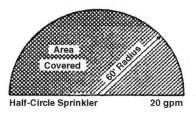

Half-Circle Sprinkler   **20 gpm**

Figure 2.12

Pressure Variation vs. Sprinkler Discharge

The relationship of pressure variation to discharge rate is given below. It is understood that these are based on calculated values which may differ from actual test performance data given in sprinkler catalogs.

Variations in Base Pressure	Variations in Sprinkler Flow (gpm)
20%	10%
30%	16%
40%	23%
50%	29%

As indicated, the sprinkler flow rate is approximately one-half of the base pressure. Actually, in the field it is the water application rate that is the most significant factor.

Example

Assume the sprinkler at Fairway 1 was running at the base pressure of 100 psi and discharging 90 gpm. Determine the flow from the sprinkler (same model) at the end of the line having a base pressure of 80 psi (20% pressure variation).

Solution

20% pressure variation; 10% discharge variation

> i.e., Sprinkler No. 1 = 90 gpm
> Sprinkler No. 2 = 81 gpm $(90 \times 0.10 = -9$ gpm$)$

In practice, a variation of 9 gpm is acceptable.

Nozzle Discharge Rates (see Table 2.4)

Formula: $gpm = P \times D^2 \times 29.82 \times C$
Nozzle $C = 0.90$ to 0.95

Example

Calculate the discharge rate (gpm) for a 0.25-in.-diameter nozzle ($C = 0.95$) operating at 100 psig

Solution

$$gpm = 100 \times 0.25^2 \times 29.82 \times 0.95$$
$$= 10 \times 0.0625 \times 29.82 \times 0.95$$
$$= 17.8$$

Table 2.4

psig	\multicolumn{9}{c}{Nozzle Discharge Rate (gpm)}								
	3/16 in. 0.1875	7/32 in. 0.28175	1/4 in. 0.25	9/32 in. 0.28125	5/16 in. 0.3125	11/32 in. 0.34375	3/8 in. 0.375	13/32 in. 0.40625	17/16 in. 0.4375
20	4.5	6.1	8.0	10.1	12.4	15.1	18.0	21.0	24.3
30	5.5	7.5	9.8	12.3	15.3	18.5	22.0	25.6	29.9
40	6.3	8.5	11.2	14.2	17.6	21.3	25.3	29.8	34.7
50	7.0	9.6	12.5	15.8	19.3	23.8	28.4	33.3	38.6
60	7.7	10.5	13.8	17.4	21.4	26.0	31.1	36.5	42.2
70	8.4	11.4	14.9	18.9	23.2	28.1	33.6	39.4	45.6
80	9.0	12.2	15.9	20.2	24.9	30.2	35.8	42.1	48.8
90	9.5	13.0	16.9	21.4	26.4	32.0	38.0	44.7	51.7
100	10.0	13.6	17.8	22.6	27.8	33.7	40.1	47.1	54.6

Precipitation Rate
Matched Precipitation Rates

With smaller turf sprinklers that have matched precipitation rates for part circles, the part- and full-circle heads *can* operate on the same control and for the same period of time.

Example

Refer to pages 367 and 368—Hunter sprinklers

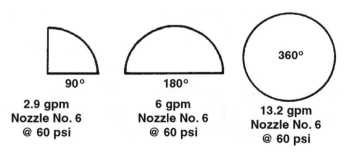

Figure 2.13

Note: The quarter-circle will make 4 sweeps (back and forth), the half-circle two sweeps, while the full-circle completes one revolution. This will give approximately uniform application of water over the entire area, since the quarter-circle = 4 sweeps × 2.9 = 11.6 gpm; half-circle 2 sweeps × 6 = 12 gpm; and the full-circle = 1 sweep × 13.2 = 13.2 gpm.

Non-Matched Precipitation Rates

With larger turf sprinklers that have the same flow, the part- and full-circles *cannot* be operated on the same control. Each must operate on separate controls, with the running time set to give uniform application rates.

Example

Refer to page 371—Toro Sprinklers, Series 690.

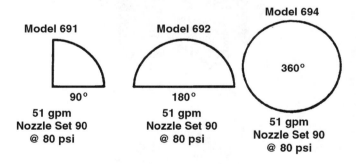

Figure 2.14

Note: The quarter-, half- and full-circle sprinklers have equivalent charge rates. While the full-circle completes one revolution applying 51 gpm over the area; the half-circle makes two sweeps applying 102 gpm and the quarter-circle will make four sweeps applying 204 gpm on the area covered. For complete information concerning sprinkler equipment, obtain catalogs from manufacturers.

Water Management and Irrigation Efficiency Recommendations

It has been found that the water application pattern of a sprinkler strongly influences uniformity results. In the past, irrigation system designers have used their experience and certain rules of thumb to determine sprinkler spacing. The information contained in this book provides the designer with additional, necessary detail to allow the optimal selection of products for the application.

The objective of the irrigation system designer is to provide an efficient sprinkler system that replenishes moisture (lost through evapotranspiration) with a minimal amount of water.

evapotranspiration (ET) = water used by the plant + water lost through evaporation (E)

Studies over the past decades have shown that water loss through evaporation is usually 5–8% of the amount applied by the sprinklers. The amount of water lost through the spray process is usually less than 2%, but can increase drastically in high wind conditions.

In general, it has been found that miscellaneous evaporation losses are negligible when compared to the effects of non-uniform sprinkler application of water.

Wind Considerations

Sprinkler spray ejected high into the air will be subjected to greater wind speeds and greater spray distortion than spray near the surface. For optimal performance in windy conditions, select sprinklers with:

1. The smallest practical radius of throw. The small curtain of water keeps the droplets close to the ground.
2. The lowest possible trajectory angle. Studies have shown that the ideal angle for a sprinkler jet is 32° above the horizontal when tested indoors, or under calm conditions. However, this is not practical since winds as low as 5 mph are experienced in virtually every irrigation site. It has been found that sprinklers with trajectory angles of 22° to 27° give op-

timal radius while minimizing water application distortion.

In applications where wind speeds exceed 5 mph, sprinklers should be spaced closer to each other.

Slope Considerations

Sprinklers should always be installed perpendicular to the slope to be irrigated. The system precipitation rate should not exceed the infiltration rate of the soil, or water will run off and possibly create erosion problems.

Sprinkler spacing *across the slope* should be reduced whenever the slope exceeds 10%. The reduction should be 1% for every 1% of slope increase over 10%. Spacing *up and down the slope* need only be adjusted at the crown and foot of the slope, keeping in mind that the sprinkler radius is shorter upslope and greater downslope.

Miscellaneous

There are many other factors which affect irrigation uniformity and system efficiency, such as soil type, relative humidity, solar radiation (affected by geographical location), altitude, etc.

Additional information is available in Buckner's "Product Catalog," "Product Specification" sheets, and "Design Manual."

Theoretical and research data may also be available from the Center for Irrigation Technology, California State University, Fresno, California, 93740. The information contained in this publication is intended to aid the designer in utilizing concepts in water management and product application. The test results contained within were performed indoors by an independent laboratory and may not be representative of actual field conditions.

Solve for Sprinkler Selection
Factors to Consider

Before selecting one model (only), assuming that a certain make was not preselected, review the following factors to consider:

- All of the models listed adequately meet the design requirements. The final choice should take into account the sprinkler manufacturing company and/or dealership in your area having a good reputation for fast and reliable service when it is most needed.
- All sprinklers have a limited range of operating pressures. Exceeding either the upper or lower limit must be avoided. Good practice dictates that a mid-range pressure be selected as first choice.
- It must be clearly understood that the sprinkler performance values listed in the sprinkler catalogs were derived under ideal no-wind conditions. In practice, however, the sprinklers are operating under far from ideal conditions. The actual amount of water that will fall at a given distance from the sprinkler is only a rough estimate at best. It is impossible to predict any difference that will occur 3 ft or 4 ft either way added or subtracted along the radius of throw. Therefore, in this case, 180 ft ± 6 ft should have no bearing whatsoever when it comes to selecting a fairway sprinkler head.

Other general factors to consider:

- *Do not mix* different types of sprinkler heads on the same circuit!
- Application rate must be nearly the same. With widely varying application rates between types of heads connected to the same circuit, part of the area will either be underwatered or overwatered in order to satisfy the conditions on the remainder.
- Even with the same type of sprinkler on a circuit, if needed, half-circle models must discharge 1/2 the gpm as that of the full-circle; quarter-circle models at 1/4 the gpm of the full-circle heads, etc. If not selected in this manner, then they must be placed on separate circuits.
- Operating pressures must be nearly the same for all sprinklers on the circuit. All sprinklers have an ideal operating pressure. If sprinklers are operated way under or way over pressure, performance will be greatly affected.

- Within the circuit of compatible sprinklers, the operating pressure at the various heads must be within certain limits for proper performance. Pressure variations between extreme heads on the circuit should be within *20% variation* for good design and proper operation of the irrigation system.

Solving for a Tentative Sprinkler Selection

A tentative choice of the fairway sprinkler can be made at this time. However, the final choice should be delayed until the final readjustments and compromises have been considered.

Example

Select a sprinkler model for a fully automatic, single-row irrigation system.

Conditions

1. *Water usage rate* = 1 in./wk.
2. The superintendent concluded that 3/8 in. of water per week at the edge of the fairways would be adequate to sustain the turfgrass where the traffic is usually the least.
3. The *average fairway width* = 126 ft
4. *Sprinkler effective coverage* of 70% = *desired effective coverage* of 3/8 in. *Note:* Review Table 2.2 on page 36.
5. *Sprinkler diameter* 126 ft/0.70 = 180 ft (± 6 ft)

Procedure

Sample pages from sprinkler equipment catalogs that contain the automatic (pop-up) models that meet the specified diameter of 180 ft ± 6 ft can be found in Appendix 5. Their performance characteristics are summarized in the following table for ease of comparison.

PRODUCT TECHNICAL DATA

MODEL: 10092

Nozzles: 16-12 (V slot)
Trajectory angle: 24 degrees

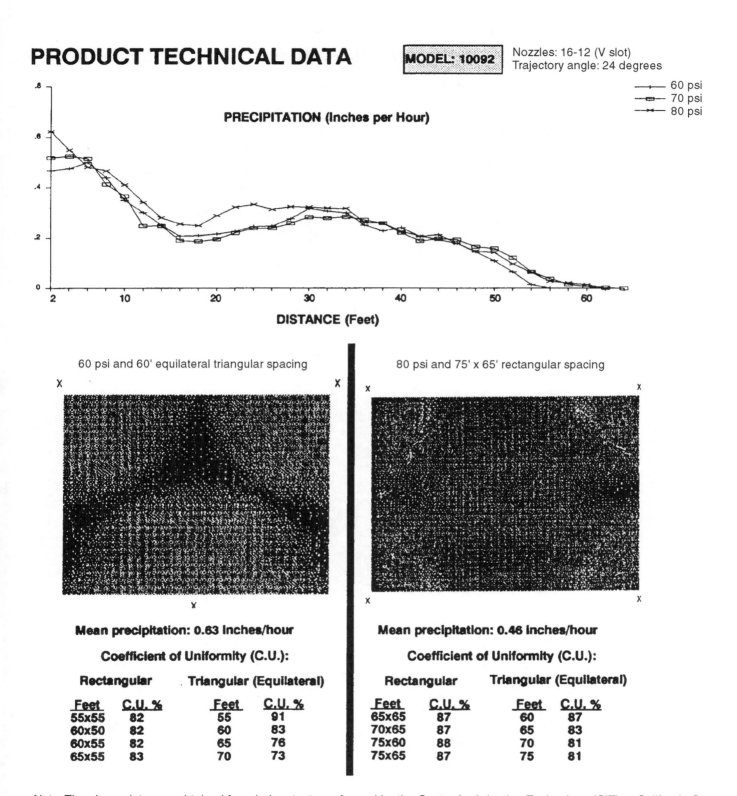

PRECIPITATION (inches per Hour)

- 60 psi
- 70 psi
- 80 psi

DISTANCE (Feet)

60 psi and 60' equilateral triangular spacing

80 psi and 75' x 65' rectangular spacing

Mean precipitation: 0.63 inches/hour

Coefficient of Uniformity (C.U.):

Rectangular		Triangular (Equilateral)	
Feet	**C.U. %**	**Feet**	**C.U. %**
55x55	82	55	91
60x50	82	60	83
60x55	82	65	76
65x55	83	70	73

Mean precipitation: 0.46 inches/hour

Coefficient of Uniformity (C.U.):

Rectangular		Triangular (Equilateral)	
Feet	**C.U. %**	**Feet**	**C.U. %**
65x65	87	60	87
70x65	87	65	83
75x60	88	70	81
75x65	87	75	81

Note: The above data was obtained from indoor tests performed by the Center for Irrigation Technology (CIT) at California State University, Fresno. No considerations have been made for varying soil and environmental conditions. The above "densograms" portray the intensity of water precipitation within the sprinklers (noted as "X"). Sprinkler radii may be slightly different from other published information due to normal manufacturing variations.

Figure 2.15

STUDENT PROBLEM

Extra spaces are provided for additional models to be listed from the given pages above and/or other sources.

Sprinkler Model Performance Table

Sprinkler Make and Model	Diameter (ft)	gpm	Base Pressure (psi)	Nozzle Size
R.B. 91 DR & 95DR	178	55	80, also others	26 × 15
Buckner 10110	178	48	80, also others	code-24-26
Weather-matic K-90	184	46	80, also others	3/8 × 7/32
Toro 670	192	57	70, also others	set-91-7/16

Solving for Sprinkler Spacing (Overlapping)—Single-Row System

Relating how far a sprinkler will throw water under ideal conditions is of little value. The worth of a sprinkler is judged by how far it can be spaced and still do an efficient job under normal conditions.

It is obvious that the uniformity of water distribution over an area is impossible with sprinklers spaced as shown below.

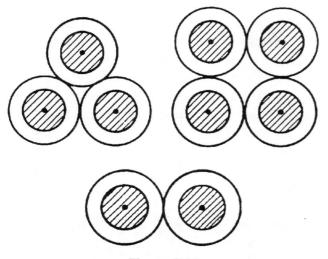

Figure 2.16

The distribution curve of a typical sprinkler shows the greatest amount of water is applied near the head location and diminishes to zero at the limit of throw.

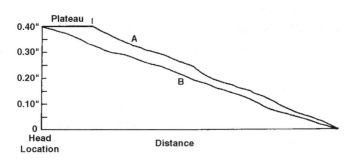

Figure 2.17. Typical distribution curves

Usually there is somewhat of a plateau (curve A) before the descent begins, and generally the longer the plateau the further apart heads may be spaced.

Curve B illustrates the distribution profile where the descent starts at the sprinkler head location.

Resulting Distribution with Overlap

The water applied by sprinkler No. 2 to Area 1 is approximately equal to Area 2. Thus, the double application of water to Area 1 brings the overall distribution to the turf area to a nearly uniform rate, as desired (shown by Line B).

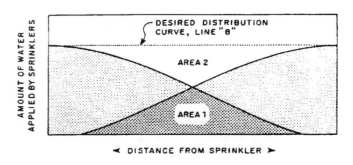

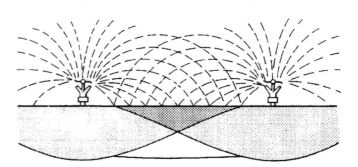

Figure 2.18

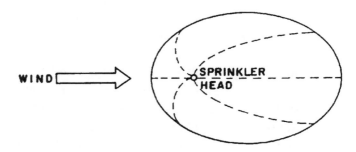

Figure 2.19. Wind vs. spacing.

The precipitation rate will build when operating against the wind and decrease when it is running downwind. Instead of having nice circles, an egg-shaped pattern results.

Maximum Spacing vs. Wind Speed

Wind Speed	Single-Row
No wind	60–65%
Up to 4 mph	55–60%
Up to 8 mph	45–50%
Above 8 mph	25–30%

Important: Good procedure dictates getting and following spacing recommendations from the manufacturer, as each sprinkler model may vary.

Wind Spacing

For wind speeds above 8 mph, the recommended spacing of 30% by the manufacturing company should be carefully considered. For example, using 200-ft-diameter sprinklers would mean 60-ft spacing ($0.3 \times 200 = 60$ ft) between sprinklers.

Operating the system without wind would cause an overlap of 40 ft from each side, giving triple coverage, with overwatering as shown here.

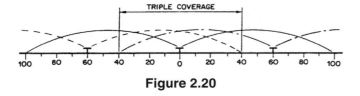

Figure 2.20

SUMMARY

The recommended spacing for a single-row system is *head-to-head* (50% spacing) or *plateau-to-plateau* (60% spacing), which allows for winds of 4 to 8 mph. *Note:* Overlapping is not a cure-all for uniformity of water distribution.

When a single row of sprinklers is used with proper *spacing,* it is a reasonable and correct assumption that there will be more water applied down the center of the row, where *overlap* is the greatest, than in the area away from the center. (See Figure 2.21.)

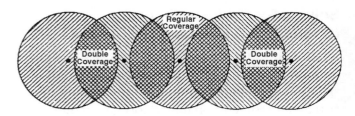

Figure 2.21. Sprinkler, constant speed.

The *head* is gear-driven, having the drive assembly arranged so that opposing 60° *arcs* operate at half speed. This allows the sprinkler to water twice as long in these two arcs in the areas of non-overlap in the pattern. This feature enables the same amount of water to be applied throughout the wetted area. However, in the single coverage zone a wedge-shaped pattern will occur. Running at half speed, the uniformity of application will vary from zero at the end of the throw and ascend to double that of the overlapped area close to the sprinkler head.

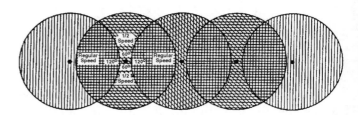

Figure 2.22. Sprinkler, 2-speed.

Solving for the Total Number of Sprinklers (Sets) to Cover All Fairways

The total number of sets to cover all fairways can be found by using a compass and/or a circle template (set to scale) to represent the spray diameter of the sprinkler previously selected.

The decisions rest with the designer to properly space and to cover the fairways accordingly. For the first draft *do not* change the circle size regardless of the variations of fairway width or shape. This will show where future corrective changes, if any, should be made.

Before starting, review the following procedure and examples. Upon completing the coverage of all the fairways it is a matter of counting the total.

Procedure

Starting at the greens, draw in the first sprinkler(s). *Do not* allow the circle to touch the putting surface. Proceed down the fairway by properly spacing and centering the fairway within the circle. In other words, maintain an equal distance on either side of the circle, as shown.

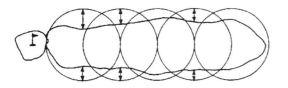

Figure 2.23.

Another procedure for finding the number of sets to cover a fairway (single-row) is by using the formula:

$$\text{number of sets (for 1 fairway)} = \frac{\text{length of fairway (ft)}}{\text{sprinkler effective diameter (ft)}}$$

Examples

Problem

Find the number of sets required for a fairway 150 ft wide and 900 ft long. The effective sprinkler diameter is 150 ft.

Solution

Number of sets = 900/500 = 6 sets on this fairway.

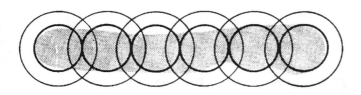

Figure 2.24

Problem

Same as above, except fairway length = 1000 ft.

Solution

Fractions less than 1/2 are dropped. The sprinklers would be spaced as follows:

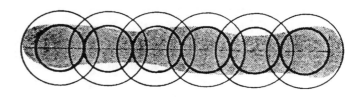

Figure 2.25

Problem

Same as above, except fairway length = 1100 ft.

Solution

Fractions less than 1/2 are added. The sprinklers would be spaced as follows:

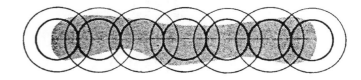

Figure 2.26

Odd-shaped fairways will require a combination of patterns to effectively cover the area to be irrigated, as illustrated in Figure 2.27.

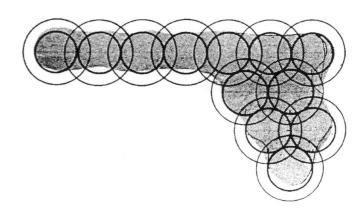

Figure 2.27

Location of First Fairway Sprinkler from Green

In spacing sprinklers on the fairway, it is advisable to start at the green, locating the first sprinkler out from the front edge of the putting surface approximately 10 ft further than the radius of throw of the sprinkler, and spacing back to the tee. By placing the first sprinkler out from the green in this manner, water from the fairway sprinkler is assured not to fall on the green, which could result in overwatering of the front of the green.

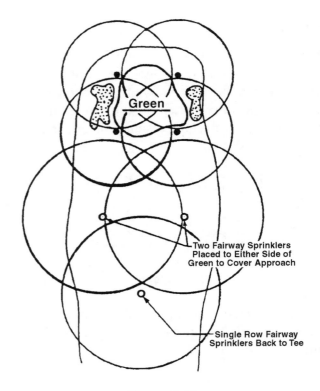

Figure 2.29

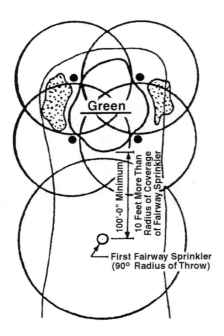

Figure 2.28

With the growing use of golf carts and their travel around the approach areas to the greens, it is becoming more of a trend to provide better coverage on the approach area. This can be effectively accomplished by placing two fairway sprinklers at the green—one to either side of the center of the green—and then continue on down the fairway with a single row. (See Figure 2.29.)

Area in Front of Tee Not Watered

In some areas, it is common to omit watering on the first 75 yards or so of the fairway, off the tee, and

thus rather than water the complete fairway from green to tee, fairway sprinklers are stopped at the end of the mowed area, in front of the tee. By so doing, we therefore reduce the total number of sprinkler heads and control valves (if an automatic system) and also may eliminate a certain amount of piping, trenching, wire, controller stations, etc.

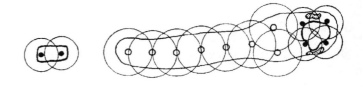

Figure 2.30

Coverage of Wide Fairway Areas

In the design of some golf courses, the mowed fairway area is increased in width at the landing area off the tee. It may be that it is desired to cover this area and thus would require going to a double row at this point, or placing additional smaller sprinklers to either side of the single row for proper coverage.

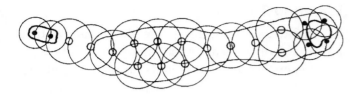

Figure 2.31

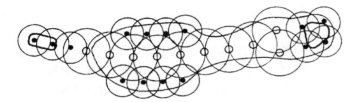

Figure 2.32

Note: If smaller sprinkler heads are used, as shown above, it is recommended that the designer operate either 2, 3, or 4 heads at the same time. In any case, their operating pressure (psi) and total (gpm) should be equivalent to one large fairway head.

Solving for Rate of Water Application (Inches Per Hour)

The water application rate for a single sprinkler *cannot be calculated* since the uniformity of water distribution ranges from a given amount at the sprinkler to nothing at the end of the throw.

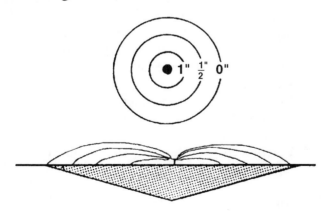

Figure 2.33. Single sprinkler: typical water distribution pattern.

The survival of a grass plant is directly related to the actual amount of water available to it—not on some calculated average figure.

In the single-row system a high degree of uniformity is obtained in the overlapped area. The water application rate for this area can be calculated by the following method:

$$\text{in./hr} = \frac{96.3 \times \text{gpm (1 sprinkler)}}{\text{spacing (\% diameter)} \times 80\% \text{ diameter}}$$

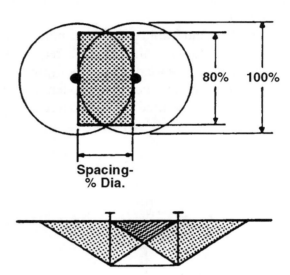

Figure 2.34. Single-row system: overlapping sprinklers.

Sample Problem

Calculate the water application rate.

Given: sprinkler discharge rate = 50 gpm
sprinkler diameter = 180 ft
sprinkler spacing = 50%

$$\text{rate of water application} = \frac{96.3 \times 50}{90 \times 144} = 0.37 \text{ in./hr}$$

Solving for the Number of Sprinklers to Operate at One Time

The price tag for an irrigation system designed to operate all of the fairway sprinklers at the same time is prohibitive, to say the least.

Example

Assume the following conditions are given:

total number of fairway sprinklers = 90
total gpm/sprinkler = 75

Simple math shows 6750 gpm would be needed to operate all the sprinklers at the same time. The economics of installing a pumping station and huge pipe sizes to handle such a volume of water could not be justified.

Figure 2.35 and Table 2.5 tend to indicate that as the water duration per night and the frequency of watering increase, the total gallons per minute decreases. In turn, the number of sprinklers required to operate at the same time likewise decreases.

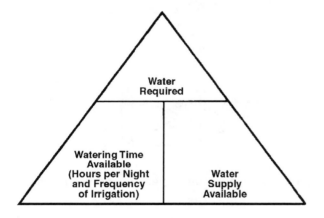

Figure 2.35

Usually the water requirement (inches of water per week required to maintain good growth in the worst seasonal average) is determined, and variation occurs in time and supply.

With the water requirement static:

Water supply *must increase* if watering time per night is decreased or watering frequency is reduced.

Watering time per night or frequency of irrigation *must increase* if water supply is reduced.

If either the watering time (per night) or frequency of irrigation or water supply are decreased, the water applied cannot match the water required. (See Table 2.5.)

Watering Time (Fairways)

In order to avoid interference with play and to take advantage of the least-wind conditions, the irrigation system is usually operated during the nighttime.

The allowable time for irrigating *fairways* is generally limited to about *6 to 8 hours* per night. The watering time for greens and tees is then added.

SOLVING FOR NUMBER OF SPRINKLERS TO BE RUN AT THE SAME TIME AND TOTAL GALLONS PER MINUTE

The following method can be used where the water supply is not fixed by any preexisting factors or conditions.

Water Usage Method (Evapotranspiration)

The water usage rate (in./wk) given in the example problem is based on the evapotranspiration rate for central Massachusetts. To solve a similar problem, it is essential to use the water usage rate for your local area.

Procedure

Fairways (single-row system):

1. Determine the total number of sprinklers needed to cover all fairways (by count).
2. Determine the water application rate (in./hr) for the sprinkler model and spacing pattern selected.
3. Determine the gpm/sprinkler.
4. Determine the water usage rate (in./wk) for the *local* area.
5. Determine the maximum watering time (hr/night).

Note: In order to avoid interference with play and to take advantage of the least-wind conditions, the irrigation system is usually operated during the nighttime.

The allowable time for irrigating *fairways* is generally limited to about *6 to 8 hours* per night. The watering time for greens and tees is then added.

Table 2.5. Required Supply Capacity (gpm) for 20-Ac Increment Area

Watering Time Available	Inches of Water Per Week Required				
	1.00	1.25	1.50	1.75	2.00
Once a week watering					
8 hours per night	1132	1414	1697	1980	2263
10 hours per night	905	1131	1358	1584	1810
12 hours per night	754	943	1132	1320	1509
Twice a week watering					
8 hours per night	565	707	849	990	1131
10 hours per night	453	566	679	792	905
12 hours per night	377	471	566	660	754
Three times a week					
8 hours per night	377	471	566	660	754
10 hours per night	302	377	453	528	603
12 hours per night	252	314	377	440	503
Every other day					
8 hours per night	323	404	485	566	647
10 hours per night	259	323	388	453	517
12 hours per night	216	269	323	377	431
Every day watering					
8 hours per night	162	202	242	283	323
10 hours per night	129	162	194	226	259
12 hours per night	108	135	162	189	216

Example

Solve for the total number of sprinklers to be run at the same time and the total gpm needed for an 18-hole golf course.

Design Determinations

1. Total number of sprinklers = 90 heads
2. Water application rate = 0.35 in./hr
3. gpm per sprinkler = 45 gpm
4. water usage rate =1.0 in./week (assume central Massachusetts). See Table 11.2 on page 260.
5. Watering time per night = 6 hr

Solution

First Step

$$min / night = \frac{\text{water usage per wk}}{\text{application rate}}$$

$$= \frac{\text{hr / wk} \times 60}{7}$$

$$\frac{1.0}{0.35} = \frac{2.86 \times 60}{7}$$

= 25 min / night, or 50 min every other night

Solve for sprinkler operating time, in minutes per night.

Second Step

Construct a worksheet using the headings (columns) as shown in Table 2.6. Increase the number of sprinklers to operate at one time until the operating time (hr/night) is reduced to the maximum allowed or below.

If the fairway watering time was deemed too long, since the time for watering greens and tees must be added, then operating more sprinklers would reduce the watering time but increase the gpm.

Table 2.6

Total No. Sprinklers	No. to Operate at One Time	No. Moves to Irrigate All Fairways	Total Operating Time (min/night)	Operating Time (hr/night)	gpm
90	1	90	90 × 25 = 2250	38.50	45
90	2	45	45 × 25 = 1125	19.25	90
90	3	30	30 × 25 = 750	12.50	135
90	4	23	23 × 25 = 575	9.60	180
90	5	18	18 × 25 = 450	7.50	225
90	6	15	15 × 25 = 375	6.25	270[a]
90	7	13	13 × 25 = 325	5.40	315
90	8	11	11 × 25 = 275	4.6	360
90	9	10	10 × 25 = 250	4.2	405

[a] Although the maximum irrigation time of 6 hr/night is exceeded by 15 min, this may be acceptable. If not, consider other alternatives, such as, select a sprinkler with a higher demand to increase the water application rate and in turn reduce the irrigation time.

Review

Operating 6 sprinklers × 15 moves can complete the irrigation cycle in 6 hr and 15 min using 270 gpm.

ACKNOWLEDGMENTS

1. "Design Information for Large Turf Systems." Manual and Handouts. Toro Manufacturing Company.

2. "Irrigation Equipment." Catalogs and Handouts. Buckner Sprinkler Manufacturing Company.

3. "Sprinkler Irrigation Equipment." Catalogs and Handouts. Rainbird Sprinkler Manufacturing Company.

4. "Sprinkler Irrigation Equipment." Catalogs and Handouts. Weather-Matic Sprinkler Manufacturing Company.

Chapter 3

Greens and Tees

GOLF COURSE GREENS

General

For an ideal round of par 72, about 36 shots are made on the green and about 18 shots to the green. The putting green figures into about three-quarters of the total strokes. It is little wonder why golfers demand excellent turf and playing surfaces on these tremendously important areas.

The superintendent should have a good understanding of green construction, which is basic to good irrigation practices.

Green Construction

Some major objectives of green construction are:

- Good drainage. (1) To allow play to continue uninterrupted from puddles of water or from soggy greens. (2) To provide good aeration that will encourage deep roots and will aid in preventing diseases.
- Resistance to compaction from heavy traffic of players. (1) To allow rapid infiltration and percolation of water. (2) To help prevent a reduction of non-capillary pore space.
- Maintain proper resiliency. To be firm enough to minimize pitting from golf balls, but not overly firm to cause them to bounce off.
- Rarely will sites selected for greens possess the natural soil conditions needed to meet these objectives. Consequently, it is com-

mon practice to build greens to specifications.

Specifications*

The specifications for a Method of Putting Green Construction as reported by the USGA Green Section Staff in the USGA *Journal of Turf Management,* September, 1960 still remain basically the same. (See Figure 3.1.)

PLANNING IRRIGATION FOR GREENS

General Comments and Suggestions

The appearance and condition of the greens are often used as a measuring stick for judging the success or failure of a golf course superintendent. Since the superintendent's bread and butter may depend on this factor, he must develop a love for each blade of grass and give it his most tender care.

Water plays a key role in growing and maintaining a healthy, vigorous plant. Therefore, special emphasis should be placed on the watering system for greens.

The size and shape of the green will govern the size of the sprinklers, their location, spacing, and number.

Points to consider:

- Provide a quick-coupling valve at each green for hand watering and syringing of automatic

---

* Complete information and details for putting green construction may be obtained by writing to: Golf House, P.O. Box 708, Far Hills, NJ 07931-0708.

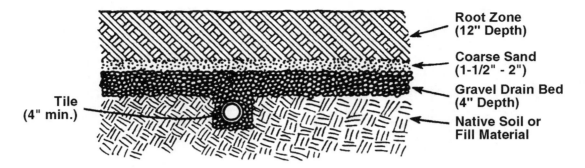

**Root Zone
(12" Depth)**

**Coarse Sand
(1-1/2" - 2")**

**Gravel Drain Bed
(4" Depth)**

**Native Soil or
Fill Material**

**Tile
(4" min.)**

Figure 3.1. Profile of a properly constructed green.

systems. Be sure it is installed before the remote control shut-off valve.

Consider the installation of a small-sized pump or pressure tank to accommodate the quick-coupling valves. This permits the use of one outlet that is independent of the large-sized fairway pumping system. In other words, it will not be necessary to start the large pump just for one outlet.

• The nearest fairway sprinkler should be placed at such a distance that it does not add water onto the green.

• Individually controlled sprinklers (valve at each sprinkler) are preferable to one control valve for all sprinklers.

When one valve control is used, the water beyond the valve will drain from the sprinkler at the lowest elevation. Consider drains or corrective measures to avoid water accumulation.

• Generally, the greens require a smaller-sized sprinkler than the fairway. A model having approximately the same pressure requirements should be selected. The gpm requirement of the sprinkler will determine the number of sprinklers that will operate at one time.

For example: The fairway was designed for 7 sprinklers at 50 gpm, or a total of 350 gpm. The green sprinklers' discharge rate can be selected on the basis of the number of greens to be done at the same time. For 4 greens (3 sprinklers per green) 350/12 = 29 gpm or, for 5 greens, 350/15 = 23 gpm per sprinkler, etc.

• Where the discrepancy of operating pressures between the green and fairway sprinklers is too great, consider operating the greens from

their own pump and separate controllers. This means irrigating the greens and fairways independently of one another and at separate times. However, utilize the fairway pipe system to convey the water to the green.

• Spacing sprinklers—order of preference:
— Square spacing:

Has the best water distribution uniformity (using plateau-to-plateau spacing), offers the advantage of squaring off corners, especially suitable for oblong-shaped greens.

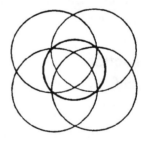

Figure 3.2

— Triangular spacing:

Good water distribution uniformity. Can be spaced further apart than a square pattern. Suitable for round-shaped greens.

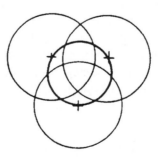

Figure 3.3

— Single sprinkler (permanent):

Extra large sprinkler diameter is required to uniformly cover the green with water. On established greens, it is necessary to tear up a section for pipe installation. Most affected by wind.

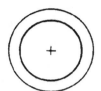

Figure 3.4

— Single sprinkler (portable):

Allows for flexibility in adjusting to changing wind direction and to give special attention to troublesome spots on the green. Inconvenient to operate.

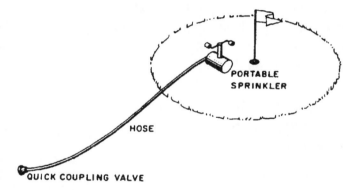

Figure 3.5

Full- and Part-Circle Sprinklers Around a Green

Usually full-circle sprinklers are used on the greens around the perimeter to:

- Give good coverage of approach and collar.
- Make these areas easier to maintain—especially important for areas where golf carts normally travel around the greens.
- Make greens look much larger—with area around green well watered.

Usually part-circle sprinklers are only used for trouble or special requirement areas: to avoid overthrow on buildings, walks, or streets, when they lie close to a green, or, very rarely, when bunkers are not properly drained so that water in them would be a problem.

Be sure to select proper nozzle sizes for part-circle sprinklers in order to give approximately the same application rate as full-circle heads. If not, the part-circle sprinklers should be put on a separate circuit from that of the full-circle heads, in order to control proper application from these heads.

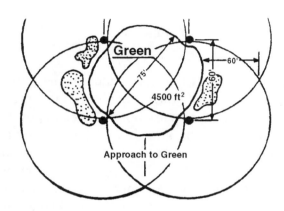

Figure 3.6

Example

4,500 ft² green (38 ft radius) = 76 ft average distance across green (shown in Figure 3.6):

75% of 76 ft = 57 ft sprinkler radius desired.
41 rotor 7/32 in. × 11/64 in. at 60 psi—16.7 gpm
119 diameter = 60 ft radius
4 sprinklers spaced 60 ft apart (50% spacing)

Oddly Shaped Greens

Use the actual shape of the green to advantage in order to get proper coverage for oddly shaped greens. This can save in the number of sprinkler heads required and give better coverage of the green. (See Figure 3.7.)

Water Distribution Uniformity

At the University of Massachusetts, the late Professor Dickinson, who started the first known turfgrass program for golf course superintendents, just

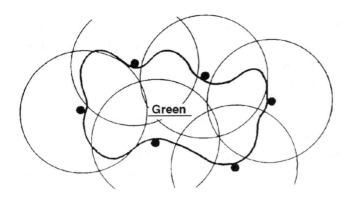

Figure 3.7

A throw of 45 ft beyond the edge is necessary to effectively water the approach to the green.

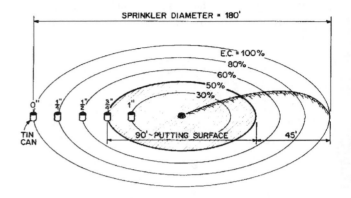

Figure 3.8

couldn't overstress the importance of the putting green to the game of golf. After all, it plays a role in 3/4 of shots, i.e., for a par-4 hole, it takes 1 approach shot, 1 shot onto the green and 2 putts. Growing and maintaining excellent quality turfgrass requires good irrigation practices. Now, if an irrigation system were designed as a number in the past were, using the "center sod-cup method" (sprinkler located in the center of the green) the sprinkler selected must provide uniform water distribution over the entire green.

The same rationale as outlined for selecting the effective coverage for fairways applies to greens using the center sod-cup method. As explained, a compromise must be made between the two extremes. The suggested first choice for greens is to use a sprinkler effective coverage of 50%.

Example

Sprinkler selection for the "center sod-cup method" (not recommended)

> average green width = 90 ft
> sprinkler effective coverage (EC) = 50%
> sprinkler diameter = 90 ft/0.50 = 180 ft

As shown Table 2.1 on page 35, for an EC of 50% the uniformity of application would decrease from 100% near the sprinkler to 75% at the edge of the putting surface (a variation of only 25% throughout). This has proven to be adequate for maintaining good turfgrass.

Solving for Green Sprinkler Selection

The selection of the sprinkler model(s) to adequately water a green is based on the size and shape, method used, the pattern and spacing. Each green must be considered individually.

Center Sod-Cup Method
(Review page 54)
Procedure

Step 1. Solve for average green width

Step 2. Effective coverage = Use 50%

Step 3. Sprinkler diameter = $\dfrac{\text{average green width}}{\text{effective coverage}}$

Example

Assume average green width = 80 ft

      i.e., sprinkler diameter = 160 ft – 80/0.5

Perimeter Method (Two-Row)
General Information

There are two basic patterns used in turf work: equal leg triangular and square spacing.

The best possible spacing is a square pattern with an overlapping of plateaus. Where heads are spaced further apart, the greater the variation in distribution. Spacing the heads closer together (or overlapping) causes a worse distribution profile.

Below are illustrations of *in-pattern coverage of sprinkler heads* (square pattern 60 ft apart) using 10 ft bands of radius with the average *precipitation rate* increased in each succeeding band from the outside to the center. The *sprinkler's* profile is shown to the left.

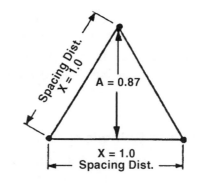

Figure 3.10

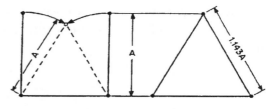

Figure 3.11

To form an equilateral triangle having the same distance A as compared to the square pattern, the spacing must be increased accordingly, as shown in Figure 3.11.

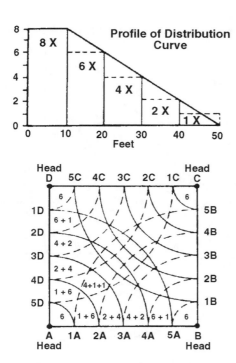

50-ft Radius, Head Spaced 60 ft Apart

Figure 3.9. Square spacing.

Equilateral Triangle

Triangular Spacing

The relationship of the height (line A) of an equilateral triangle to the length of the sides (line X) is 0.87 to 1.0. (See Figure 3.10.)

On this basis, the distance of A will be 87% of the length of X. Therefore, the spacing or distance between sprinklers on the line can be easily calculated. If distance A is known, length X = A × 0.87. (See Figure 3.11.)

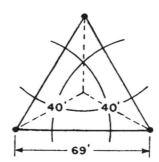

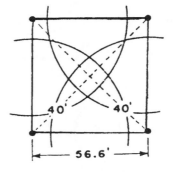

Figure 3.12

A geometric analysis of the comparative sizes of a square and a triangle which have a distance of 40 ft from their exact center to the corner location show why the square spacing must be approximately 18% closer.

Green Sprinkler Selection

The suggested procedure for selecting green sprinklers is by following steps 1 through 5 as outlined and explained. Unusually large or oddly shaped greens must be considered individually. It may be necessary to use either larger-diameter sprinkler heads, more heads, and/or locate the heads inside the putting surface.

Procedure

Step 1. Solve for Average Green Width

Note: Take the average of all the green widths that have relatively the same diameters (only). As stated above, exceptionally large and oddly shaped greens must be designed accordingly.

Step 2. Select Sprinkler Pattern—Square and/or Triangular Spacing

The equilateral triangle is well-suited for roundish but not for oblong-shaped greens, as shown below, whereas the square pattern could be used for either case.

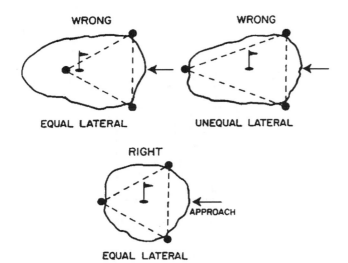

Figure 3.13

Step 3. Solve for Sprinkler Diameter
  a. Square Spacing

Procedure

The sprinkler *radius* should cover 75% to 80% of the average green width.

Example

Sprinkler Diameter = 0.80×80 ft $\times 2 = 128$ ft

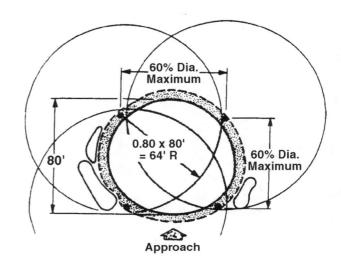

Figure 3.14

b. Triangular Spacing

Example

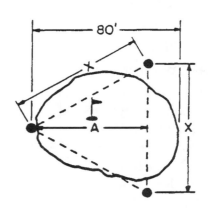

Figure 3.15

Procedure

Requires 3 steps

1. Solve for length of line A, same as above – A = 80 ft × 0.8 = 64 ft
2. Solve for sprinkler spacing X.
   X = 64/0.87 = 73 ft
3. Solve for sprinkler diameter
   See page 107, spacing for winds up to 8 mph = 55%
   Sprinkler diameter = 73/0.55 = 133 ft

Note: For all practical purposes, a difference of 5 feet between the sprinkler diameters for the square vs. triangular spacing can be considered insignificant. Actually, the numbers could be rounded off to 130 ft or so and the sprinkler model selected would be adequate for either pattern.

Step 4. Solve for Green Sprinkler Performance Factors

First Approach:

If possible, match the green sprinklers to the fairway sprinkler previously selected.

Performance factors:

1. Gallons Per Minute—Use about 1/4 (square spacing) or 1/3 (triangular spacing) of the fairway sprinkler.
2. Pressure—Use the same operating pressure (psi) as the fairway head.
3. Diameter—Use the method where the sprinkler radius = 75% to 80% of the average green diameter.

Note: Operating the 4 or 3 green sprinklers at the same time would be equivalent to operating one fairway sprinkler.

Second Approach: If possible match the number of greens that will be irrigated at the same time to utilize the total gallons per minute (gpm) and fairway design factors.

Example

1. Fairway Sprinkler Selection

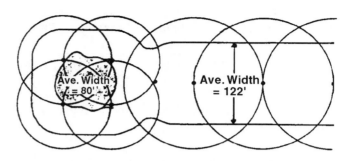

Figure 3.16

Procedure

$$\text{sprinkler diameter} = \frac{\text{average width}}{\text{EC}}$$

$$= \frac{122 \text{ ft}}{0.70} = 174 \text{ ft}$$

Specifications for the sprinkler model selected are shown in the table below.

Name and Model	Total Diameter (ft)	gpm	Precipitation (in./hr)	Base Pressure (psi)	Nozzle Size (in.)
10110	173	45	0.35	70	3/8 × 7/32

Following is a summary of fairway sprinkler design factors.

Operating time—min/night	=	25
Total number of sprinklers	=	90
No. heads on at same time	=	6
No. of moves per night	=	15
gpm per sprinkler	=	45
Total gpm (6 × 45)	=	270
Operating time—hr/night	=	6.25 (tentative)

2. Green Sprinkler Model Selection—Square Spacing

Procedure

sprinkler radius = 80 ft × 0.80 = 64 ft
sprinkler diameter = 64 ft ÷ 0.50 = 128

Specifications for the sprinkler model selected are shown in the following table.

Name and Model	Total Diameter (ft)	gpm	Precipitation (in./hr)	Base Pressure (psi)	Nozzle Size (in.)
10092	127	22.9	0.54	70	1/4 × 13/64

i.e., 270/22.9 = 11.8 or 12 heads and/or 3 greens on at the same time

Step 5. Solve for sprinkler spacing—See page 44.

Square pattern = 50% spacing for winds up to 8 mph

Step 6. Solve for *total* number of sprinklers needed to irrigate all the greens. Draw circles to scale on the map of the golf course, then count.

Suggestions for placement of Green 8 sprinklers:

1. Cut out templates to scale and fit to green. Avoid placement of sprinklers in the green or too far from the green. Also, avoid bunkers.

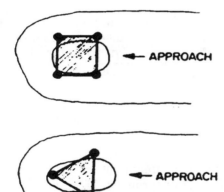

Figure 3.17

Note: Place two heads to the approach side of the green.

Step 7. Solve for water application rate.

$$\text{in./hr} = \frac{\text{gpm (one head only)} \times 96.3}{S \times L}$$

where: S = spacing of sprinklers on line
L = distance (ft) between lines

For example:

$$\text{in./hr} = \frac{23 \times 96.3}{64 \times 64} = \frac{2215}{4096} = 0.54 \text{ in.}$$

Step 8. Solve for operating time—min/night. *Note:* Use the same design factors and procedure as for fairways.

For example:

$$\text{min / night} = \frac{\text{water usage (in./wk)}}{\text{appl. rate (in./hr)}} = \frac{\text{hr / wk}}{7} \times 60$$

$$= \frac{1.0 \text{ in.}}{0.54 \text{ in.}} = \frac{1.85}{7} \times 60$$

$$= 16 \text{ min / night or 32 min}$$

$$\text{every other night}$$

Step 9. Solve for operating time (hr) to complete the irrigation cycle for all greens.

For example:

$$\text{moves} = \frac{\text{total no. greens}}{\text{greens on at same time}} = \frac{18 \text{ (assumed)}}{3} = 6$$

$$= \frac{6 \times 16 \text{ min}}{60} = \frac{96}{60} = 1.6 \text{ hr / night}$$

Total irrigation time (hr/night):

fairways (6 heads at 270 gpm) = 6.25 hr/night
greens (12 heads at 270 gpm) = <u>1.6 hr/night</u>
7.85 hr/night

Step 10. Sprinkler Selection—Adjustments and Compromises.

Important: At this stage of the irrigation system design process all options remain open for changes regarding sprinkler model selection, including fairway heads, number to operate at the same time, total gpm, etc.

The final decisions should be on hold until a careful review and analysis of each scenario is completed.

For example:

To reduce the operating time (hr/night), consider the following:

1. Operate 8 fairway heads—increase the total to 360 gpm.
2. Operate 360 gpm/22.9 gpm = 16 sprinklers and/or 4 greens.

For example operating time (hr/night):

$$\text{fairways (8 heads at 360 gpm)} = 4.6 \text{ hr}$$

$$\frac{18 \text{ greens}}{4 \text{ greens}} = \frac{5 \text{ moves} \times 17 \text{ min}}{60 \text{ min}} = \underline{1.4 \text{ hr}}$$

$$\text{Total} = 6.0 \text{ hr}$$

Sprinkler Spacing Patterns

There are two basic types of sprinkler spacing patterns. Their shape and dimensions will depend upon the diameters of coverage (D) by the individual sprinklers and the wind velocities apparent at the time of sprinkling. Listed below is a description of each one of the basic patterns and one modified pattern.

Square Pattern

Below is a sample pattern where the dimensions S and L are equal to each other (S = L). The square pattern is used when the sprinkled area is itself a square or nearly square, and requires sprinklers to be placed in 90° corners and along each boundary. The distance across the diagonal of the pattern is approximately 1.4 times the distance S or L. Because of this the maximum value of S and L must be reduced, to avoid weak spots in the center of the pattern. The following chart will detail recommended square pattern spacing ranges for each wind velocity range.

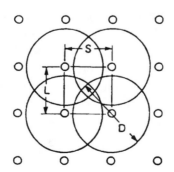

Figure 3.18

Average Square Spacing Ranges

For a Wind Velocity of	Use a Maximum Spacing(s) of
0 to 3 mph	55% of diameter D
4 to 7 mph	50% of diameter D
8 to 12 mph	45% of diameter D

Triangular Pattern

Figure 3.19 depicts a sample *equilateral triangle* pattern. The three heads which comprise the pattern are all equal distances from each other. The height of the pattern L is 0.86 times the distance(s). The triangular pattern is generally used in areas which have irregular boundaries or boundaries which do not require part-circle sprinklers along its line. Since the triangular pattern allows each row of sprinklers to be offset with respect to the position of adjacent sprinklers, it is possible to space the sprinklers at greater distances than possible when using a square pattern. The effect of spacing sprinklers at great distance from one another is to be able to cover the same or larger areas with fewer heads. The following chart lists the recommended triangular spacing ranges.

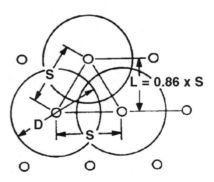

Figure 3.19

Average Equilateral Triangle Spacing Ranges

For a Wind Velocity of	Use a Maximum Spacing(s) of
0 to 3 mph	60% of diameter D
4 to 7 mph	55% of diameter D
8 to 12 mph	50% of diameter D

Precipitation Rates

General

The precipitation rate for a given pattern of similar sprinklers is an indication of approximately how much water is being applied in the area of the pattern. The average precipitation rate (PR) is expressed in the units in./hr. The equation or formula used to calculate the precipitation rate is:

$$PR = \frac{96.3 \times \text{gpm (applied by sprinklers to area)}}{S \times L \text{ (spacing of sprinklers, in ft)}}$$

where: 96.3 = constant (in./ft²/hr)
 gpm = gallons per minute applied to area by sprinklers
 S = spacing between sprinklers
 L = spacing between rows

The constant 96.3 is derived as follows:

$$1 \text{ gal} = 231 \text{ in.}^3 \text{ and } 1 \text{ ft}^2 = 144 \text{ in.}^2$$

$$\frac{231 \text{ in.}^3/\text{gal}}{144 \text{ in.}^2/\text{ft}^2} = 1.604 \text{ in./ft}^2 / \min$$

$$1.604 \times 60 \min = 96.3 \text{ in./ft}^2/\text{hr}$$

The accuracy or validity of the PR calculation depends greatly on the proportional distribution by each sprinkler to the area of the pattern. The ideal situation is where each sprinkler contributes exactly the same proportional quantity (gpm) to the pattern, that is:

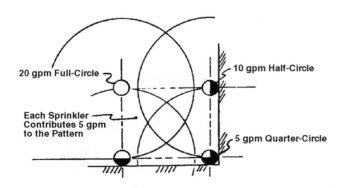

Figure 3.20

If there is a large difference in proportional discharges, the calculated (PR) will be high because some of the sprinklers will be overwatering while the others will be supplying only adequate amounts of water to the pattern.

If each sprinkler is providing approximately the proper proportion, the average PR may be approximated as:

$$PR = \frac{96.3 \times \text{(equivalent full-circle sprinkler in gpm)}}{S \times L \text{ (spacing of sprinklers, in ft)}}$$

Tees

General practice for watering tees is to place full-circle sprinklers on the tees for proper coverage. Commonly, the same sprinkler that is used around the greens is used to cover the tees.

On small tees, the sprinkler may be larger than required to just cover the tee, and it will give good coverage to the area around the tee. This reduces the number of different model sprinklers that are used on the system—which is an advantage from a maintenance standpoint.

Tees can be each on a separate station of a controller, or several can be connected to the same station of a controller, as desired.

Tee Sprinkler Selection

First choice is to select the same green sprinkler model as for greens if reasonably adequate coverage can be attained. Otherwise, follow the same selection procedure as outlined for greens. (See Figure 3.21.)

On a very small tee of 300 ft² or less, one sprinkler may be placed to the side of the tee. Reasonable coverage can be expected provided that the longest dimension of the tee fits within 60% of the sprinkler's throw diameter.

The sprinkler(s) should be placed on the side of the prevailing wind.

For wide tees it will require full- or part-circle heads on each side of the tee. (See Figure 3.22.)

**Square, Short Tee Covered
From Each End**

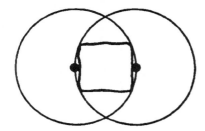

**Larger Tee Requires
Several Sprinklers**

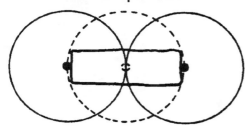

Figure 3.21. Typical tee designs

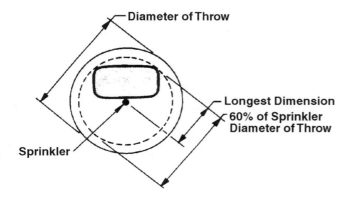

Figure 3.22. Single sprinkler covering a small tee.

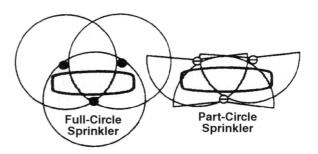

Figure 3.23. Multiple sprinklers on large tee.

ACKNOWLEDGMENTS

1. "The Specifications for a Method of Putting Green Construction." USGA Green Section.
2. "General Design Information." Booklet. Rainbird Sprinkler Manufacturing Company.

Chapter 4

Pipe Layout

FAIRWAY PIPE LAYOUT

Example

On the "problem" golf course shown, plan the pipe layout for a single-row system having the pipelines placed in the center of each fairway.

The water supply will be taken from the pond, which is at an elevation of 40 ft.

Suggested Procedure

First, draw in all essential pipelines—in this case, down the center of each fairway from the first to the last head only.

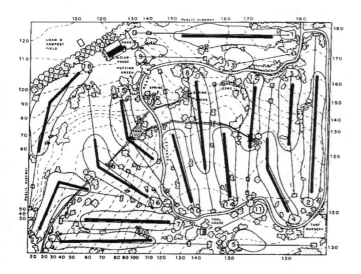

Figure 4.1

Next, consider the possible locations of the main pipeline from the water source that will tie in all the lateral lines. One possibility is presented as shown: the *dead end system.*

A study of the contour lines reveals that in this case the main pipeline, as each of the laterals, runs *uphill.* As had been stated previously, the pressure will drop 1 psi for each 2.31 ft rise in elevation. *Note:* The elevation difference from the main pipeline, at Fairway 12 up to Green 1 is about 90 ft (175 ft – 85 ft) or 39 psi.

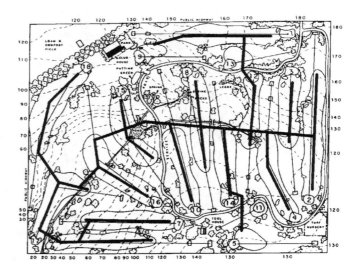

Figure 4.2

Or, consider placing the location of the main pipeline as shown.

Notice: For the most part, the lateral lines now run downhill. Various means are available to assist in controlling pressure gains in downhill laterals.

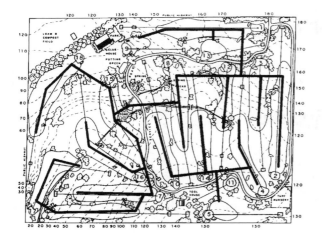

Figure 4.3

Next, consider completing loops in a *loop system,* as shown.

Obviously, the piping layout is considerably overdesigned. A number of pipe sections can be eliminated without impairing the integrity of the system.

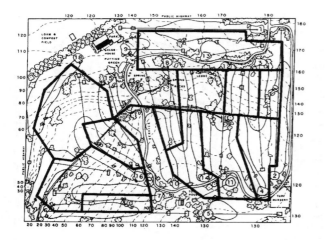

Figure 4.4

Next, modify the previous piping as shown here. By all means, try other possibilities that may come to mind.

Important: It should be understood that all of the pipe layouts considered would adequately do the job. Therefore, each layout should be carefully analyzed and compared. The one that offers the most benefits should be selected.

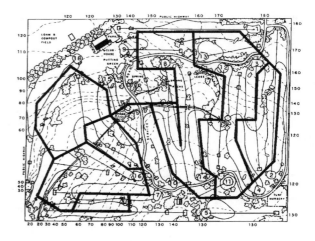

Figure 4.5

GREEN PIPE LAYOUT

General Design Information

Greens are the most critical areas of a golf course. They require the most exacting control of all maintenance procedures. Special attention and care must be given to the irrigation installation and operating practices as well.

Points to Consider

The composite illustration below contains a number of factors to consider related to green irrigation system design.

- You must have an accurate drawing or equivalent showing the size and shape of the putting surface to locate sprinklers effectively.
- Selection of the proper sprinkler and spacing and location of these heads to give complete coverage and uniform distribution is most important.
- A minimum of 4 sprinklers are required for a green to get good coverage, unless the green is extremely small—3000 ft² or less.
- Select a sprinkler with a radius of throw of approximately 75% to 80% of the average distance across the green.
- Maximum spacing of sprinklers around greens should be 60% of the diameter of coverage of the sprinkler. If windy conditions are present, spacing should be reduced to near 50% of the diameter.

- The feeder line should run to the side or back of the green. Avoid locating control valves in the approach area.
- Shown are two automatic controllers: (a) a valve-in-line *block system* designed to operate two or more heads per valve; (b) a *valve-in-head system*. Each head can be controlled individually and/or grouped together.
- The intersecting points of adjoining sprinkler patterns should be well off the putting surface (from 20 ft to 30 ft minimum). Otherwise, the wind distortion could cause dry spots.

The following illustrations show and discuss various methods of piping and controlling the green sprinkler heads.

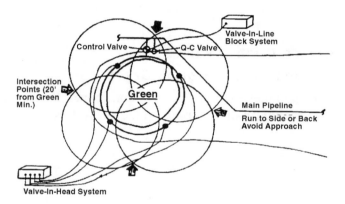

Figure 4.6

Center Sod-Cup

This method is the least expensive to install. However, it does contain some weaknesses, such as poor water uniformity, effect of wind, and having to dig up a green in case of pipe or sprinkler failure.

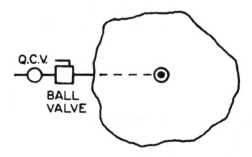

Figure 4.7

Valve-in-Head

An individual control valve for each sprinkler head at a green can be installed, which gives maximum flexibility in water control of the green.

However, to design or install this way requires a large number of additional controller stations, a larger number of control valves, a much greater amount of wire, etc. Each valve would have to be on a separate station of a controller to be effective.

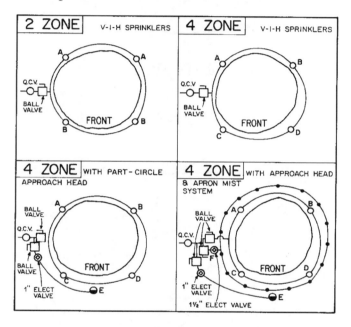

Figure 4.8

Two Circuits Per Green

In using one control valve for all sprinklers on a green there are two common methods of piping. One is to divide the supply line in each direction around the green to pick up sprinkler heads to each side of the green. (See Figure 4.9.)

Control Valve at Each Sprinkler

In areas where there is a prevailing wind during the watering period, it is desirable to control the green sprinklers on two separate circuits—one circuit on the windward side and the other on the leeward side. Each control valve is connected to a separate station of the controller so that maximum control of the watering of the green can be accomplished. (See Figure 4.10.)

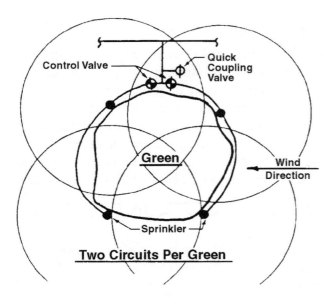

Figure 4.9

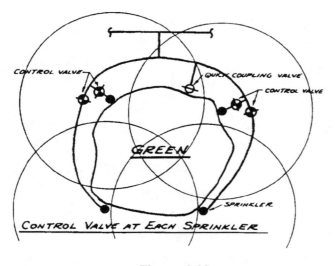

Figure 4.10

Collar Sprinklers Around Green

On greens that are built up and have rather high banks on the collars—and because of golf cart traffic around greens—we see a trend toward special consideration of the collars and area around the green.

This can be handled by placing a second row of sprinklers completely around the green, except for the approach area. (See Figure 4.11.)

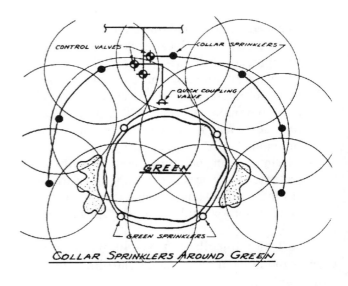

Figure 4.11

In-Line Control Valves
Divided Circuit Around Green

The other method, which is very common, is to circle the green with pipe to pick up the sprinklers.

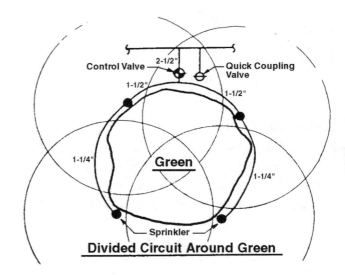

Figure 4.12

Looped Circuit Around Green

This method helps to equalize the operating pressure to all the sprinkler heads around the green. (See Figure 4.13.)

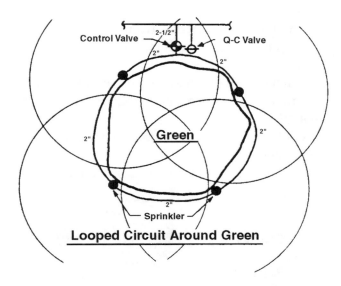

Figure 4.13

Full- and Part-Circle Sprinklers Around a Green

Be sure to select proper nozzle sizes for part-circle sprinklers to give approximately the same application rate as full-circle heads. *If not,* the part-circle sprinklers should be put on a separate circuit from that of the full-circle heads in order to control proper application from these heads.

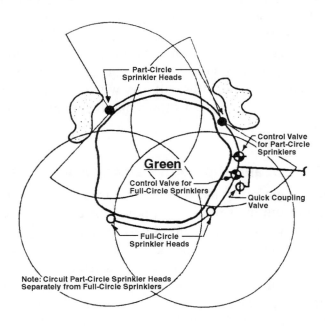

Figure 4.14

PIPING TEE INSTALLATIONS

Figure 4.15 shows various piping tee installations, including ball valves and quick-coupling valves.

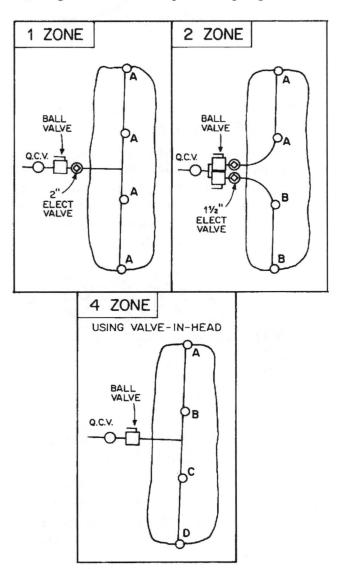

Figure 4.15. Piping tee installations.

Installation of Stop-a-Matic® Valve

In the design or installation of the sprinklers and piping around the green, be aware of the need for Stop-a-matic® valves under any heads at a lower elevation than the others, which may cause a problem of draining the piping each time the control valve is closed. (See Figure 4.16.)

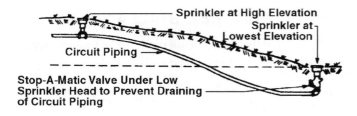

Figure 4.16

Auxiliary Quick-Coupling Valves on Automatic Fairway Systems

It may be desirable, on automatic fairway systems, to install three or four quick-coupling valves down a fairway for use in watering trees, shrubs or rough areas. The addition of these valves can be of great help in the maintenance of the course and add little to the overall cost of the system.

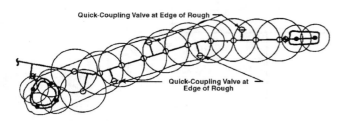

Figure 4.17

SOLVING FOR MAIN AND LATERAL PIPELINE LAYOUT

After the sprinkler selection, and the patterns and placement for the fairways, greens, and tees are finalized, the next step is to plan a network of pipelines that allow water to flow from the supply to every sprinkler.

Since no two systems are exactly alike, each system must be treated as a unique case and acted upon accordingly. The following procedure and examples illustrate the type of reasoning that may be used to derive a final piping layout.

Procedure: Single-Row System

1. Draw the essential pipelines connecting the sprinklers along the center of the fairways (only).

2. Draw the main pipeline from the water source (pumping station) to connect all the fairway pipelines.
3. From the main or fairway pipelines, extend laterals to the greens, tees, and other areas to be irrigated.
4. Avoid unnecessary trenching and the use of large pipelines.
5. Avoid trenching through greens and tees.

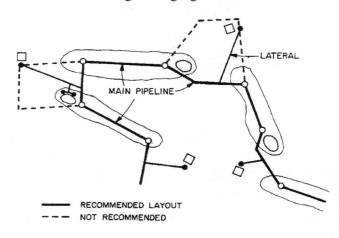

Figure 4.18

Piping Layout for Greens and Tees

As stated at the offset, there are no standard rules when it comes to planning a piping layout for a golf course. In general, after the fairway layout has been finalized, the next step is to extend laterals from the main lines to the greens and tees. The final decision for accomplishing this will vary with the individual designer. Figure 4.19 shows one of the possibilities.

Control and Piping of Fairway Sprinklers

On single-row fairway systems, it is most economical to run the supply main down the center of the fairway and place a control valve in the swing joint riser piping to each sprinkler head. Control valves can be connected individually to a separate station of the controller—giving maximum flexibility—or several can be connected to one station of a controller as desired. If several valves are connected to the same stations, caution must be taken *not* to tie a head in a very low, normally wet area together with a head on a high, normally dry area. (See Figures 4.20 through 4.25.)

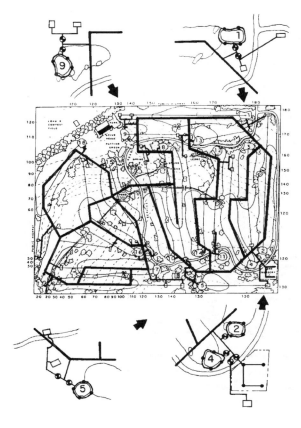

Figure 4.19

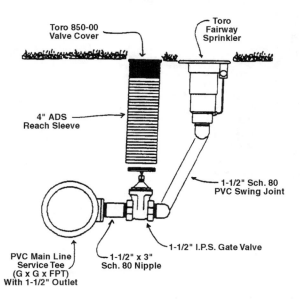

Figure 4.20. Fairway sprinkler detail (on main line). Not to scale.

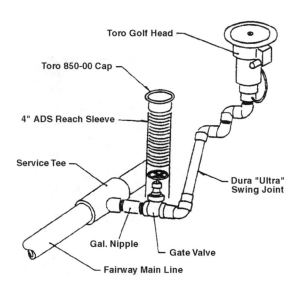

Figure 4.21. Typical dura-swing joint.

HOW TO DETERMINE THE REQUIRED THRUST-BLOCK AREA

Thrust blocks are made of concrete, usually placed between the fitting and the trench wall. It is important to place the concrete so that it extends to an *undisturbed* (freshly cut) trench wall—soft stuff will not hold. (See Figure 2.26.)

The concrete mix used should be of adequate strength and will often be specified by the engineer. A typical mix consists of one part cement, two and one-half parts sand, and five parts gravel. (Washed gravel should be used.)

First let's assume that the block is to hold against the horizontal thrust at a 90° bend, for 10-in. Class 150 pipe. Let's further assume that the pipeline will be pressure tested at 200 lb.

The soil, for this example, is sand.

1. Check Table 4.1. Find 10-in. Class 150 pipe size—moving to the right (to the 90° bend column) we find 15,200 lb for each 100 psi of water pressure. Since we are working on a total test pressure of 200 lb, double 15,200, making the total 30,400 lb.
2. Determine the bearing capacity of sand. Table 4.2 provides bearing load data that is often used. If the bearing capacity of sand is 2,000 lb/ft^2, the total force (30,400 lb) divided by

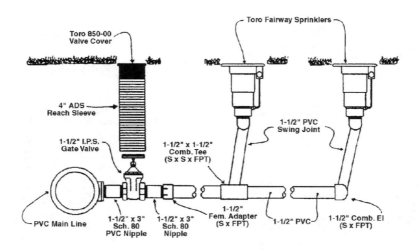

Figure 4.22. Two-row fairway sprinkler. Not to scale.

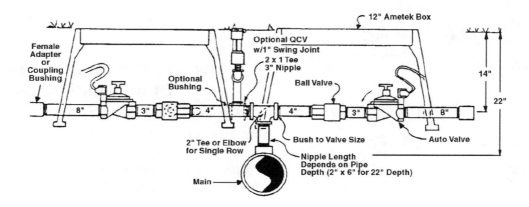

Figure 4.23. Two-row fairway assembly with quick-coupling valve.

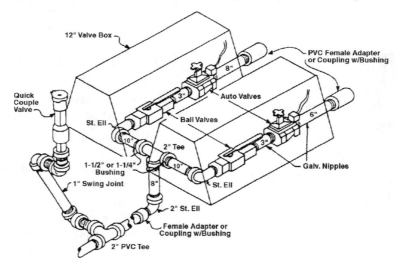

Figure 4.24. Typical double-valve green assembly.

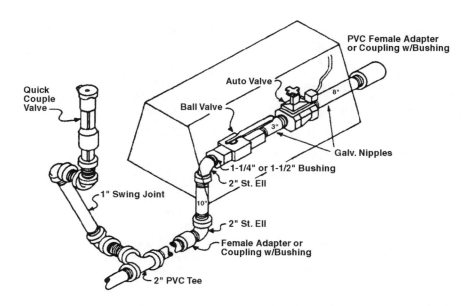

Figure 4.25. Typical tee assembly.

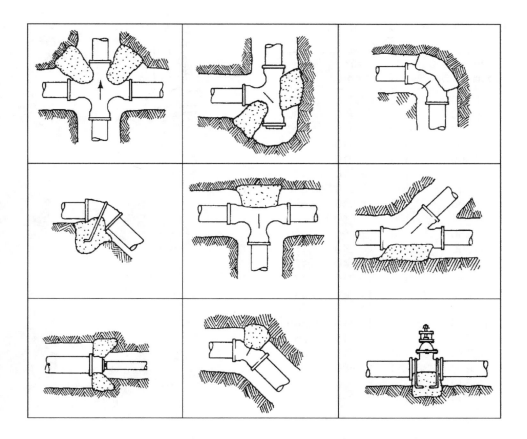

Figure 4.27. Typical thrust-block details for ring-tite pipe.

2,000 gives the ft^2 of area needed (15.2). A bearing area of 4 ft × 4 ft (16 ft^2) is indicated.

Other Thrust and Anchorage Problems

Thrusts in Soft, Unstable Soils

In soft, unstable soils, thrusts are resisted by running corrosion-resistant tie-rods to solid foundations or by removing the soft material and replacing it with ballast of sufficient size and weight to resist thrusts.

Upward Thrusts at Fittings

Where a fitting is used to make a vertical bend, anchor the fitting to a concrete thrust-block designed to key in to undisturbed soil and to have enough weight to resist upward and outward thrust. Since the newly placed backfill may not have sufficient holding power. (See Tables 4.1 and 4.2 and Figure 4.27.)

Table 4.1. Thrust at Fittings in Pounds at 100 psi of Water Pressure.

Transite Pipe Size and Class	Tees and Dead Ends	90° Bend	45° Bend	22.5° Bend
4"—100	1,720	2,440	1,320	660
150	1,850	2,610	1,420	720
200	1,850	2,610	1,420	720
6"—100	3,800	5,370	2,910	1,470
150	3,800	5,370	2,910	1,470
200	3,800	5,370	2,910	1,470
8"—100	6,580	9,300	5,040	2,550
150	6,580	9,300	5,040	2,550
200	6,580	9,300	5,040	2,550
10"—100	9,380	13,270	7,190	3,640
150	10,750	15,200	8,240	4,170
200	10,750	15,200	8,240	4,170
12"—100	13,330	18,860	10,240	5,170
150	15,310	21,640	11,720	5,940
200	15,310	21,640	11,720	5,940
14"—100	17,930	25,360	13,740	6,960
150	20,770	29,360	15,910	8,060
200	20,770	29,360	15,910	8,060
16"—100	23,210	32,820	17,880	9,000
150	26,880	38,010	20,590	10,430
200	26,880	38,010	20,590	10,430

For Class 100 pipe—add 50 for test—total 150. Table shows 13,270 lb for Class 100 pipe at 100 lb pressure—so increasing 13,270 half as much again to obtain thrust at 150 lb, the answer would be 19,905 lb total.

Table 4.2. Safe Bearing Loads

* Johns-Manville assumes no responsibility for the following bearing load data, which is often used. The engineer is responsible for determining safe bearing loads and when a doubt exists, soil bearing tests may be specified. The bearing loads given are for horizontal thrusts when depth of cover exceeds 2 ft.

Soil	Safe Bearing Load (lb/ft^2)
Muck,[a] peat, etc.	0
Soft clay	1,000
Sand	2,000
Sand and gravel	3,000
Sand and gravel cemented with clay	4,000
Hard shale	10,000

[a] In muck or peat, all thrusts are resisted by piles or tie rods to solid foundations or by removal of muck or peat and replacement with ballast of sufficient stability to resist thrusts.

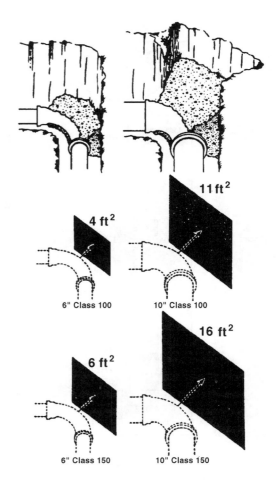

Figure 2.27. A comparison of thrust-block areas.

Backfilling and Tamping

Backfilling usually follows pipe installation as closely as possible. This protects pipe from falling boulders, eliminates the possibility of lifting the pipe due to flooding of open trench, avoids shifting pipe out of line by cave-ins and, in cold weather, lessens the possibility of backfill material becoming frozen.

Keep in mind that the purpose of backfilling is not only to protect the pipe by covering it, but to provide firm, continuous support that will prevent the pipe from settling or resting on the couplings. The essentials of a first-class backfilling job can be summarized as follows:

- Provide a continuous bed by carefully consolidating approved material under pipe and couplings and between the run of pipe and the trench walls.
- Provide a cushion on top by hand-placing approved material to at least 12 in. over the pipe. The balance can then be backfilled by machine.

Obtaining Proper Backfill Material

The initial backfill material used should be free of large stones, dry lumps and, in cold weather, free of frozen lumps. Dry or frozen lumps will slump down when they become wet or thaw out, leaving the pipe with insufficient support or causing dangerous settling. Generally, a material that is slightly damp will pack more solidly under the pipe.

Placing and Tamping Under the Pipe

A good tamping job can be done more easily than a poor job. The key to it is in the amount of soil that is thrown in to be tamped under.

This initial backfill is always placed by hand. It should be shoveled in evenly along both sides of the pipe, making a layer about 4 in. thick—but not more. Then the tamping bar is used to tamp this soil firmly under the pipe. If more than 4 in. of soil is shoveled in before tamping, the soil can bridge and fail to go under the pipe (Figure 4.28a). This may look airtight from the surface, but it is easy to see that, when it occurs, support under the pipe is not provided.

Figure 4.28a. Wrong! Too much soil. Tamp bar compacts it improperly—and leaves a void under the pipe.

Next, another 4-in. layer is shoveled in and tamped. This is repeated until the pipe is firmly bedded in compact soil. (See Figure 4.28b.)

Figure 4.28b. Right—4 in. layers of soil—which tamp bar can force under pipe—thus ensuring a firm bed.

Using the Flat Tamper

The next step is to provide lateral support by tamping soil firmly between trench-walls and the pipe. This too is done in 4 in. layers, but a flat tamper is used. (The tamper used under the pipe will not work properly at the sides.)

Hand placing and flat tamping are continued to a point halfway up the pipe wall. (See Figure 4.28c.) When this point is reached, the tamping can be stopped.

Figure 4.28c. When pipe firmly bedded—use fist tamper at sides. Continue fist tamping, in 4 in. layers halfway up pipe.

Completing the Initial Backfill

Backfilling the approved material is continued, by hand, until there is a protective cushion of at least 12 in. of cover over the pipe. This completes what is called the *initial* backfill—the thoroughly tamped soil which provides a continuous, supporting bed for the pipeline (as shown in Figures 4.29a and b).

Figure 4.29a

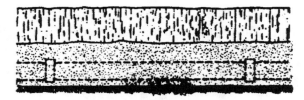

Figure 4.29b

Even in fairly stormy areas, suitable material for initial backfill can often be shaved from the sides of the trench.

Where a clay soil condition is encountered, the pipe should be enveloped in a minimum of 4 in. of sand, then the initial backfill completed to at least 12 in. above the pipe with selected material.

In cold weather, unfrozen material can often be found below the frozen crust of the excavated earth. If the exposed material becomes completely frozen, then dry sand or other unfrozen material should be hauled in for fill.

In bogs and marshes, the excavated material is usually little more than vegetable matter, and should not be used for bedding purposes. In such cases gravel or crushed stone should be hauled in for fill.

Hand Tamping

About Tamping Bars

Two types of tamping bars are required for a complete job. First a bar with a narrow head, or blade, is used to tamp under the pipe and couplings. Then a bar with a flat head is used to compact soil at the sides.

For packing soil under pipe and couplings, tamping bars should not be improvised from two-by-fours, shovel blades, pick handles and the like. A good tamping bar should be long enough to handle easily, heavy enough to do the work but not wear out the worker, and the right size and shape to pack

soil properly. The tamping bars illustrated will do a good job, can be handled easily, and are quickly made up in the shop. (See Figure 4.30.)

Flat tampers are standard equipment. Either hand or pneumatic tampers may be used.

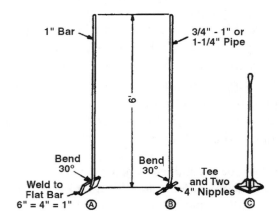

Figure 4.30. Tamping bars of the type shown at A and B are used for tamping under pipe and couplings. The fist tamper C is used at the sides, between pipe and wall of the trench. (Do not attempt to use the fist tamper in place of A or B—it will not do the work properly.)

MAINTENANCE AND REPAIRS

Particularly for a pressurized golf course irrigation system, it may be necessary to drain the entire system whenever it becomes necessary to repair or replace a sprinkler, valve, or broken pipe. Needless to say, this becomes a time-consuming and costly practice to completely drain and repressurize the system. However, installing a shut-off valve (ball valve) before each sprinkler as well as at strategic locations that would isolate sections, such as one or two fairways, would be recommended. (See Figures 4.31 and 4.32.)

Irrigation System Drainage

In the "hard-freeze" areas where the pipelines are installed above the frost line, the complete removal of all the water from the *entire* system is a *must* in order to avoid costly damage.

Since the mid-sixties when the use of a high-volume air compressor was demonstrated and found capable of "blowing" water out of the system, there

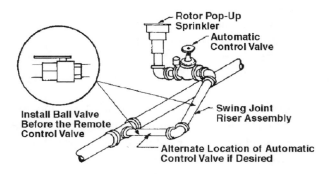

Figure 4.31. Sprinkler isolation.

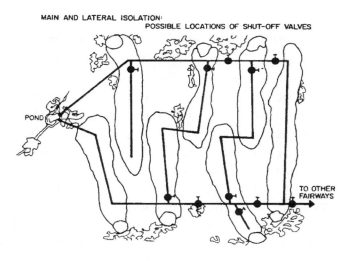

Figure 4.32. Main and lateral isolation: possible locations of shut-off valves.

has been a gradual change from laying the pipelines to grade to permit drainage by gravity flow to installing the pipelines 2–2.5 ft below the surface following the contour of the land.

The use of the air compressor for blowing out the system *does not* eliminate the need for drain valves. On the contrary, drain valves are recommended, especially on lines that can be conveniently drained to a lower grade such as a pond, stream, culvert, or a sump. It is considered good practice to remove as much water as possible from the system by gravity flow drain outlets and to use air pressure to blow out remote control valves along with the pipelines which do not readily lend themselves to draining by drain valves.

Blowing Out the System

Because the system design, pipe layouts, and installations may differ on golf courses, there are no standard rules for removing water by air pressure. However, the following checklist is offered for consideration.

The key to success for blowing out the system is the capacity of the air compressor and the maximum line pressure. A compressor having a *capacity of 250–300 cfm* and a maximum *line pressure of 60 psi* is recommended for pipe sizes between 4–8 inches.

If pockets of water are present behind pockets of air in the line, the air escapes very rapidly through the swing joint and out the sprinkler nozzle. At line pressures above 60 psi, the velocity of air, due to its low density and viscosity, can increase by as much as 5-fold at 100 psi. When this is followed by a slug of water, the velocity immediately drops back, causing an excessive pressure surge (water hammer effect) capable of stripping the threaded nipple of a schedule 80 PVC pipe connected to the sprinkler, causing the swing joint to fail.

Before rushing out to rent an air compressor and starting to blow out the system, it would be time well spent in thoroughly studying the design plan of your irrigation system (as installed). Then, prepare a list of steps giving the procedure to be followed.

To assist in preparing the list of steps, it has been suggested that you picture yourself as entering the pipeline at the source and having the task of pushing the water out to the top of the ground as you move through the pipelines. Careful attention must be given to the route that must be followed to push the water out of each dead end. Note especially branch tees that feed two fairways and feed lines that split and go in two directions as loops. Remember, the same routes you used will be followed by a wall of air when you blow out the system.

Steps

1. Shut off the mainline water supply.
2. Open up all the drain valves. Be sure that the lines being drained are vented.
3. To remove water from pipelines that do not have drains, a sprinkler can be inserted or opened at the

high and low points. This will allow air to replace water as it flows out of the lower head.

On lines where the water is prevented from flowing back through the remote control valve because the diaphragm spring is closed, use the air compressor to push the water out of the valve along the flow line and out of the sprinkler head.

4. Once the water is *completely* drained out of the pipelines, close all of the drain valves.

5. Connect the portable air compressor through a quick-coupler valve, or equivalent, at the pumphouse near the supply source. The system as a whole can be blown out, or any section of the system that can be isolated can be blown out separately.

6. Be sure to allow the air compressor to build pressure in order to release a large volume of air to push out the water. Operate the compressor continuously—evacuation procedure is complete to prevent air pockets in the lines. Do not exceed a line pressure of 50–60 psi.

7. Manually run the controller through all stations so that every remote control valve is actuated.

8. Repeat the above procedure until air and no more water is coming out of the sprinkler heads.

9. Starting again at the pumphouse and working out to the end of each lateral, crack the drain valves slightly and blow out any water that may have collected. Wait a minute and repeat. Then, close the drain valves, which will prevent surface water from entering, as well as gravel from sumps. This practice eliminates the chore of closing valves before starting up the system again.

AIR ENTRAPMENT

Introduction

Air in pipelines can cause serious operating and testing difficulties: (1) air may reduce the carrying capacity of the line because of reduced cross-sectional area and the fluctuation of flow caused by air movement, (2) serious surge pressures may be generated from these flow fluctuations, and (3) release or venting of entrapped air can cause water hammer pressures sufficient to rupture the pipeline.

Sources of Air in Pipelines

The source of air in a pipeline may be attributed to one or more of the following sources:

1. Entrapment of air during filling, either initially or when refilled after drainage.
2. Entrapment of air at the point where water enters the line, i.e., the pump inlet or gravity inlet.
3. Release of dissolved air from the fluid due to changes in temperature and pressure.
4. The intake of air through air release valves or vacuum or siphon breakers. This happens when the pipeline pressure drops below atmospheric pressure, such as during a negative surge or when a portion of the line may be acting as a siphon.

Problems Associated with Air Entrapment

Air pockets would normally be found at the high points of the pipeline during low flow rates. As the discharge is increased these pockets are forced along the line by the moving water and may become lodged at high points, where they reduce the area available for flow. Then the working pressure at the pump must be increased to maintain the same discharge. Because of these problems associated with air in lines it is desirable to remove it. However, this is not always a simple operation. In fact, additional problems may be encountered when an air-water mixture is vented through an air release valve. When lodged, compressed air reaches a vent, the vent provides very little resistance to the passage of air and it is quickly released from the system. In turn, the *speed* of the water column that is behind the air pocket is greatly increased. When the water column reaches the restricted area of the small vent, a sudden change in velocity occurs and water hammer surges are generated.

Control of Air

The most critical period during the life of a pipeline is initial filling. The tendency is to entrap air at the high points, as illustrated.

1. Pipelines should be laid to a grade which results in a minimum number of high points, based on the terrain and economic considerations. Abrupt transitions and sharp peaks should be avoided.

2. Automatic air release valves with a bleed-off ports should be installed at all high point locations or where air pockets would be expected to accumulate. Provisions should be made for disposal of excess water discharging from the bleed-off port.

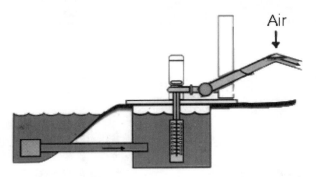

Figure 4.33

Air Relief Valve

Air relief valves are designed to exhaust air under various pressure conditions while restricting the flow of liquid.

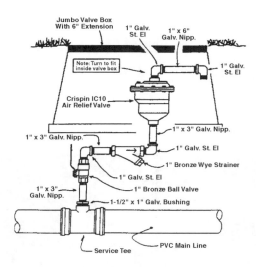

Figure 4.34. Air relief valve assembly. Not to scale.

Air Release and Vacuum Breaker Valve

Model 33A eliminates air and prevents vacuum formations in pipelines. During normal pipeline operation, air accumulation and buoyancy cause a float ball to lower or lift. As water level lowers inside the valve, small amounts of accumulated air are released through the small orifice. Once air is released, the float poppet system closes drip-tight.

Air release and vacuum breaker valves are typically installed at high points in the pipelines for air release, or at the anticipated pipeline vacuum occurrence locations.

Along uniform grade line piping, install Model 33A at regular intervals (approximately 1/2 mile).

Figure 4.35. Air release and vacuum breaker valve.

Ring-Tite PVC Typical Installation Procedure

The following figures show the typical installation procedure for PVC pipe.

Figure 4.36

Figure 4.37

Figure 4.40

Set ring in groove with marked edge facing toward end of bell.

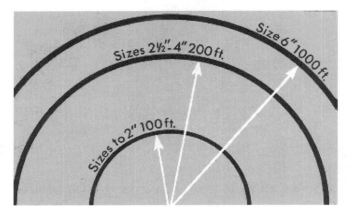

Figure 4.38

Good procedure dictates checking with the manufacturer for detailed instructions.

Figure 4.41

Lubricate pipe end. A light film of lubricant is enough.

Figure 4.39

Clean bell and end. Be sure no dirt or other foreign matter can lodge between the ring and the bell or pipe end.

Figure 4.42

Push pipe end into position so that the reference mark on the pipe end is either in line with or within the bell.

Figure 4.43

1. The bell is an integral part of the pipe—no solvent weld problems.

2. The rubber ring provides the seal—no solvent weld problems.

3. No expansion-contraction problems due to solvent weld joints. Elongation and shortening of lengths are compensated for at the joints. No need to "snake" the pipe in the trench.

4. All joint elements factory made—no field-mixing or application of cement.

5. No waiting for solvent cement to dry—units can be handled right away.

6. No waiting for solvent cement to dry—ready right away for test and full pressure.

7. The bell wall is thicker than the pipe wall—extra strength at the joint.

8. Higher levels of strength and durability proved by test.

No solvent weld problems.

1. Lubricate pipe end.

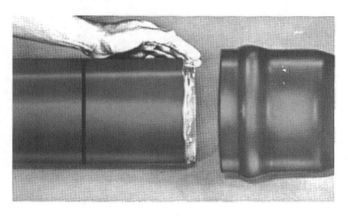

Figure 4.44

2. Push pipe in until reference mark is hidden within the bell.

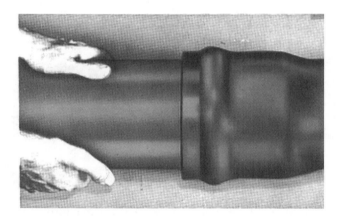

Figure 4.45

Make Connections to Transite Pipe and Other Pipe Materials

As illustrated in Figures 4.47 through 4.50, J-M Ring-Tite PVC Pipe is available with adaptors that provide for simple, trouble-free connections to Transite pipe, cast iron, steel and other pipe materials.

Figure 4.46

Simple to Make Joints with the Double-Bell Coupling

1. Push a double-bell coupling onto the spigot end of a length of pipe, as shown.

Figure 4.47

2. Then push the spigot end of the next length of pipe into the coupling bell and adjust positions of spigot ends and coupling so that the reference marks on each pipe are hidden under the coupling end. The double-bell coupling may also be used to join PVC to the same size of IPS steel pipe.

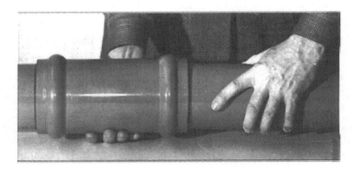

Figure 4.48

In the same way, the double-bell coupling may be used to install new pipe lengths into an existing line. After the replacement pipe has been cut to the proper length, mount a double-bell coupling on each end. Push each coupling all the way onto the pipe end and position the new length in the line. Then slide each coupling into place, bridging the ends and aligning the coupling over the reference marks.

Tap with Saddle Assemblies

J-M Ring-Tite PVC Pipe may be easily tapped wet or dry with a tapping tool, as shown, and saddle assembling. Use only tapping tools intended for tapping PVC plastic pipe.

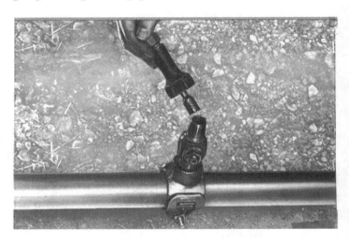

Figure 4.49

Typical Solvent Welding Procedure for Class I PVC Pipe

Caution: The procedure varies with the class of pipe and solvents. Obtain and follow manufacturer's instructions.

Note: Condition pipe and fittings to the same temperature.

1. Cut pipe square to desired length using hand saw and miter box or mechanical cut-off saw.

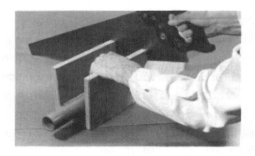

Figure 4.50

2. A tube cutter designed for plastic may be used as an alternate cutting method. (It is essential to remove raised bead left on outside of pipe.)

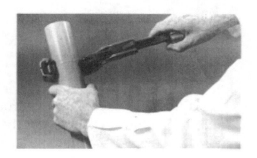

Figure 4.51

3. Chamfer end of pipe to approximately the dimensions illustrated below.

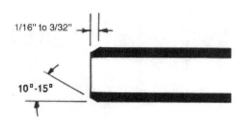

Figure 4.52

4. Clean and dry pipe and fitting socket of all dirt, moisture, and grease. Use a clean dry rag.

Figure 4.53

5. Check dry fit of pipe in fitting. Pipe should enter fitting socket about 1/3 to 3/4 depth.

Figure 4.54

6. First dissolve inside socket surface by brushing with #P70 primer. Be sure to use brush at least one half the size of pipe diameter. Use a scrubbing motion to assure penetration. Repeated applications may be necessary.

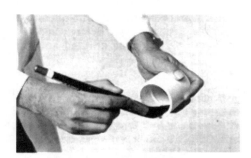

Figure 4.55

7. Next, dissolve surface of male end of pipe, to be inserted into socket, to depth of fitting socket by brushing on liberal coat of #P70 primer. Be sure entire surface is well dissolved.

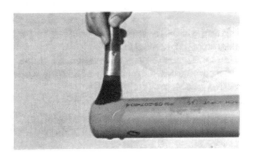

Figure 4.56

8. Again brush inside socket surface with #P70 primer. Then, without delay, apply #711 (#710 for pipe 2 in. and smaller) filler cement *liberally* to male end of pipe. The amount should be more than sufficient to fill any gap.

Figure 4.57

9. Also apply #711 (#710 for pipe 2 in. and smaller) filler cement lightly to inside of socket using straight outward strokes to keep excess filler solvent out of socket. This is to prevent solvent damage to pipe. Time is important at this stage. *Note:* The solvent cement should be applied deliberately but without delay, and it may be necessary for two men to work together when cementing larger pipe and fittings.

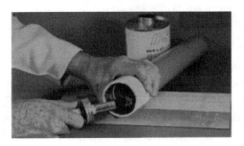

Figure 4.58

10. While both the inside socket surface and the outside surface of the male end of the pipe are *soft* and *wet* with solvent cement, forcefully bottom the male end of the pipe in the socket, giving the male end a one-quarter turn if possible. *The pipe must go to the bottom of the socket.*

Hold the joint together until both soft surfaces are firmly gripped (usually less than 30 seconds).

Figure 4.59

11. After assembly, wipe excess cement from the pipe at the end of the fitting socket. A properly made joint will normally show a bead around its entire perimeter. Any gaps at this point may indicate a defective assembly job, due to insufficient cement or the use of light-bodied cement on large diameters where heavy-bodied cement should have been used.

Figure 4.60

12. Do not disturb the joint until initial setup of the cement occurs. See Table 4.3.
13. Allow the joint to cure for an adequate time before pressure testing. See Table 4.4 for recommended times.

The following cure schedules are suggested as guides. They are based on laboratory test data, and

Table 4.3

Set Time—Handle the newly assembled joints carefully until the cement has gone through the set period. Recommended set time is related to temperature as follows:

30 min minimum at 60°F to 100°F
1 hr minimum at 40°F to 60°F
2 hr minimum at 20°F to 40°F
4 hr minimum at 0°F to 20°F

should not be taken to be the recommendations of all cement manufacturers. Individual manufacturer's recommendations for their particular cement should be followed. These cure schedules are based on laboratory test data obtained on *net fit joints* (net fit = in a dry fit the pipe bottom snugly in the fitting socket without meeting interference).

14. Short-cuts—*Do not take short-cuts.* Experience has shown that short-cuts from the instructions given above are the cause of most field failures. Don't take a chance.

ACKNOWLEDGMENTS

1. "Architect-Engineer Turf Sprinkler Manual." Rainbird Sprinkler Manufacturing Company.
2. "Design Information for Large Turf Irrigation Systems." Toro Sprinkler Manufacturing Company.
3. "Golf Course Irrigation System Design." Manual and Handouts. Buckner Sprinkler Manufacturing Company.
4. "Pipe Installation Guide." Booklet. Johns-Manville Pipe Manufacturing Company.
5. "Surge Control Valve." Catalog. CLA-VAL Automatic Control Valves.

Table 4.4

Temperature Range During Cure Period	Test Pressures for Pipe Sizes 0.5 in. to 1.25 in.		Test Pressures for Pipe Sizes 1.05 in. to 3 in.		Test Pressures for Pipe Sizes 3.5 in. to 8 in.	
	Up to 180 psi	Above 180 to 370 psi	Up to 180 psi	Above 180 to 315 psi	Up to 180 psi	Above 180 to 315 psi
60°–100°F	1 hr	6 hr	2 hr	12 hr	6 hr	24 hr
40°–60°F	2 hr	12 hr	4 hr	24 hr	12 hr	48 hr
10°–40°F	8 hr	48 hr	16 hr	96 hr	48 hr	8 days

Chapter 5

Pipe Size Selection

PRINCIPLES OF PIPE SELECTION

It is mandatory to have a good understanding of the following pipe materials and basic terminology in order to select the type of pipe and size of pipe and to use the friction loss tables.

Pipe Materials

Currently, two types of plastic pipes are used for practically all golf course irrigation systems. They are polyvinyl chloride (PVC) and polyethylene (PE).

PVC pipe is usually manufactured as rigid pipe in lengths of 20 ft. However, it is made as flexible pipe in long coils when plasticizers are added during the manufacturing process. PVC pipe has a high tensile strength and is resistant to nearly all types of deterioration.

PE pipe is a flexible pipe that is manufactured in long coils and has a lower tensile strength. It is commonly installed by pulling the pipe below the surface instead of digging a trench to lay the pipe.

PE pipe is more subject to "creep," which means that when pipes full of water freeze they will expand and stretch the pipe wall, which does not return to its original size. After a number of cycles of freezing and thawing the pipe wall becomes much thinner and weaker. PE pipe cannot be solvent welded and in most cases will require the use of insert fittings and clamps.

Basic Terminology
C-Value

C-value = coefficient of flow. This is based on the inside wall surface *roughness* or *smoothness*. The higher the C-value, the smoother or less rough the interior pipe wall.

The various types of pipes were tested and rated in terms of a numerical value using 15-year-old cast iron (CI) pipe as the "standard" (life of CI = 30 years) which was assigned a C-value = 100.

Note: The inside pipe wall will rust or tuberculate with time; therefore new cast iron pipe has a higher C-value (C = 110).

Pipe Class

Pipe class = pressure rating (psi). For example, Class 80, 100, 160, 200, etc., indicates the safe operating pressure of the pipe.

The pressure class provides for a safety factor of about 2:1 over the hydrostatic test. This may not be adequate for irrigation systems with long runs or quick-closing valves. It is often recommended to select a pipe with a rating somewhat higher than the maximum design working pressure.

Standard Dimension Ratio

All plastic pipe has pressure ratings, but pipe that is made with a standard dimension ratio (SDR-PR) has the same pressure rating for all diameters of pipe. This pipe is given a pressure class designation, which is generally listed in the friction loss table titles.

Pipe Schedule

By comparison, plastic pipe manufactured with schedule dimensions such as Schedule 40 or Schedule 80 has a different pressure rating for all diameters of pipe even though they are all made of the same material. The pressure ratings are not listed in the friction tables.

Table 5.1 shows a comparison of wall thickness for some SDR-PR class pipe and schedule pipe:

Table 5.2 shows a comparison of wall thickness and pressure rating of Pressure Class 200 vs. Schedule 40 pipe.

Notice the following

The wall thickness of SDR-PR pipe increases faster than Schedule 40 as the nominal pipe diameter increases.

The pressure rating of SDR-PR pipe does not change, whereas the pressure rating of schedule pipe does.

Schedule 40 maintains a higher pressure rating as compared to Class 200 up to a nominal pipe size of 4 in. or less.

Pressure Variations

Pressure directly influences the distance a sprinkler can throw water and how much water the sprinkler will use. The lower the pressure, the less distance (area) the sprinkler can cover. The greater the pres-

Table 5.1. Standard Pipe Dimensions. All Dimensions in Inches Rigid PVC Plastic Pipe.

Nominal Pipe Diameter (in.)	Outside Diameter	CL 160 SDR 26 i.d.	Wall	CL 200 SDR 21 i.d.	Wall	CL 315 SDR 13.5 i.d.	Wall	SCH 40 Plastic i.d.	Wall	SCH 80 Plastic i.d.	Wall
0.5	0.840				0.716	0.062	0.622	0.109	0.546	0.147	
0.75	1.050		0.930	0.060	0.894	0.078	0.824	0.113	0.742	0.154	
1	1.315	1.195	0.060	1.189	0.063	1.121	0.097	1.049	0.133	0.957	0.179
1.25	1.660	1.532	0.064	1.502	0.079	1.414	0.123	1.380	0.140	1.278	0.191
1.5	1.900	1.754	0.073	1.720	0.090	1.618	0.141	1.610	0.145	1.500	0.200
2	2.375	2.193	0.091	2.149	0.113	2.023	0.176	2.067	0.154	1.939	0.218
2.5	2.875	2.655	0.110	2.601	0.137	2.449	0.213	2.469	0.203	2.323	0.276
3	3.500	3.230	0.135	3.166	0.167	2.982	0.259	3.068	0.216	2.900	0.300
4	4.500	4.154	0.173	4.072	0.214	3.834	0.333	4.026	0.237	3.826	0.337
6	6.625	6.115	0.255	5.993	0.316	5.643	0.491	6.065	0.280	5.761	0.432
8	8.625	7.961	0.332	7.805	0.410						
10	10.750	9.924	0.413	9.728	0.511						
12	12.750	11.770	0.490	11.538	0.606						

Table 5.2. Wall Thickness and Pressure Rating for Pressure Class 200 vs. Schedule 40 Pipe.

Nominal Pipe Diameter (in.)	Pressure Class 200 Wall Thickness (in.)	Pressure Rating (psi)	Schedule 40 Wall Thickness (in.)	Pressure Increase (%)	Rating[a] (psi)
2	0.113	200	0.154	136	270
4	0.214	200	0.237	111	220
6	0.316	200	0.280	0.89	180

[a] Pressure rating based on wall thickness.

sure, the further the distance (area) the sprinkler can cover. When designing an irrigation system, the pressure needed or available is a prerequisite to selection of the sprinklers to be used.

Pressure can be created in two ways: by using a pump, or by using elevation. In a pump system, mechanical energy is used to create pressure. With elevation, 0.433 psi of pressure is created for each foot of elevation the water is stacked up. The pressure for an irrigation system will be created in one of these two ways. With a pump, the designer can pick the pressure to produce to match the system design. With a gravity (elevation) or municipal water system, the designer needs to work within the constraints of the pressure available.

The pressure variations found throughout an irrigation system are caused by: (1) elevation differences, and (2) friction head that occurs when water is running in the pipelines. These are actually independent of each other and must be calculated individually.

The pressure effects caused by elevation can be modified by use of pressure regulators, booster pumps, pressure tanks, and water storage ponds advantageously located. Pressure variations caused by friction head can be controlled somewhat by correctly sizing the pipes, valves, and fittings.

Friction Head (Pressure Loss)

Since water is not merely lifted but flows from one point to another, other factors enter into the picture. The movement of any substance from one point to another generates friction, which is a pressure loss due to the water rubbing against the pipe surfaces. When pumping, the frictional resistance to flow must be overcome.

Friction varies with the following factors:

1. increases with an increase in velocity
2. decreases with an increase in pipe diameter
3. increases with the roughness of pipe
4. increases with length of pipe

Friction losses will occur regardless of the direction of flow, from straight up to straight down or any angle in between.

Since friction varies due to many factors, direct management by gauges is recommended after the system has been installed. This is done by subtracting the static head from the gauge readings multiplied by 2.31, which gives the friction loss for the pipe valves and fittings. However, when planning a system, use the manufacturer's "friction tables" for the various pipeline components.

Water Hammer or Surge

Water hammer will occur whenever the movement of water in a pipe is stopped either in part or completely. The worst possible condition occurs when it is abruptly stopped, for example, when a valve suddenly closes, during sudden pump or power failure, or when filling the system. This would also happen when water in the main line moves back toward the pump before the check valve closes.

When the flow of water is stopped instantaneously, the water rebounds, setting up a pressure wave that rushes through the pipe. This pressure wave travels back down the pipe until it becomes equal to the pressure moving forward, then it rebounds back up the pipe again. This bouncing back and forth continues until the pressure wave or surge dissipates itself. The same principle applies as dropping a golf ball to the pavement. The ball will rebound with the greatest force at the first bounce and will continue to bounce lower and lower until it has no energy left.

The pressure wave that develops depends on the velocity of flow and the rate at which the flow is stopped. Under certain conditions, it can reach proportions that will burst any pipe. In other words, the higher the velocity, the higher the pressure from water hammer. The pressure due to water hammer is over and above the working pressure of the system; therefore, the two pressures must be added.

For example, the theoretical pressure rise due to water hammer in PVC Schedule 40 plastic pipe carrying 150 gpm in a 2.5-in. pipe at velocity of 10 fps is about 190 psi. Therefore, assume working pressure = 100 psi + water hammer 190 psi = peak pressure of 290. Higher flow rates would increase the velocity, and consequently a greater water hammer would result. For this reason it is suggested to select pipe sizes which will not exceed a maximum velocity of over 5 fps.

Note: Additional information concerning water hammer (surge), its causes, control methods and calculations can be found on pages 88–93.

Estimate Potential Mainline Surge Pressure

Surge pressures which may damage the mainline piping system should be considered when designing your sprinkler system.

Surge pressure or *water hammer* occurs in a mainline pipe when the flow of water in that pipe is suddenly stopped. The magnitude of the surge pressure depends upon several factors:

1. The initial velocity of flow in fps, and/or the quantity of water flowing through the pipe.
2. The length of time it took to stop the flow of water in the pipe.
3. The length of the mainline pipe between the point where the flow first stopped and the first entrance connection into the source of water.

The total pressure subjected to the mainline pipe during a surge condition is equal to:

$$P_{total} = P_o + P_s$$

where: P_{total} = total system pressure during a surge (psi)

P_o = the operating pressure at the time of the surge (psi)

P_s = the surge pressure: an increase in pressure over and above the existing operating pressure at the time of the surge (psi)

$$P_s = \frac{V \times L \times 0.07}{t}$$

V = original velocity of flow at time of surge (fps)

L = the length of the straight mainline pipe which extends between the water source and the point in the mainline (valve or pump location) where the flow was stopped (ft)

t = the approximate time it required to stop the flow of water, i.e., time to close the valve (sec)

Example 1

Sudden closure of a control valve at approximately 5 fps flow velocity.

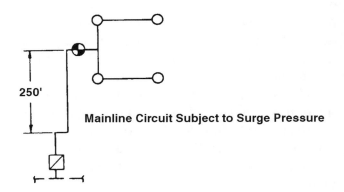

Mainline Circuit Subject to Surge Pressure

250'

Figure 5.1

If the existing pressure in the main line is 100 psi in the Class 160 PVC plastic pipe main line, a 2-sec valve closing time will create approximately the following surge pressure condition:

$$P_s = \frac{V \times L \times 0.07}{t}$$
$$= \frac{5 \text{ fps} \times 250 \text{ ft} \times 0.07}{2 \text{ sec minimum}} = 43.8$$

$$P_{total} = P_o + P_s = 100 \text{ psi} + 43.8 \text{ psi} = 743.8 \text{ psi}$$

Example 2

Sudden closure of a valve at 5 fps velocity. If the valve suddenly closes in approximately 0.5 sec the surge pressure and total system pressure will be:

$$P_s = \frac{5 \text{ fps} \times 250 \text{ ft} \times 0.07}{0.5 \text{ sec minimum}}$$
$$P_{total} = 100 \text{ psi} + 175 \text{ psi} = 275 \text{ psi}$$

Example 3

Valve closure at 10 fps water flow velocity.

$$= \frac{10\ \text{fps} \times 250\ \text{ft} \times 0.07}{0.5\ \text{sec minimum}} = 350\ \text{psi}$$

$$P_{total} = 100\ \text{psi} + 350\ \text{psi} = 450\ \text{psi}$$

In Example 1 the surge pressure is quite low because the initial velocities were held near 5 fps and because the closing time of the valve is reasonably slow. Notice that the total system pressure (P_{total}) is less than the working pressure rating of the plastic pipe (i.e., 143.8 psi is less than 160 psi allowable working pressure.)

In Example 2 the surge pressure is high because the closing time of the valve is short. The total system pressure of 275 psi is greater than the allowable working pressure of the pipe. However, it is somewhat less than the 450 psi burst pressure rating of the pipe, and should not cause any damage.

Example 3 depicts the results of using a quick-closing valve at a 10 fps velocity of flow. The total pressure during surge approaches the burst pressure rating of the pipe (450 psi). The pipe may not burst when subjected to the high pressure initially, but may burst during subsequent surge conditions.

Summary

A good guideline to follow is to try to keep the total pressure during a surge equal to or less than the nominal operating pressure of the pipe. If this is not possible, the total system pressure during the surge should be less than 75% of the burst pressure rating of the plastic pipe. The burst pressure rating of plastic (PVC) pipe is approximately 2.5 to 2.8 times the nominal operating pressure rating.

Another useful guideline to use is to maintain flow velocities as near 5 fps as possible. If velocities above 5 to 5.5 fps must be used, a slow-closing valve such as the Rain Bird Model EF or The valve should be used.

Valve Operation

Unfortunately most valves (including ball, gate, globe, and butterfly valves) do not cut flow proportionate to the valve stem travel.

Figure 5.2 illustrates how the valve closing time could affect the maximum pressures developed. For

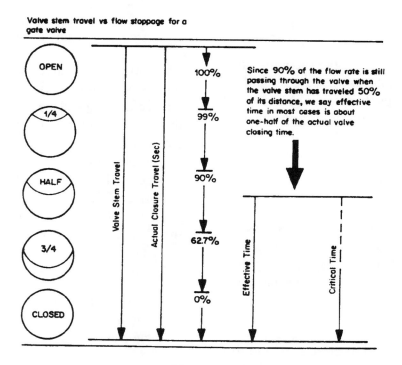

Figure 5.2

gate valves the final 5% of the valve stem travel is most critical. It is extremely important, therefore, to base timing of valve closing on the "effective closing time" of the particular valve in question. Generally it is about one-half of the actual closing time.

Caution in Filling Empty Pipelines

In filling a long piping system, the flow should be controlled with a gate valve being about 0.25 of the operating capacity. When the lines have filled, the valve should be opened slowly until full operating capacity and pressure are attained.

The average water velocity when filling a pipeline should be 1 fps or less and definitely should not exceed 2 fps. Table 5.3 has been provided to relate this velocity filling rate to an equivalent volume flow rate in gpm.

Table 5.3. Filling Rates in gpm Equivalent to Filling Velocities of 1 fps for Pipes Flowing Full

Nominal Pipe Size	Flow Rate (gpm)
4	40
6	88
8	157
10	245
12	353
14	480

Keeping filling rates within the recommended 1 fps cannot be overemphasized. Care should be taken not to choose pumps that would be oversized for a particular filling job without being able to control or throttle the pump output flow rate.

Other Control Considerations
Surge Relief Valves

In cases where sharp changes in flow velocities occur the most economical solution may be to install a pressure relief valve. This valve opens at a preset pressure and discharges the fluid to relieve the surge.

Surge Tanks

Surge tanks can be installed to effectively control both positive and negative surges. In general, they act as temporary storage for excess liquid that has been diverted from the main flow to prevent over-pressures, or as supplies of fluid to be added in the case of negative pressures.

Check Valves

Install check valves designed to close at zero velocity for relatively long pipelines that slope up and away from the pump, as well as for shorter lines that slope sharply up from the pump.

Valve Closing Speed Characteristics

Valve closing speed is critical to the life of an irrigation system. Fast closures create tremendous surge pressures and water hammer. Surge pressure (or water hammer) in main line or lateral piping occurs when the flow of water is suddenly stopped. The magnitude of surge pressure depends upon several factors:

1. The initial velocity and the quantity of water flowing through the pipe.
2. The length of time it takes to stop the flow of water in the pipe.
3. The length of the piping between the water source and the point where water flow is stopped (usually a valve location).

The definition of valve closing speed has been misunderstood for many years. Many have believed that valve closing speed was the time it takes from when the irrigation controller issues a signal until the valve actually closes. Valve closing speed is actually a measurement from the time flow rate starts to drop until zero flow. The charts in this book are representative of this latter definition.

This subject has received much attention from the Center for Irrigation Technology at California State University as a result of field requests (Figure 5.3). Their findings were summarized as follows:

> The valve testing was initiated by concerns from the field that water hammer (surge) was aggravated by a fast valve closure speed and thus could be contributing to observed PVC fitting failure.

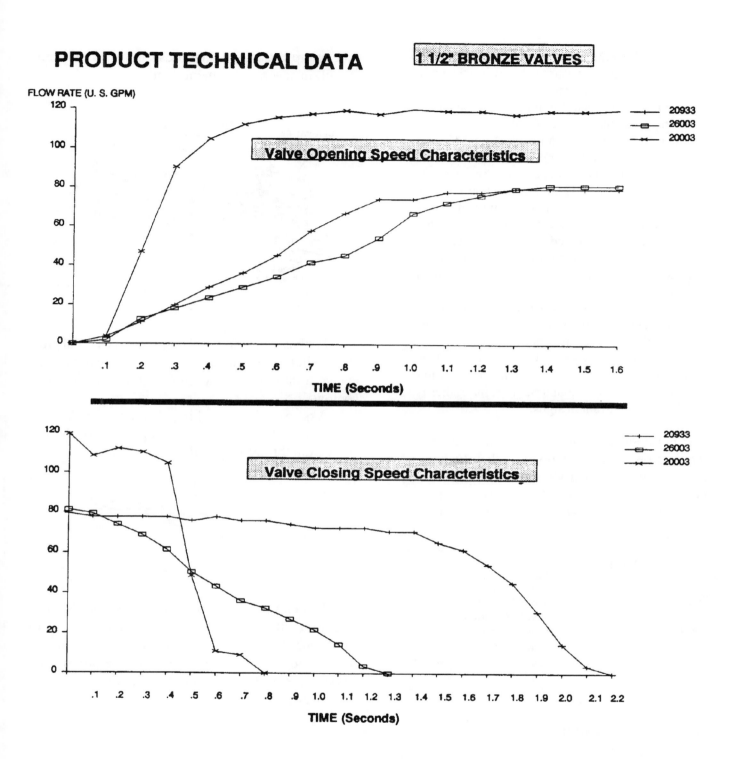

Figure 5.3. Product technical data. *Note:* The above data was obtained from tests performed by the Center for Irrigation Technology (CIT) at California State University, Fresno. Actual product performance may be slightly different from the information contained herewith due to normal manufacturing variations.

The three main factors that influence water hammer (surge) are water velocity, pipe length, and valve closing speed. Water velocity is a function of flow rate and pipe size. Pipe length is defined as the distance from the valve to the water source (pump) and back. Valve closing speed has typically been measured from the time the solenoid is deactivated (power off) until flow ceases. This elapsed time can run from 2 to 5 seconds or more for typical irrigation valves. However, test results indicate that the actual closing speed, when measured from the moment flow rates start to decay until flow ceases, may only be four- to six-tenths of a second...

The actual closing speed, being much faster than previously thought, can significantly increase the intensity of water hammer.

The testing of several commercial valves indicates this phenomenon is not found only with one particular manufacturer, but rather is inherent in current designs.

It is important to note that tests by the Center for Irrigation Technology revealed that Buckner 1.5-in. bronze valves greatly outperformed the typical industry closing speeds of 0.4 to 0.6 sec. The closing speeds for Buckner valves are summarized as follows:

Model	Closing Speed (sec)	Description
20933	2.2	diaphragm valve with "3-way" solenoid
26003	1.3	piston valve with "3-way" solenoid
20003	0.8	diaphragm valve with "2-way" solenoid

Surge Control Valve

A *surge control valve* is a device used to protect pumping equipment, and pipelines from dangerous pressure surges (water hammer) caused by rapid changes of flow velocity within a pipeline.

When pumping systems are started and stopped gradually, harmful surges do not occur. However, should a power failure occur, the abrupt stopping of the pump can cause dangerous surges, which could result in severe equipment damage.

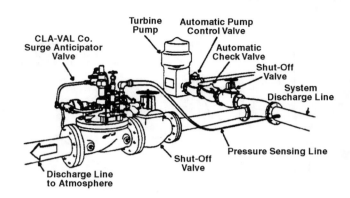

Figure 5.4

For basic surge control problems, a Cla-Val pressure relief valve can be used to relieve the high surge wave to atmosphere, as shown in the illustrations below.

A train engine is pushing a train up a grade. This compares with a pump delivering water to a higher elevation through a pipeline.

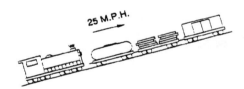

Figure 5.5a

The engine has stopped abruptly, but the momentum will continue to carry the train forward up the grade. The same condition will exist at the pump discharge upon a power failure to the pump. The momentum of the water column continues to carry it forward, creating a low pressure at the pump discharge.

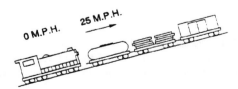

Figure 5.5b

The train has stopped after expending its energy in moving up the grade. The column of water in a pipeline has likewise come to rest.

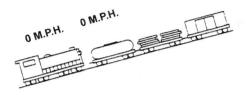

Figure 5.5c

The train has now started to coast backward and down the track and is about to collide with the engine. The same is true in the pipeline. The water column also reverses and flows back down the pipeline toward the pump with approximately the same velocity as before the pump failed.

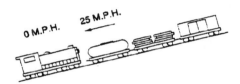

Figure 5.5d

The backward-moving train collides with the stationary engine, with disastrous results. Correspondingly, when water returning down a pipeline is stopped by a closed check valve, a high surge pressure develops. The result of this sudden stop can be damaging to the pipeline and/or the pumping equipment. The magnitude of this high pressure surge is directly proportional to the rate of change of velocity of the fluid.

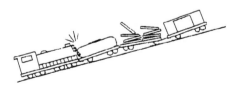

Figure 5.5e

This illustration shows the returning train being switched to a sidetrack, thus averting a collision with the engine. A Cla-Val surge control valve, when installed in a system, acts in the same manner; the high surge is directed or sidetracked to atmosphere, protecting the pump and system.

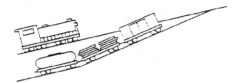

Figure 5.5f

The creation of a low pressure condition at the pump discharge is a result of the abrupt stopping of the pump and the column of water rushing away from it. The surge control valve senses this low pressure and opens in anticipation of the high wave which will be returning. Thus the returning high wave does not build up against a closed check valve, but is sidetracked out of the system, eliminating any possible damage.

SOLVING FOR PIPE SIZE SELECTION

The most important factors that determine the proper pipe sizes are the layout of the pipelines and the number of heads that will be running at the same time per line.

Example

Assume a green was designed for four heads (square spacing). The designer may choose to operate just one head at a time and install a pipe size to meet the demands of one sprinkler. This practice would reduce the pipe size and, in turn, lower the cost of the system. On the other hand, he may decide to operate two or four heads at the same time and install the proper pipe sizes accordingly.

Important: Once the pipe sizes are selected and installed the operator *must* run the system as designed.

Pipe Sizing Design Factors

1. Sprinkler models' gpm/head and operating pressure.
2. Number of heads that will run per zone and/or pipeline.
3. Length (ft) of each pipe section.

4. Operating pressure: To determine *class* of pipe.
5. Piping layout—dead-end and/or loops.
6. Piping layout—valve-in-head vs. block system.

Note: When in doubt, select the next size larger.

Pipe Sizing—Limiting Factors

1. Limit the friction loss from the source to the end of each leg of the pipeline to 20% of the operating pressure.
2. Limit the velocity to a maximum of 5 ft/sec.

Pipe Sizing Methods

1. Trial and error method
2. Pressure variation method

Comparative Pipe Capacities

Pipe carrying capacities can be compared by the area (square inches) of the orifice of the pipe, as shown.

$$\text{2-in. pipe} = 3.14 \times 1^2 = 3.14 \text{ in.}^2$$
$$\text{4-in. pipe} = 3.14 \times 2^2 = 12.56 \text{ in.}^2$$
$$\text{8-in. pipe} = 3.14 \times 4^2 = 50.24 \text{ in.}^2$$

This shows that the orifice area of an 8-in. pipe = four 4-in. pipes and sixteen 2-in. pipes.

Dead-End and Loop Piping

Note: The following problems illustrate the procedure for selecting pipe sizes for the dead-end and loop systems. Hydraulically speaking, either method would perform adequately providing the proper pipe sizes, etc., were selected and used. The benefits and/or disadvantages of each should be carefully considered before making a final choice for an actual installation.

EXAMPLE PROBLEMS

Main and/or Lateral—Loop System
Dead-End System

Problem 1

Assume the dead-end system in Figure 5.6 is designed to run two 50-gpm sprinklers per fairway spaced 100 ft apart. Solve for the pipe size(s) and total friction loss in each leg. Use Class 160 PVC plastic pipe. (Appendix 4.)

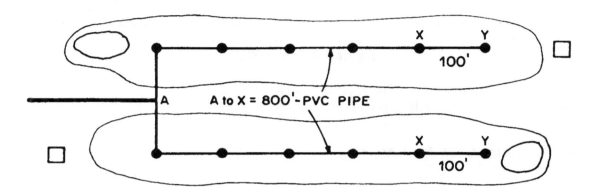

Figure 5.6. Dead-end system.

Procedure

Section	gpm	Pipe Size (in.)	Velocity (ft/sec)	Friction (psi/100 ft)	Distance per 100 ft	Total Friction (psi)
Source	100	2.5	5.78	1.86	8.00	14.88
to X		3*	3.91	0.72	8.00	5.76
X to End	50	2*	4.24	1.31	1.00	1.31
		2.5	2.89	0.52	1.00	0.52
					Total	*7.07

*Recommended choices

Problem 2

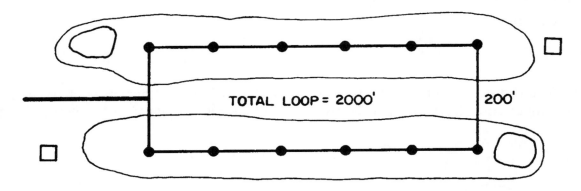

Figure 5.7. Loop system.

Solve for pipe size and friction loss for the loop as shown above.
    Note: Use the same design conditions and type of pipe—see Problem 1.

Calculating Pressure Loss for Loops

Procedure

The friction loss is based on the following:

1. One-half the total gallons per minute (gpm) supply.
2. One-half the total distance around the loop.

Section	gpm	Pipe Size (in.)	Distance (ft)	Friction Loss per 100 ft	Distance per 100 ft	Total Loss (psi)
Loop	100	2.5	1000	1.86	10.00	18.60
	100	3	1000	0.72	10.00	7.20*

*Recommended choice

Loop System

The loop system offers some advantages over the dead-end and should be used where the length of pipeline (trenching) is relatively short between laterals to complete the loop.

Advantages

1. *May* reduce pipe sizes
2. Equalizes the pressure loss on flat land. (*Note:* On hillsides the pressure in the lower leg will be higher. It is directly related to elevation difference).
3. The total sprinklers can be operated anywhere on the loop, i.e., 4 heads in any fairway.

Pipe Size Selection—Dead-End vs. Loop

Problem 1a

Select the pipe size needed to carry 350 gpm from a pump located at the pond to Fairway 10 (solid line) dead-end.

Problem 1b

Assume that all conditions are the same except the loop is completed by adding 500 ft of pipeline. Use Class 160 PVC pipe and Friction Loss Table in Appendix 4.

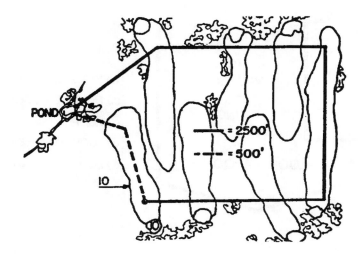

Figure 5.8

Procedure

Trial and Error Method

Problem 1a.

Section	gpm	Pipe Size (in.)	Dist. (ft)	Friction (psi/100 ft)	Total (psi)	Velocity (ft/sec)	Comments
Pond to No. 10	350	4	2500	2.30	57.5	8.3	
	350	5	2500	0.77	19.3	5.4	
	350	6	2500	0.33	8.3	3.8	Best

Problem 1b.

Section	gpm	Pipe Size (in.)	Dist. (ft)	Friction (psi/100 ft)	Total (psi)	Velocity (ft/sec)	Comments
Loop	175	3	1500	2.00	30.0	6.8	
	175	4	1500	0.59	8.9	4.2	Best
	175	5	1500	0.20	3.0	2.7	

Note: In actual practice the pipe size selected is generally based on the velocity of 5 ft/sec or less.

STUDENT PROBLEM

Select the pipe size for each section of pipeline. Use Class 160 PVC pipe. Use the same procedure as shown in the previous sample problem.

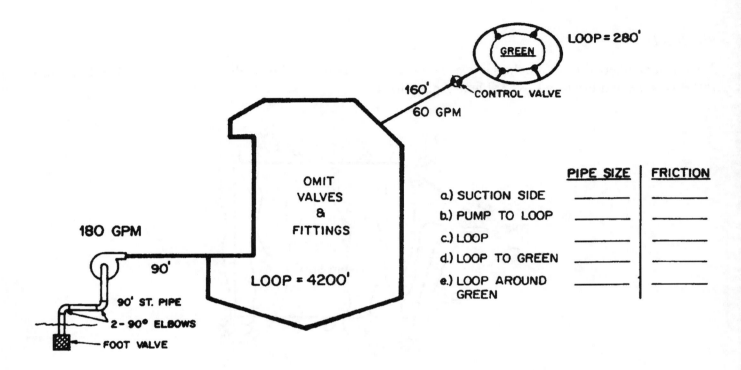

Figure 5.9

LOOP SYSTEM ANALYSIS

If the point of demand is of an unequal distance from the point of supply, but the pipe size along both paths is the same, calculate the friction loss by figuring one-half the gallonage flow at one-half the distance of the loop.

Example

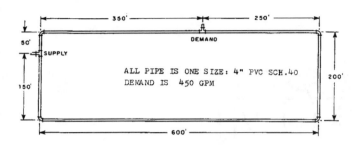

DEMAND

SUPPLY

ALL PIPE IS ONE SIZE: 4" PVC SCH.40
DEMAND IS 450 GPM

Figure 5.10

450 gpm ÷ 2 = 225 gpm
50 + 350 + 250 + 200 + 600 + 150 = 1600 ft of
    pipe
1600 ÷ 2 = 800 ft of pipe per leg
Friction Loss = 1.04 psi per 100 ft of 4-in. PVC
    Schedule 40
at 225 gpm × 800 ft = 8.32 psi
(friction loss in the loop of 8.32 is *maximum*)

This is considered a "rule of thumb" and does create a safety factor as shown below in constructing the gallonage flow and friction losses per leg of the loop. To find the gpm in each leg of the loop, use the Friction Loss Table for 4-in. PVC Schedule 40 plastic pipe in Appendix 4.

	Size	Distance	gpm	Loss
Leg A	4 in.	400 ft	X	8.32
Leg B	4 in.	1200 ft	450–X	8.32
	Totals	1600 ft	450	8.32

Leg A 8.32 2.08 loss = 325 gpm (Enter 4-in. Sch. 40 PVC)

Leg B 8.32 0.69 loss = <u>180 gpm</u> (Find 2.1 friction loss Read gpm Flow)

Total = 505 gpm (55 gpm over actual demand)

If necessary, the 180/505 ratio can be applied to the actual demand and losses figured more accurately:

180/505 X 450/X = 160.4 gpm Leg B
325/505 X 450/X = <u>289.6</u> gpm Leg A
                    450   gpm

Recalculating the loss:

	Size (in.)	Distance (ft)	gpm	Loss/ 100 ft	Total Loss
Leg A	4	400	289.6	1.66	6.6 psi
Leg B	4	1200	160.4	0.55	6.6 psi

The *actual* friction loss in the 100 ft of pipe is 1.7 psi less than that figured using the *quick estimating* method.

EXAMPLE PROBLEM

Outlet at Unequal Distance From Supply

Solve for the flow (gpm) and friction loss (psi) that will occur in Leg A (solid line) and Leg B (dash line) as shown. *Note:* Use the format and procedure outlined on previous page.

Conditions

a. Pipe size = 4-in. PVC Class 160
b. Flow to Loop = 350 gpm
c. Friction Loss Table (see Appendix 4)

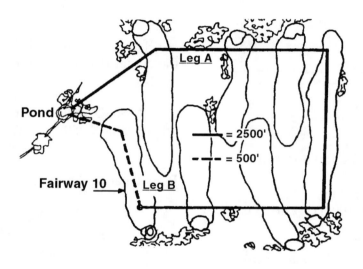

Figure 5.11

Procedure

1. Determine the *maximum* friction loss in the loop.

> Loop Flow = 350/2 = 175 gpm
> Distance = 3000/2 = 1500 ft
> Friction Loss = 0.60 psi/100 ft × 15 = 9.0 psi (max.)

2. Solve for *maximum* loss in each leg.

	Size (in.)	Distance (ft)	gpm	Loss (psi)
Leg A	4	2500	X	9.0
Leg B	4	500	350-X	9.0

3. Find friction loss per 100 ft.

> Leg A   9.0/25 = 0.36 psi/100 ft = 135 gpm   (Enter 4 in. Col. Class 160 PVC)
> (Find 0.36 F.L. Read 135 gpm)
>
> Leg B   9.0/5 = 1.8 psi/100 ft =   <u>320</u> gpm
> Total =   455 gpm (105 gpm over actual demand)

4. Calculate *actual flow* per leg and *actual demand.*

Leg A = 135/455 × X/350 = 103.8 gpm
Leg B = 320/455 × X/350 = <u>246.2</u> gpm
Total = 350.0 gpm

5. Recalculate the friction loss per leg.

	Size (in.)	Distance (ft)	gpm	Loss/100 ft	Total Loss (psi)
Leg A	4	2500	104	0.23	5.75
Leg B	4	500	246	1.15	5.75

Note: The *actual* friction loss in the loop is 3.25 psi less than that figured using the quick estimating method.

PIPE LAYOUT

Pipe sizes can be reduced substantially by running fewer sprinklers at the same time per zone and/or fairways, etc.

Figures 5.12 through 5.17 show that the same area can be irrigated properly by a number of different piping layouts. The layouts can vary with respect to the location of the water source, obstacles, and other landscape features.

The piping layout shown in Figure 5.12 is based on a golf course divided into 5 zones and *one* 50-gpm sprinkler per zone. Note the gallons per minute carried by each pipeline.

The system shown in Figure 5.13 is designed to operate a maximum of *five* 50-gpm sprinklers per zone.

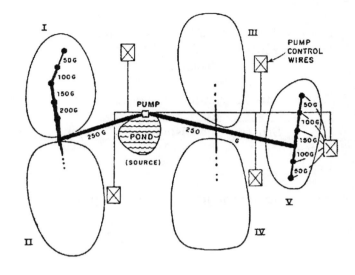

Figure 5.13

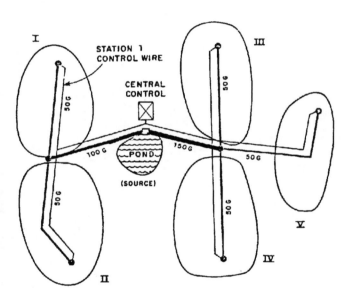

Figure 5.12

STUDENT PROBLEMS

Problem 1. Assume that the layout of the main and lateral pipelines from the source was planned as shown in Figure 5.14. The system is designed to operate a total of five 50-gpm sprinklers in the zone and a *maximum* of *two* that would run in any one of the fairways.

   Determine the carrying capacity of each section of the lateral pipelines. Indicate on the sketch.

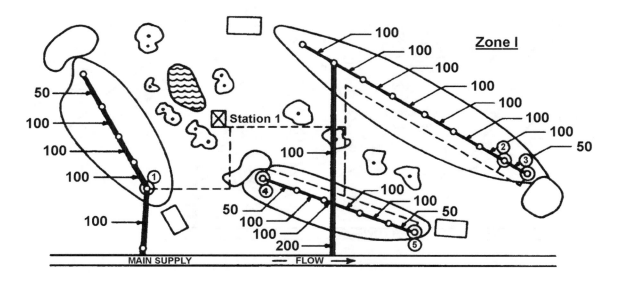

Figure 5.14

Problem 2. Assume that the main and lateral pipelines from the source was planned as shown in Figure 5.15. The system was designed to operate a maximum of five 50-gpm sprinklers in any of the fairways.

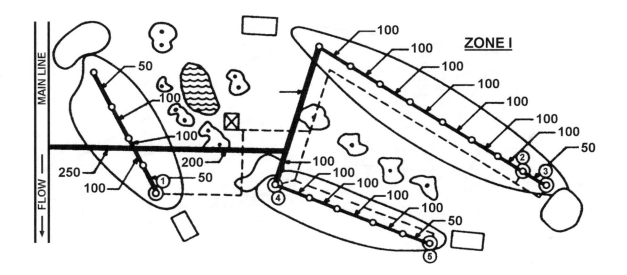

Figure 5.15

Problem 3. Assume that this watering system was designed to run *five* 50-gpm sprinklers in any one of the fairways. Indicate the flow capacities of each section of the pipelines.

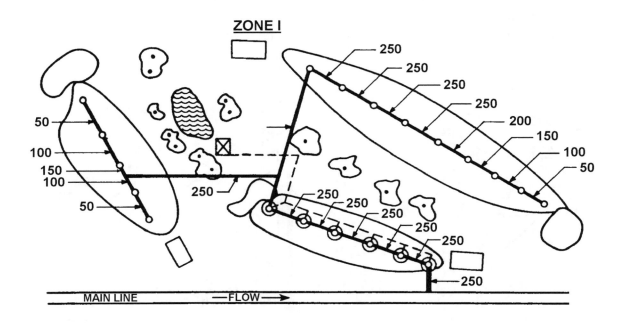

Figure 5.16

Problem 4. The design conditions for the pipe layout loop system shown in Figure 5.17 are the same as Problem 3. Indicate the gpm flow to the loop and throughout the loop.

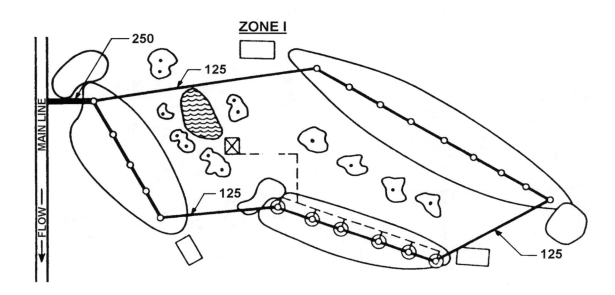

Figure 5.17

Dead-End Pipeline or System

The dead-end piping system is a one-way flow from the source to a given point or to the last outlet on the pipeline.

Example 1

Select the Pipe Size for the main pipeline (only).

Design Conditions (see Figure 5.18):

1. For this problem, the friction loss for the main pipe-line is to be limited to 10 psi or 23 ft.
2. System is designed to run all 7 sprinklers at the same time. Sprinkler demand is 50 gpm per sprinkler.
3. Type of pipe is PVC Class 160 (neglect friction losses for valves and fittings).
4. Land surface distance = 1500 ft.

Procedure

Trial and error method. Find the friction losses for various pipe sizes and make the selection on the basis of the above limitations. See Appendix 4.

Pipe Size Choices

Size	gpm	Velocity (ft/sec)	Loss/ 100 ft	Distance (ft)	Total Loss
4	350	8.27	2.15	1500	32.25
5	350	5.4l	0.77	1500	11.55*
6	350	3.81	0.33	1500	4.95

*First choice—select 5-in. pipe even though the recommended limit is exceeded slightly.

Alternative—Use 500 ft of 6-in. pipe = 1.65 psi
1000 ft of 5-in. pipe = 7.70 psi
Total Friction Loss = 9.35 psi

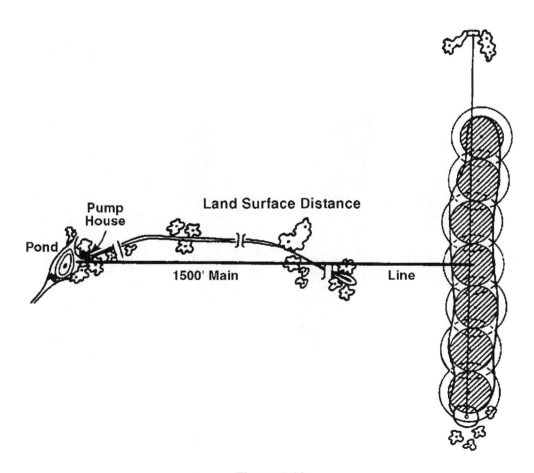

Figure 5.18

Example 2

Select the main pipeline sizes for an irrigation system designed to operate a *maximum of two sprinklers in any fairway.* (Neglect losses in fittings, valves and joints.)

Conditions

1. Use same limitations as for Problem 1.
2. Seven sprinklers at 50 gpm/sprinkler to be operated at the same time.

- 1 sprinkler operating in Fairway 8
- 2 sprinklers operating in Fairway 12
- 2 sprinklers operating in Fairway 3
- 2 sprinklers operating in Fairway 12

Procedure

Trial and Error Method

Possible Choices

Section	gpm	Pipe Size (in.)	Velocity (ft/sec)	Friction psi/ 100 ft	Distance per 100 ft	Total psi
Pump to	350	6	3.81	0.33	5.00	1.65
8th	350	5	5.41	0.77	5.00	3.85*
8 to 12th	300	5	4.64	0.58	4.50	2.61*
	300	4	7.09**	1.61	4.50	7.25
12 to 3rd	200	4	4.72	0.76	3.50	2.66*
	200	3	7.82**	2.59	3.50	9.07
3 - 2nd	100	3	3.91	0.72	2.00	1.44*
		2.5	5.78	1.86	2.00	3.72
					*Total	10.56

*First choices
**Velocity too high

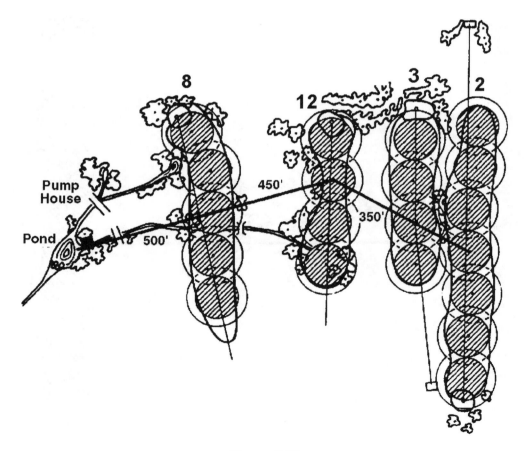

Figure 5.19

SAMPLE PROBLEM

On the dead-end pipe layout below record the gallons per minute (gpm) to be carried by each pipe section indicated by an arrow.

Design Conditions

Fairways: Complete irrigation of all fairways.

1. Operate five sprinklers at the same time—maximum of one sprinkler in any of the fairways shown.
2. Sprinkler model = 90 gpm
3. Operating Pressure = 80 psi

Greens/Nursery: Next, irrigate all greens.

1. Operate any six greens, at the same time, of the seven greens shown.
2. Total gpm (piped as shown) = 75 gpm/green
3. Operating pressure = 80 psi.

Tees: After greens, irrigate tees.

1. Operate six sprinklers at the same time.
2. Sprinkler model = 15 gpm, 80 psi.

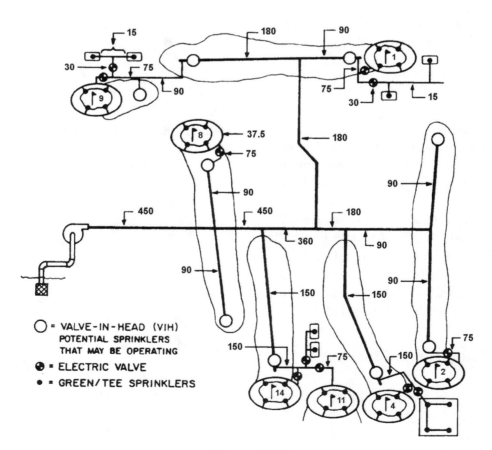

Figure 5.20

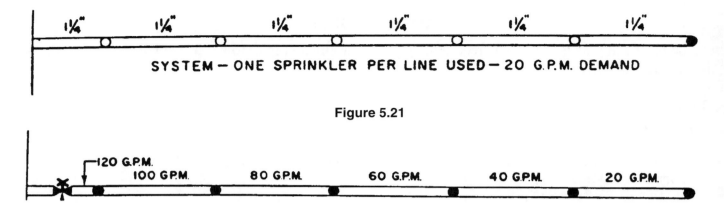

Figure 5.21

Figure 5.22

VALVE-IN-HEAD SYSTEM VS. BLOCK SYSTEM

Example Problems
Valve-In-Head System

The piping is sectioned where *usually* one sprinkler runs on any one lateral. The pipe size can be constant all the way. Pressure losses are figured for one gallonage demand for the length of pipeline. (See Figure 5.21.)

Block System

The pipeline consists of a number of sprinklers (block) that are controlled by one valve. The pipe sizing can be telescoped down as the demand decreases from each head along the line. (See Figure 5.22.)

If the cost is a consideration of the *designer,* the material cost of a *valve-in-head system* is usually less than a *block system*. It has the added advantage of allowing a greater degree of versatility and results in a more efficient irrigation system than a block system does. (See Figures 5.23 and 5.24.)

Pressure Variation Method for Pipeline Sizing

Ideally, the pressure (friction) loss in the pipelines should be negligible. However, to comply with this objective would require excessively large pipe sizes that cannot be justified. On the other hand, a pres-

sure drop that exceeds the lower limit of the required sprinkler operating pressure must be avoided.

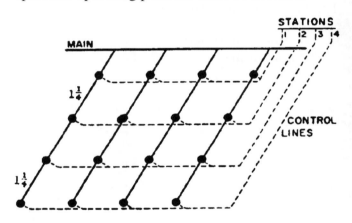

Figure 5.23. Valve-in-head design.

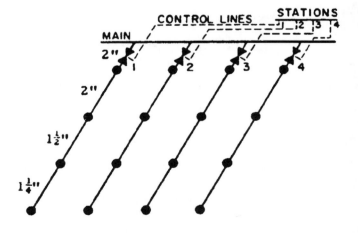

Figure 5.24. Block design.

Consequently, a general "rule of thumb" is used which limits the friction loss from the source (pump) to the end of *each* leg to approximately 20% of the system operating pressure. A Pressure Variation of 20% = A Discharge Rate of Variation of 10%.

Example

Determine the pressure (psi) and discharge (gpm) as shown in Figure 5.25.

Assume:

- Operating Pressure = 80 psi
- Allowable Friction Loss = (80 × 0.20) = 16 psi
- Sprinkler Discharge Rate = 50 gpm
- Allowable Pressure Variation = 20%
- Discharge Variation = 10%
- gpm = 50 – (50 × 0.10) = –5 gpm

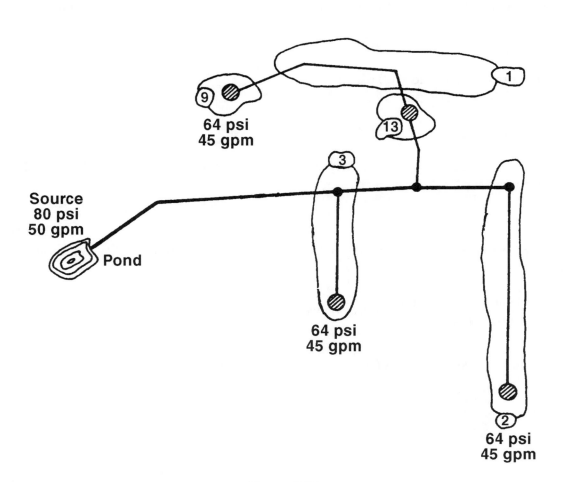

Figure 5.25

SAMPLE PROBLEM

Pipe Sizing

The procedure for solving pipe sizes and allowable friction losses by the pressure variation method is outlined on page 114.

To obtain a pressure balance as close as possible at the end of each leg it may be necessary to manipulate pipe sizes.

Operating Conditions (see Figure 5.26)

- Run 9 sprinklers at the same time—
- Operating pressure of each head = 60 psi
- Discharge rate of each head = 20 gpm
- Pipe lengths are as shown
- Pipe—Class 160 PVC plastic

Procedure

Problem

Select pipe sizes which will give a minimum of variation of pressure and discharge between any two sprinkler heads on the system.

Known:

- Operating pressure of each head = 60 psi
- Discharge of each head = 20 gpm
- Pipe lengths as shown
- Pipe is PVC 160 plastic.

If the loss in each circuit is approximately equal, then each sprinkler will operate at about the same pressure and discharge.

Procedure

Design to make pipe selections automatically balanced and to come within the 20% variation. Compute the friction factor.

$$\frac{\text{sprinkler nozzle pressure} \times \% \text{ variation}}{\text{length of circuit (in 100's of ft)}} = \text{friction factor}$$

Enter chart and select size which is *less than or equal to the friction factor, and also less than a velocity of 5 ft/sec.*

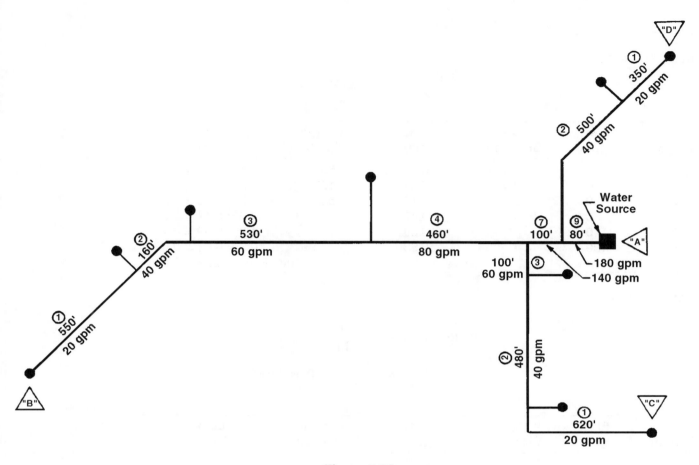

Figure 5.26

Example

Circuit "A" to "B"

$$\frac{70 \text{ psi} \times 0.2}{18.8} = 0.745 \text{ psi} / 100 \text{ ft}$$

Section	Flow gpm	Pipe Size (in.)	Loss (psi)	Length Pipe (ft)	Loss Totals
1	20	1.5	0.71	550	5.5 × 0.71 = 3.9
2	40	2.5	0.34	160	1.6 × 0.34 = 2.6
3	60	2.5	0.72	530	5.3 × 0.72 = 3.8
4	80	3	0.47	460	4.6 × 0.47 = 2.2
7	140	4*	0.39	100	1.0 × 0.39 = 0.4
9	180	4	0.63	80	0.8 × 0.63 = 0.5
					Branch Total Friction Loss 13.4

BRANCH "A" to "C"

$$\text{friction factor} = \frac{70 \text{ psi} \times 0.2}{13.8}$$

$$= \frac{14}{13.8} = 1.0 \text{ psi} / 100 \text{ ft}$$

Section	Flow gpm	Pipe Size (in.)	Loss (psi)	Length Pipe (ft)	Loss Totals
1	20	1.5	0.71	620	6.2 × 0.71 = 4.4
2	40	2	0.86	480	4.8 × 0.86 = 4.1
3	60	2.5	0.72	100	1.0 × 0.72 = 0.7
7	140	40*	0.39	100	1.0 × 0.39 = 0.4
9	180	4	0.63	80	0.8 × 0.63 = 0.5
					Branch Total Friction Loss 10.1

BRANCH "A" to "D"

$$\text{friction factor} = \frac{70 \text{ psi} \times 0.2}{9.3}$$

$$= \frac{14}{9.3} = 1.5 \text{ psi} / 100 \text{ ft}$$

Section	Flow gpm	Pipe Size (in.)	Loss (psi)	Length Pipe (ft)	Loss Totals
1	20	1.5	1.37	350	3.5 × 1.37 = 4.8
2	40	2	0.86	500	5.0 × 0.86 = 4.3
9	180	4	0.63	80	0.8 × 0.63 = 0.5
					Branch Total Friction Loss 9.6

*3.5 in. PVC Plastic Pipe Is Not Available.

Note: Some pipe sections are mutually common to two or more branches. The final pipe size for the mutually common sections is based upon the largest size computed for that section in any calculation.

STUDENT PROBLEM

Pressure Variation Method
Pipe Sizing

Fill in the blank spaces in the tables on the following page. Follow the same format as the sample problem.

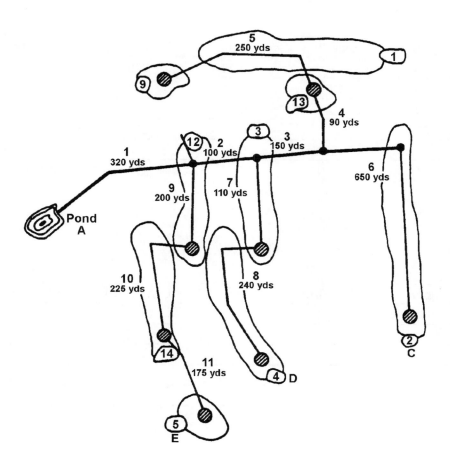

Figure 5.27

Design Conditions

- Operating Pressure = 80 psi
- gpm Per Sprinkler = 50 gpm
- Operate 8 Heads as shown
- Pipe Lengths as Shown
- Pump House at Pond
- Pipe—PVC Class 160
- Allowable Friction Loss = 20% of the Operating Pressure

BRANCH "A" to "B"

$$\text{friction factor} = \frac{80 \text{ psi} \times 0.2}{27.3}$$

$$= \frac{16}{27.3} \quad 0.59 \text{ psi} / 100 \text{ ft}$$

Section	gpm	Size (in.)	psi	Length (ft)	Total Loss
1	400	6	0.42	960	$9.6 \times 0.42 = 4.03$
2	250	5	0.41	300	$3.0 \times 0.41 = 1.23$
3	150	4	0.45	450*	$4.5 \times 0.45 = 2.03$
4	100	3	0.72	270	$2.7 \times 0.72 = 1.94$
5	50	2.5*	0.52	750	$7.5 \times 0.52 = 3.90$

Branch Total Friction Loss   13.13

BRANCH "A" to "C"

Friction Factor =

Section	gpm	Size (in.)	psi	Length (ft)	Total Loss

BRANCH "A" to "D"

Friction Factor =

Section	gpm	Size (in.)	psi	Length (ft)	Total Loss

BRANCH "A" to "E"

Friction Factor =

Section	gpm	Size (in.)	psi	Length (ft)	Total Loss

STUDENT PROBLEM

Pipe Sizing Choice

1. Pressure Variation Method
2. Trial and Error Method

From the dead-end piping layout shown below, fill in the blank spaces for each section of pipe listed in the table.

Use Class 200 PVC Plastic Pipe—see Appendix 4.

Omit valves and fittings

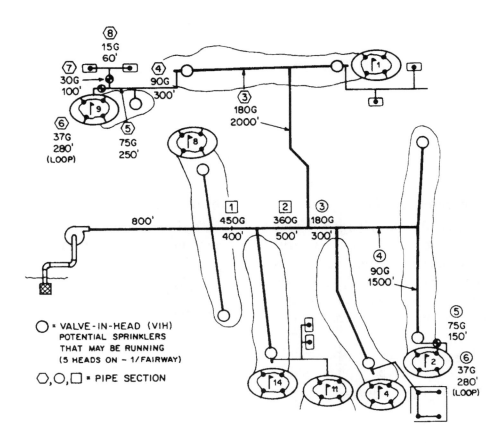

Pump—Green #2

Section	Pipe Size (in.)	gpm	Distance (ft)	Loss/100 ft	Total Loss (psi)	Velocity (ft/sec)
1		450	1200			
2		360	500			
3		180	200			
4		90	1500			
5		75	150			
6		37	280/2			

Pump—Green #9 + Tee #1

Section	Pipe Size (in.)	gpm	Distance (ft)	Loss/100 ft	Total Loss (psi)	Velocity (ft/sec)
1		450	1200			
2		360	500			
3		180	2000			
4		90	300			
5		75	400			
6		37	280/2			
7		30	100			
8		15	60			
	Total Loss (psi)					

Friction Loss in Valves and Fittings

The charts on page 118 and table on page 119 give the friction loss values for various valves and fittings. CAUTION! The numerical values are in terms of "Equivalent Length of Straight Pipe." Therefore, it is necessary to add the equivalent values of each valve and/or in the pipeline to the *actual* feet of straight pipe.

Knowing the flow (gpm) and feet of straight pipe, the friction loss for a given type of pipe is found by the same procedure used previously.

Chart Values

To find the values, refer to the example on the chart which illustrates the procedure.

Table Values

The table is self-explanatory.

EXAMPLE PROBLEMS

1. Find the equivalent length of straight pipe for each of the valves and fittings listed. Use Table 5.4.

a. Two 4-in. globe valves	220.8 ft
b. Four 3-in. gate valves	7.6 ft
c. Five 6-in. standard elbows	80 ft
d. One contraction 6-in. to 3-in.	2.8 ft
Total	311.2 ft

2. Determine the feet of *straight pipe* and the *friction loss* of the Suction Side and the Discharge Side—Use Class 125 PVC pipe. See Appendix 4.

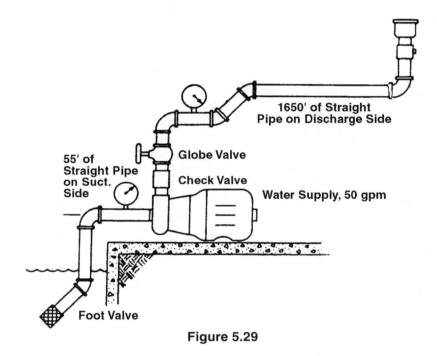

Figure 5.29

Pipe Selection Procedure

Discharge Side—Use the trial and error method.
Suction Side—Use one pipe size larger.

Suction Side
    Select Pipe Size = 2.5 in.
    Calculate St. Pipe (ft)
        Foot valve 55 ft st. pipe
        45° elbow 3 ft st. pipe
        90° elbow 6.5 ft st. pipe
        St. pipe 55 ft
            Total 119.5 ft
    Friction loss 1.3 ft
    $(1.2 \times 0.48 \times 2.31)$

Discharge Side
    Select Pipe Size = 2 in.
    Check valve 19 ft st. pipe
    Globe valve 55 ft st. pipe
    Two 90° elbows 11 ft st. pipe
    Two 45° elbows 5 ft st. pipe
    St. pipe 1650 ft
    Total 1740 ft
    Friction 48.2 ft
    $(17.4 \times 1.2 \times 2.31)$

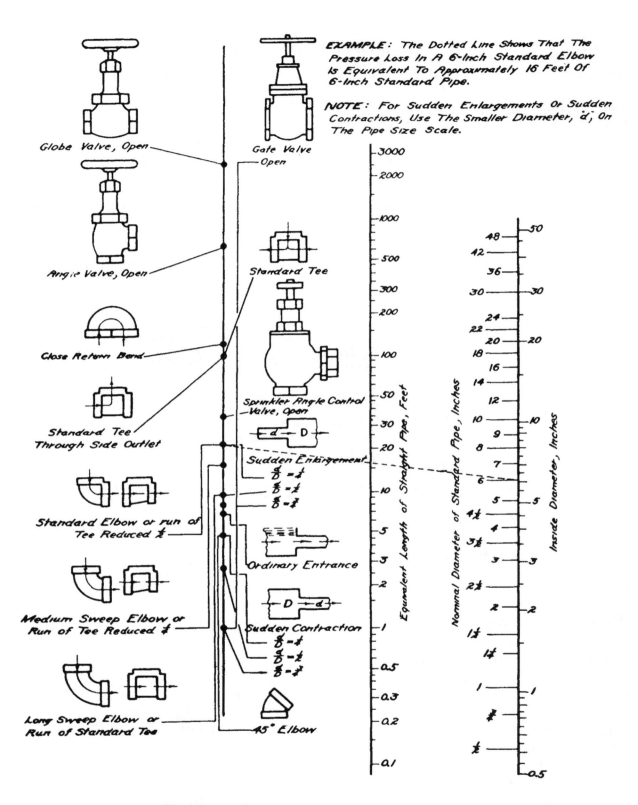

Figure 5.30. Pressure losses in valves and fittings.

Table 5.4. Friction Loss in Valves and Fittings, Equivalent Length of Straight Pipe

Size of Fittings, In.	1	1.5	2	2.5	3	4	6	8
90° elbow (standard)	2.7	4.3	5.5	6.5	8.0	11.0	16.0	21.0
45° elbow	1.3	2.0	2.5	3.0	3.8	5.0	7.5	10.0
90° elbow (long sweep)	1.7	2.7	3.5	4.2	5.2	7.0	11.0	14.0
Tee—Run Thru	1.7	2.3	3.5	4.2	5.2	7.0	11.0	14
Tee—Thru Side	5.7	9.9	12.0	13.0	17	21	33	43.0
Gate Valve (fully open)	0.6	1.0	1.5	1.7	1.9	2.5	3.5	4.5
Gate Valve (1/2 open)	14	22	27	33	41	53	100	130
Globe Valve	27.0	43	55	67	82	110	140	
Check Valve	8.0	14.0	19	23	32	43	63	90
Foot Valve		38	46	55	64	71	77	79
Enlargement 1:2		2.6	3.2	3.8	4.7	6.2	9.5	13.0
3:4		1.0	1.2	1.3	1.7	2.3	3.4	4.5
Contraction 2:1		1.5	1.8	2.2	2.8	3.6	5.6	
4:3		1.0	1.2	1.3	1.7	2.3	3.4	4.5

Valve Size Selection

In practice, the valve size is based on a pressure drop limitation, usually, 5 to 8 psi maximum. *Note:* When calculating friction losses, first determine the total loss from the beginning to the end of the line, then add the total valve friction losses.

Pressure loss tables are available for most of the valves on the market. The following represent a few typical examples.

ACKNOWLEDGMENTS

1. "Architect-Engineer Turf Sprinkler Manual." Rain Bird Sprinkler Manufacturing Company.
2. "Design Information for Large Turf Irrigation Systems." Toro Sprinkler Manufacturing Company.
3. "Golf Course Irrigation System Design." Manual and Handouts. Buckner Sprinkler Manufacturing Company.
4. "Pipe Installation Guide." Booklet. Johns-Manville Pipe Manufacturing Company.
5. "Surge Control Valve." Catalog. CLA-VAL Automatic Control Valves.

| Model | Size NPT | Flow, GPM | | | | | | | | | | | | | | |
		5	10	15	20	30	40	50	60	80	100	120	140	160	180	200
20050	¾"	Pressure Loss, PSIG														
20050	¾"	3.0	3.4	3.9	4.8	8.6	15.2									
20051	1"	3.0	3.4	3.9	4.8	8.6	12.6	18.9								
20052	1¼"		3.3	3.8	4.6	5.6	7.3	9.4	12.0	18.0						
20053	1½"			2.1	2.6	2.8	3.0	3.6	4.0	7.5	11.0	16.0	21.2			
20054	2"				1.4	1.5	1.6	1.8	2.1	2.9	3.9	5.5	7.5	10.0	12.0	15.5

Note: Valve size should be determined by selection of approximately 5 to 8 PSIG pressure loss. Operation of valves in shaded area of chart may result in erratic operation.

EAV

EGV

ELECTRIC VALVE PRESSURE LOSS - Cover vented to downstream pressure

Model No.	Size	PSI loss @ 50 GPM	PSI loss @ 100 GPM	PSI loss @ 200 GPM	PSI loss @ 300 GPM	PSI loss @ 400 GPM	PSI loss @ 500 GPM	PSI loss @ 600 GPM	PSI loss @ 700 GPM	PSI loss @ 800 GPM
EAV-300MFC	3"	4.5	5.0	6.1	8.0	15.0	19.0			
EAV-400MFC	4"		5.0	6.1	6.6	6.9	7.3	11.0	13.0	17.5
EGV-300MFC	3"	6.0	7.5	9.5	12.0	19.0				
EGV-400MFC	4"	5.0	5.5	6.1	6.5	6.8	8.3	11.0	17.5	

252-25-08   252-25-06   216-21-06   216-26-08   216-26-06   216-06-12

Friction Loss																		
	GPM flow																	
Size	5	10	15	20	30	40	50	60	70	80	100	120	150	170	180	200	250	300	350
Hydraulic																		
1"	<1	<1	1.5	2.5	5.5	9.0												
1¼"				2.0	2.7	3.7	4.8	6.0	8.0	10.5	13.0							
1½"			<1	1.5	2.5	3.0	4.5	5.0		8.0	11.5	14.0						
2"				<1	1.0	1.1	1.5	2.5		3.0	5.5	7.0	10.0	11.5	14.5			
2½"						<1	<1	1.0	1.2	1.5	1.7	2.7	3.5	4.0	5.5			
3"							<1	1.0	1.2	1.5	2.5	3.5	4.0	5.0	5.5	6.0		
Electric																		
1"	2.0	3.5	4.0	4.5	7.0	9.5												
1¼"				6.5	7.0	7.5	8.0	9.0	11.0	13.5	17.0							
1½"			5.0	5.2	5.4	5.6	5.8	6.0	11.0	13.5	16.0							
2"			2.0	2.5	3.0	3.5	4.0	4.5	7.0	8.5	11.0	13.0	15.0					
2½"					2.0	2.2	2.3	2.4	2.5	3.0	4.0	4.5	5.5	7.0				
3"							2.2	2.4	2.5	3.0	4.0	4.5	5.5	6.5	7.0	7.5		

Note: When designing a system, be sure to calculate total friction loss to ensure sufficient downstream pressure for optimum sprinkler performance.

Model 100 -20

6" Globe, Flanged

6" Angle, Flanged

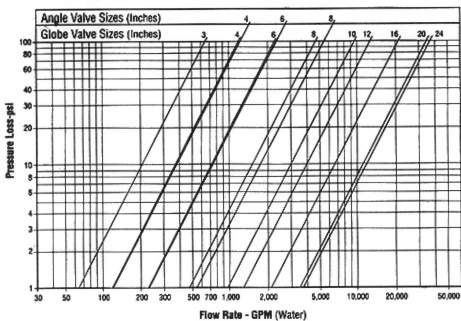

Flow Chart-Normal Flow (Based on flow through a wide open valve.)

Angle Valve Sizes (Inches)   4   6   8
Globe Valve Sizes (Inches)   3   4   6   8   10 12   16   20 24

Pressure Loss-psi

Flow Rate - GPM (Water)

Chapter 6

Two-Row System

The designer should consider using a two-row or multi-row system when he has determined the following:

1. The practical limitations of the available sprinkler spray diameter (effective coverage) is exceeded.
2. The advantages offered by the two-row system vs. the single-row system are preferred.

Practical Limitations

Example

Select two pop-up sprinkler models from available sprinkler catalogs or from the sprinkler models found in Appendix 5, and record them in the table below.

Conditions

Average fairway width = 175 ft.
Effective coverage (EC) = 70%.

Solution

Sprinkler diameter = 175 ft/0.7 = 250 ft.

Advantages

- Smaller Sprinkler Heads
  — Lower discharge rates (gpm) may result in smaller pipe sizes (depending on layout).
  — Lower precipitation rates will allow better penetration into the soil.
  — Lower operating pressure may reduce pump size and operating cost.
- Better uniformity of water distribution across the fairway.
- Less wind effect.
- Less watering of roughs.

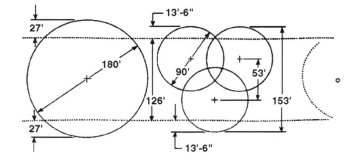

Figure 6.1

Possible Choices

Name and Model	Total Diameter (in.)	gpm	Precipitation (in./hr)	Base Pressure (psi)	Nozzle Size (in.)
	250	None	Available		
	250	None	Available		

Two-Row Square Spacing vs. Three-Row

Of utmost concern is the central portion of the fairway where the most play and traffic occur. The following illustrations show that the two-row system does provide good uniformity where needed most (center of the fairway), whereas the three-row system extends the uniformity of application across the width of the fairway.

Two-Row, One-Speed vs. Two-Speed

The same problem remains in a two-row system design that exists in a single-row system design mentioned previously. *Overlapped* areas receive more water than *non-overlapped* areas.

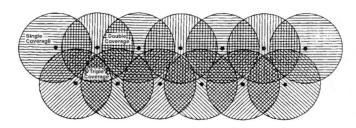

Figure 6.3

Again, a two-speed head is used; but this time it is a 180° two-speed head. Half of the *arc*—the half which is covered by *overlap* from an opposing head—runs at regular speed. The other half runs at half speed, allowing the sprinkler to apply twice the normal amount of water over the area that is not covered by *overlapping sprinklers.* However, the wedge-shaped uniformity of application will occur in the single coverage area.

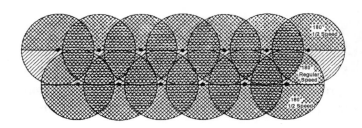

Figure 6.4

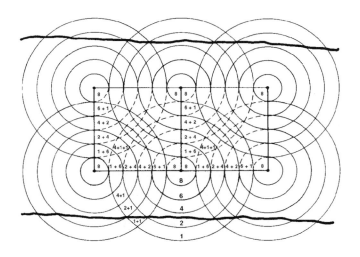

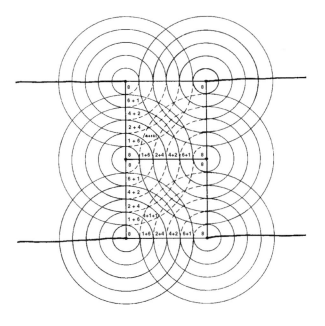

Figure 6.2

STUDENT PROBLEM

Two-Row System Design Procedure

Design a fully automatic two-row irrigation system. *Note:* All design conditions and a "problem" golf course map will be assigned.

Suggested Steps—Fairways

Step 1. Solve for average fairway width.

Step 2. Solve for location of fairway sprinklers.
(*Note:* Use *light* guidelines on the problem map. Refer to the illustration below.)

Example

Assume average fairway width = 126 ft

1. Establish center measuring line.
2. At intervals from the center line, measure out 1/4 of the average fairway width. Locate sprinkler lines as shown.

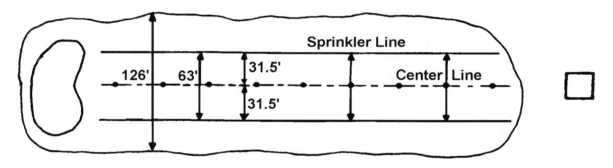

Figure 6.5

Step 3. Solve for sprinkler diameter.

Example

Solve for sprinkler diameter for the following conditions:

1. Average fairway width = 126 ft
2. Sprinkler spacing
   - Square pattern = 55%
   - Triangular pattern = 60%

Solution: Review page 55 (Equilateral Triangles).

Square Pattern

$$\text{Sprinkler diameter } = \frac{1/2 \text{ fairway width}}{\% \text{ spacing}}$$

$$= \frac{63}{0.55} = 114 \text{ ft}$$

Triangular Pattern

Caution: Do not use the "effective coverage" method as for the single-row system. Use the following procedure.

Solution

1. Sprinkler spacing =

$$\frac{1/2 \text{ average width}}{0.87} = \frac{63}{0.87} = 72 \text{ ft}$$

2. Sprinkler diameter =

$$\frac{\text{sprinkler spacing}}{\% \text{ of sprinkler diameter spacing}} = \frac{72}{0.60} = 120 \text{ ft}$$

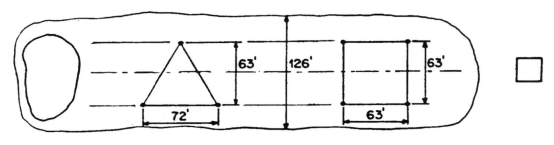

Figure 6.6

Step 4. Solve for tentative sprinkler selection (based on sprinkler diameter). Use sprinkler catalogs.

Possible Choices

Name and Model	Total Diameter (in.)	gpm	Precipitation (in./hr)	Base Pressure (psi)	Nozzle Size (in.)

Step 5. Solve for total number of sprinklers.
   To cover all fairways, draw circles to scale on map, then count.

Considerations: Two-row system sprinkler placement.

With a two-row system, it is more adaptable to start at the green with a square spacing or a rectangular spacing rather than a triangular spacing, which allows good coverage of the approach area to the green, as opposed to using the triangular spacing and having difficulty in proper coverage of this area, as shown below.

1. Rectangular Spacing

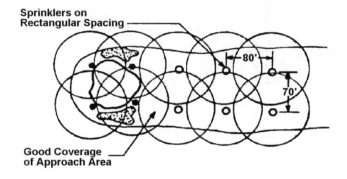

Figure 6.7

2. Triangular Spacing—Note weak coverage

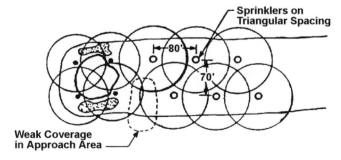

Figure 6.8

3. Possible Corrective Measures for Weak Coverage Area

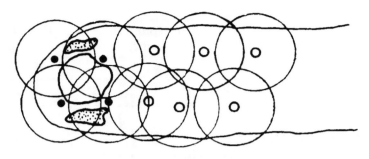

Figure 6.9

Considerations

1. Three Sprinklers for Approach to Green

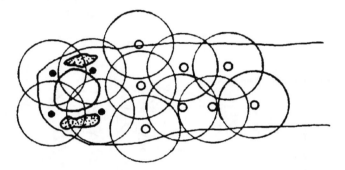

Figure 6.10

Rectangular and Triangular Fairway Sprinkler Spacing on a Dog-Leg Hole

On fairways which are dog-legs the rectangular spacing will work toward a triangular spacing around the curve, and gives very good coverage.

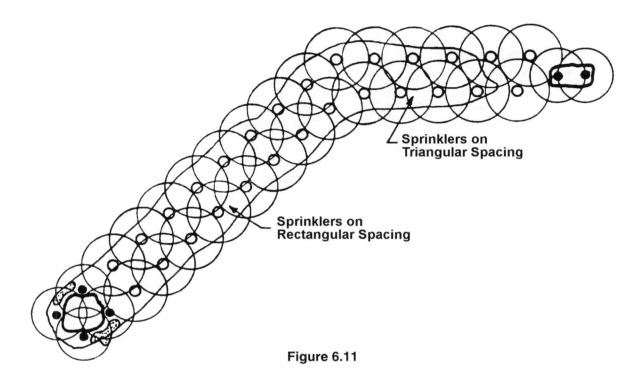

Figure 6.11

Step 6. Solve for total number of sets

Set = number of sprinklers that operate per control valve.

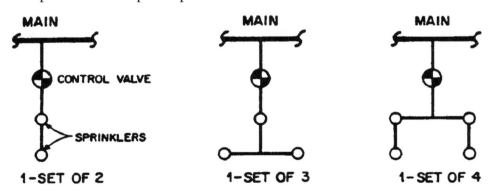

Figure 6.12

i.e., total sets $= \dfrac{\text{total fairway sprinklers}}{\text{no. sprinklers per set}}$

Step 7. Solve for water application rate.
Review two-row system.

$$\text{in./hr} = \frac{\text{gpm (one head)} \times 96.3}{S \times L}$$

where: S = spacing (ft) sprinklers on line
L = distance (ft) between lines

Step 8. Solve for operating time (min/night) and number of sets to operate at one time.
Note: Use design conditions and procedure outlined for one-row system. See pages 48 to 50.

Step 9. Solve for total gpm.

Step 10. Solve for pipe layout and pipe sizes.

Two-Row—Main and Lateral Pipeline Layout

The procedure for planning the two-row piping layout is basically the same as for the greens (two-row) etc. (See pages 91–93.) Start by drawing in the essential main pipelines, then consider various alternatives to complete the layout. (See Figure 6.13.)

Valve-in-Head (Loop within Fairway)

See page 133.

Factors to consider:

- Avoid uphill laterals in main lines. Also avoid laterals to greens and tees—*if possible.*
- Greens—Run laterals to the sides or back of greens. Avoid approach to green.
- Avoid running main pipelines through greens.

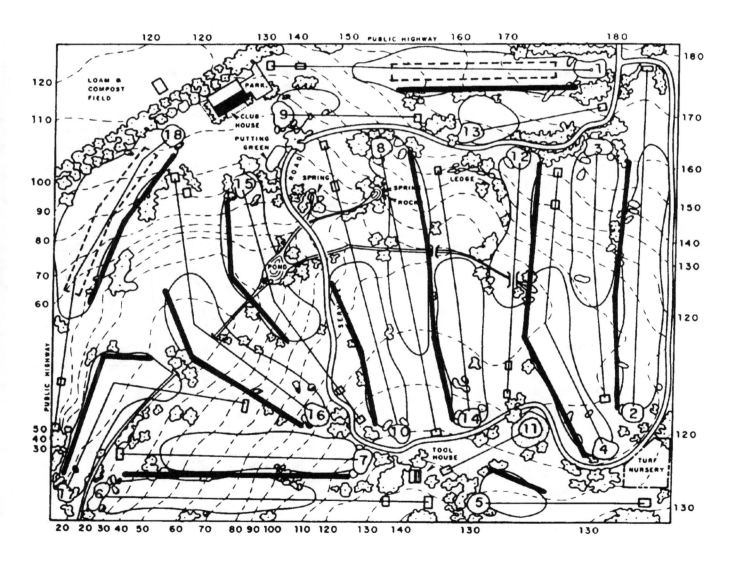

Figure 6.13

Piping—Two Sprinklers Per Valve, Triangular Spacing

The lateral to the first sprinkler is perpendicular to the supply main using a 90° standard or reducing tee joint. The section of pipe to the second sprinkler is gently bent. It can be installed by either trenching or pulling the pipe.

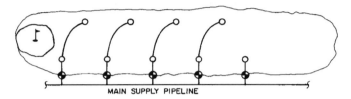

Figure 6.14

Piping—Four Sprinklers per Valve for Sloping Fairways

For sloping fairways, the piping method shown allows for ease of trenching and/or pulling pipe having the lines run in one direction along the length of the fairway. This arrangement helps to equalize the pressure and to drain the pipe lines by gravity.

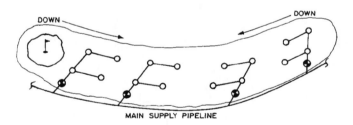

Figure 6.15

Piping—Main and Laterals

Piping—Two Sprinklers Per Valve— Parallel Fairways

If two fairways run parallel and are relatively close together, the main may be run between the two fairways and thus supply both fairways. (See Figure 6.16.)

One control valve is generally used to supply two sprinkler heads—although a separate valve for each sprinkler could be used if desirable and cost is not prohibitive.

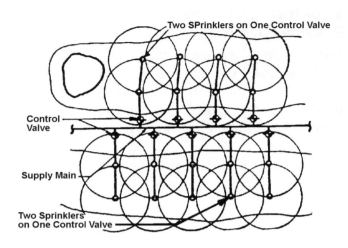

Figure 6.16

Piping—Four Sprinklers Per Valve— Parallel Fairways

If low-gallonage sprinklers are being used such that total gallonage of four sprinklers is within the capacity limits of a control valve, then four sprinklers may be supplied from one control valve. This would reduce the cost, but also reduce the flexibility of the system.

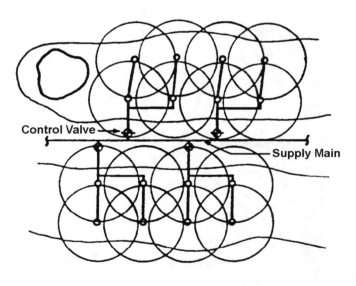

Figure 6.17

Piping—Three Sprinklers Per Valve

Figure 6.18 illustrates the two-row triangular pattern and piping layouts. It should be noted that either piping method will cause considerable trenching and/or pipe-pulling difficulties.

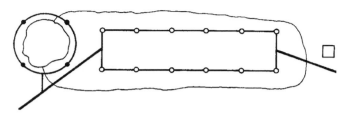

Figure 6.19

Piping—Two Sprinklers per Valve

In the design of a double-row system, it is more desirable to run the supply main along the side of the fairway—or if two fairways run parallel and are relatively close together, the main may be run between the two fairways and thus supply both fairways.

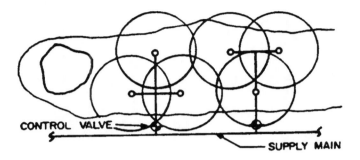

Figure 6.18

Two-Row System—Pipeline Layout

Piping

Figures 6.19 and 6.20 show the two most common methods used for a two-row fairway irrigation system.

Valve-in-Head (VIH) System

Loop

In addition to the advantages of the valve-in-head system outlined previously, the trench and installation of lateral pipelines is simplified in comparison to the block systems.

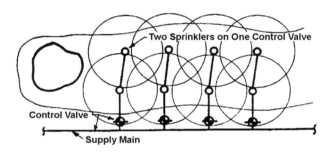

Figure 6.20

One control valve is generally used to supply two sprinkler heads—although a separate valve for each sprinkler could be used if desirable and cost is not prohibitive.

Chapter 7

Electrical Fundamentals

Introduction

Golf course irrigation control systems have evolved from the manually operated to the fully automatic. Currently there are two types available: (1) electrical, and (2) hydraulic. By far, the electricals are ranked number 1 in use. These range from the relatively simple electronic to the highly sophisticated computerized and/or radio controlled, which usually include a weather station. These require electrically operated relays, valves, and/or sprinkler heads and pump stations. In order to function properly, the components require a minimal voltage, which involves proper wire sizing.

In view of the important role of electricity related to the operation of controllers, pumps, maintenance equipment, golf carts, etc., it seemed appropriate to include a brief review of electrical principles entitled "Understanding Electricity."

A study of the following section shows that the electrical systems and water systems have many similarities and are often used to compare or explain the principles of one or the other.

Generator vs. Pump

Although the construction and operating principles are different, they do have a similar function in common. The generator creates a flow of electrons in wires, whereas the pump creates a flow of water in pipes.

Voltage vs. Pressure or Head

Voltage is a measure of electrical pressure caused by the separation of + and − charges. Pounds per square inch (psi) is a measure of hydraulic pressure caused by spinning an impeller, which forces water out of the pump and into a pipe line where a pressure buildup occurs.

Amperage vs. Gallons per Minute (gpm)

Amps is a measure of the rate of flow of electrons in a circuit and gpm is the rate of flow of water in pipelines.

Resistance (Ohms) vs. Friction Loss

Ohms is a measure of resistance to the flow of electrons in wires, and friction is a resistance to flow of water in pipes. The resistance to both will vary with the type of material, length, cross-sectional area, and temperature. Also, the electrical system and water system comply with *Ohm's Laws* of series and parallel circuits.

UNDERSTANDING ELECTRICITY

The Electron Theory

According to this theory all matter is composed of atoms. The atom operates as a miniature solar system, like the sun with the planets moving around it.

No one has ever seen an atom; therefore it is necessary to resort to symbols to represent what we think is actually happening.

The Atom

The center of the atom is composed of protons and neutrons. The protons have a positive (+) electrical charge. The neutrons are uncharged, but help increase the mass of the nucleus of the atom. (These are omitted in the sketches.)

The positive charge of each proton is of a magnitude to balance the negative (–) charge of one or more electrons. When the + and – charges are in balance externally, the atom is in a natural state, or electrically neutral.

The Atomic Structure

The nucleus and number of protons will differ in matter. Each and every material has its own unique atom which constitutes its special characteristics and identity.

The atom consists of "planetary" electrons which circulate around the nucleus at 186,000 mi/sec. These electrons cannot be readily removed from their orbit.

Also circulating about the nucleus at 186,000 mi/sec are the "free" or "gypsy" electrons which are held *loosely* in their orbit.

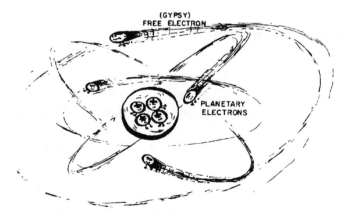

Figure 7.1. Atom with electrons and nucleus with positively charged protons.

The material used in both the production and distribution of electricity *must* contain the so-called "free" electrons. Materials such as silver, copper, aluminum, etc., fall into this category and are considered to be good conductors.

Materials that have atoms with "planetary" electrons with few if any free electrons cannot be used. Items such as rubber, plastics, glass, etc., are considered to be good insulators.

The Electric Generator

The generator is a device which can artificially unbalance the electrical charges of the atom. Nature always tries to maintain an electrical balance. The electrons of a negatively charged atom will move to one that is positively charged. This movement constitutes a miniature electrical current.

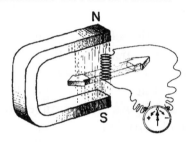

Figure 7.2

Generator Components

Regardless of size, all generators are made with the following components:

1. coil of insulated wire
2. magnet (magnetic field)
3. movement (means of moving or rotating a coil of wire through the magnetic field)

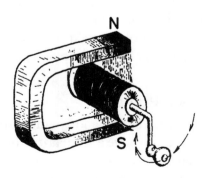

Figure 7.3

The Basis of All Electricity

Let us imagine that we can see what is happening in a tiny segment of wire as it passes through the magnetic field. The free electrons of Atom 2 are deflected away, leaving the atom positively charged, whereas Atom 3 gains electrons and becomes negatively charged. This unbalance of + and – charges is the basis of all electricity.

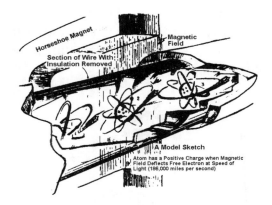

Figure 7.4

The Complete Circuit

As long as the generator is running, the electrical imbalance is maintained. However, the only way that electrons will flow is by providing a path (conductor) that is uninterrupted by an open switch, broken wire, etc., which is attached from one end of the coil of wire in the magnetic field back to the other end.

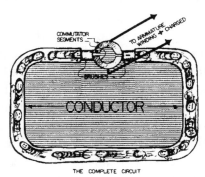

Figure 7.5

Let us again consider what will happen. As the free electrons from Atom 2 are deflected, leaving it

positively charged, the free electrons from Atom 1 move in to balance the charge, but are deflected away by the magnetic field. Meanwhile, the electrons before Atom 1 move to balance the deficit, causing a complete chain reaction along the entire circuit. The speed at which the electrons flow is 186,000 mi/ sec.

The Electromagnet

An electromagnet is created by a flow of electrons through a coil of insulated wire wrapped around an iron core.

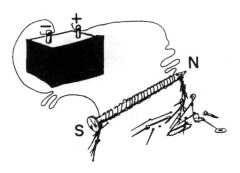

Figure 7.6. The electromagnet.

The Solenoid

Solenoids are electromagnets that are extensively used in all aspects of an automatically controlled irrigation system. They are found in all the valve-in-head (VIH) sprinkler heads, remote control valves, relay switches, computer components, etc. In addition, they are utilized in all electric motors, appliances, doorbells, etc.

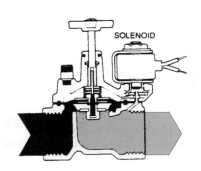

Figure 7.7

The Direct Current (DC) Generator

The automotive generator, which may still be found in some cars, has the three basic components of all generators. Electromagnets (field windings) are used in lieu of permanent magnets, and the coil of insulated wire, called an armature, is rotated in the magnetic field by the engine.

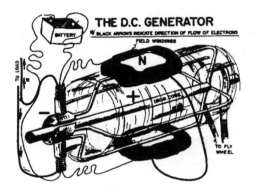

Figure 7.8. The DC generator.

Basically, when electrons flow from the battery through the field windings, an electromagnet is made. After the engine starts and the generator starts working, the magnetic field is self sustaining (trace circuits).

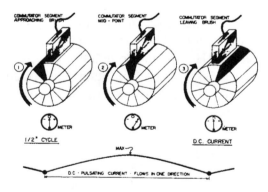

Figure 7.9

The commutator of a DC generator is divided into a number of insulated segments. The brushes are set to pick up an impulse or movement of electrons. Actually, direct current is a series of impulses or (1/2 cycle) which would appear as shown.

Figure 7.10

DC current flows in one direction through the circuit.

The Alternator—Alternating Current (AC)

Note: This illustration is used for the purpose of explanation only. No effort was made to show the actual construction nor the actual operating principles involved.

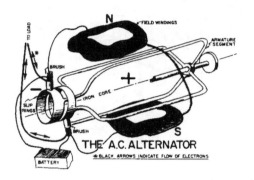

Figure 7.11. The AC alternator.

The primary difference between a DC generator and an AC alternator is that the first has a "commutator," whereas the latter has "slip-rings."

The slip-rings enable the brushes to pick the output from the armature winding through a complete revolution in the magnetic field. If viewed on an oscilloscope it would appear as a sine wave.

Alternating current means that the current flow changes direction of flow twice each cycle. In other words, if the flow went out your left hand and back on your right hand for 1/2 a cycle it would go out on your right hand and back on the left hand for the other 1/2 of the cycle.

The number of times the voltage goes through its cycle of change per unit of time is called "frequency" measured in cycles per second (cps). A new unit was adopted ca. 1928 called "hertz" (Hz) that enables us to avoid any mention of a unit of time. A 60 cycle per second voltage can now be called 60 Hz.

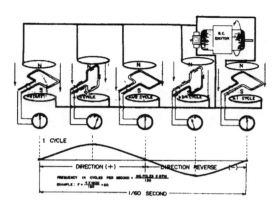

Figure 7.12

ELECTRICAL TERMS AND THEIR RELATIONSHIPS

Voltage—Symbol (E)

The voltage, also called *electromotive force* or *electrical pressure,* is perhaps the most difficult of the terms to describe. The voltage output of a generator can be changed by increasing any of the three basic components: that is, a more powerful magnetic field, more windings on the coil of insulated wire or a more rapid movement of the coil in the magnetic field. In turn, this will produce a greater potential difference between the + and – charges. The greater the potential difference, as in large generators, the greater the force or pressure available to drive the electrons along the conductor when the circuit is completed.

As an analogy, consider the effect of stretching a bow and arrow or a rubber band; the further it is extended, the greater the force upon release.

ELECTRO-MOTIVE FORCE*

MR. R. DEMONSTRATES VOLTAGE*

*(THE POTENTIAL DIFFERENCE BETWEEN PLUS AND MINUS CHARGES.)

Figure 7.13

Amperage—Electric Current

Amperage (symbol I) is the *intensity* or the *rate of flow of electrons.* To help clarify rate of flow, it is necessary to introduce the term coulomb. A coulomb is a specific quantity and is defined as 6,280,000,000,000,000,000 electrons passing a given point per second. Obviously, it would be very awkward to express the rate of flow in terms of numbers.

Ammeters are calibrated in *"amps."* It should be understood that when the above number of electrons flow past a point or through an ammeter, it can be said that the rate of flow is equal to 1 coulomb, or actually 1 amp. There is no regard for numbers or time factors.

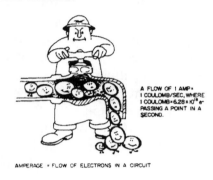

A FLOW OF 1 AMP = 1 COULOMB/SEC, WHERE 1 COULOMB = 6.28 x 10¹⁸ e⁻ PASSING A POINT IN A SECOND.

AMPERAGE = FLOW OF ELECTRONS IN A CIRCUIT

Figure 7.14

Wattage

Wattage (symbol W or P) is the term used to describe the total energy in the form of light, heat, and power. It is the product of volts × amps or (E×I).

Light is created by a flow of electrons through a tungsten filament enclosed in a argon gas–filled globe. This is necessary to prevent the tungsten from immediately burning up in the presence of oxygen.

Heat is created in a like manner, but a more rugged material like nichrome is used. This wire will not burn up in the presence of oxygen.

Power or *work* is obtained from electrical motors. Without going into a lengthy discussion on the operating principles of motors, suffice it to say that the driving force of all electric motors is based on a rotating magnetic field.

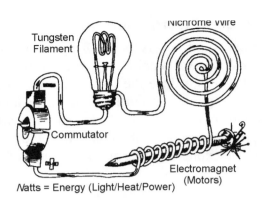

Figure 7.15

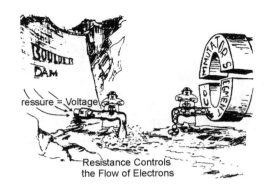

Figure 7.16

Resistance—Ohms

The resistance (symbol R) is measured by an *ohm-meter* and expressed simply as *ohms*.

The importance in the design of all electrical equipment and devices can be appreciated by considering the following illustration.

Imagine a commercial power plant having a generator that is capable of supplying the needs of an entire city. Yet, if we isolate one home and turn on one light, the resistance of this one light will govern the number of electrons that will flow from the generator through the completed circuit.

Perhaps this can be explained by the comparison shown. The billions of gallons of water held in storage by a mighty dam can supply the needs of a number of cities at the same time. However, if only one isolated faucet is turned on, the amount of water that will flow will be governed by the resistance of the faucet.

Important: The lower the resistance, the higher the electron flow.

Although tungsten is classified as *high resistance,* if it is a short length and large diameter, it could have a lower resistance than a long thin piece of copper wire. (See Figure 7.16.)

Resistance Factors

- Kind of material—The resistance will vary from good conductors to good insulators.
- Cross-sectional area—Resistance will be *inversely proportional* to its cross-sectional area.
- Length—The resistance is directly proportional to the length.
- Temperature—The higher the temperature, the higher the resistance.

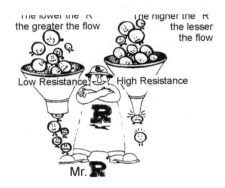

Figure 7.17

Ohm's Laws

The formulas, as shown in the formula wheel following, indicate the relationships of voltage, amperage, wattage, and resistance. Therefore, by knowing any two values, the others can be found by selecting the correct formula.

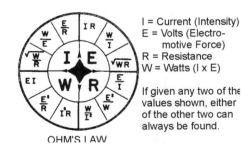

I = Current (Intensity)
E = Volts (Electro-motive Force)
R = Resistance
W = Watts (I x E)

If given any two of the values shown, either of the other two can always be found.

Figure 7.18

Another Approach

Another approach is the use of the triangles shown. However, some problems may require the use of both triangles.

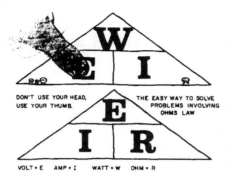

Figure 7.19

OHM'S LAWS OF SERIES AND PARALLEL CIRCUITS

This series of problems is included to mathematically show and explain the behavior of electricity when electrical devices are wired into a circuit in series and in parallel.

Most of the wiring is done *in parallel*. As can be seen by placing your thumb on any lamp(s), obstructing glow to one will not obstruct the flow to others; each will receive the full voltage. In a *series* circuit the voltage divides among the resistances. Also, turning off one lamp will break the circuit, and all lights are turned off.

Parallel Circuits

Law 1.
$$R_t = \frac{1}{\dfrac{1}{r_1} + \dfrac{1}{r_2} + \dfrac{1}{r_3} + \dfrac{1}{r_n}}$$

Law 2. I_t = sum of separate I's.
Law 3. E stays the same throughout.

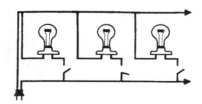

Figure 7.20. A parallel circuit.

Series Circuits

Law 1. R_t = sum of the separate R's.
Law 2. I stays the same.
Law 3. E_t divides among the resistance.

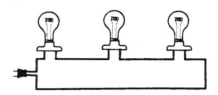

Figure 7.21. A series circuit.

Example (Parallel Circuit)

The motor nameplate shows that it is a 120/240-V motor. Terminal connections are for 120 V, as shown.

$$R_t = \frac{1}{\dfrac{1}{20} + \dfrac{1}{20}} = \frac{1}{\dfrac{2}{20}} = \frac{1 \times 20}{2} = 10$$

$$I_t = \frac{E}{R_t} = \frac{120}{10} = 12 \text{ amps}$$

$$W = EI = 120 \times 12 = 1440 \text{ watts}$$

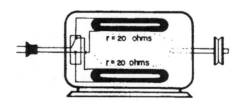

Figure 7.22. Motor wired in parallel.

Example

Assume the motor wired as shown was connected to a 240-V power line.

$$R_t = 10 \text{ ohms}$$

$$I_t = \frac{240}{10} = 24 \text{ amps}$$

$$W = 240 \times 24 = 5760 \text{ watts (winding would burn out)}$$

Example (Series Circuit)

The motor was rewired, as shown, and connected to a 120-V line.

$$R_t = 20 + 20 = 40 \text{ ohms}$$

$$I = \frac{120}{40} = 3 \text{ amps}$$

$$W = 120 \times 3 = 360 \text{ watts (too low to function)}$$

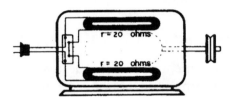

Figure 7.23. Motor wired in series.

Example

Finally, assume the motor was wired to a 240-V line.

$$R_t = 40 \text{ ohms}$$

$$I_t = \frac{240}{40} = 6 \text{ amps}$$

$$W = 240 \times 6 = 1440 \text{ watts}$$

Conclusion

The windings of the motor must be wired in parallel for 120 V and in series for 240 V.

AC TRANSFORMER

The transformer is defined as a device capable of stepping up the electrical power or stepping it down. It makes possible the transmission of electricity over long distances with relatively small wire sizes.

Note: Wire size is based on amperage carrying capacity: the higher the amps, the larger the wire size.

Transformers operate on alternating current only. It is essential to have a continuous making and breaking of a magnetic field in the primary. This is done automatically 120 times a second on AC current.

This momentary magnetic field produced in the primary causes a momentary movement of electrons in the secondary, which quickly decays.

A continuous flow of electrons (i.e., direct current) to the primary would create a momentary surge of electrons in the secondary, and then both would remain as permanent magnets, with no flow of electrons in the secondary.

Principles of Operations

The voltage is stepped up or stepped down directly proportionally to the ratio of turns on the primary side (the side connected to the power source), as compared to the turns on the secondary side.

The amperage is affected inversely proportionally to the ratio of turns.

The wattage (power) will be the same on both the primary and secondary sides of the transformer.

Transformer Equations

$$\frac{\text{volts on primary}}{\text{volts on secondary}} = \frac{\text{turns on primary}}{\text{turns on secondary}}$$

$$\frac{\text{amps on secondary}}{\text{amps on primary}} = \frac{\text{turns on primary}}{\text{turns on secondary}}$$

Example (Step-Down Transformer)

Note: The actual number of turns of wire on the primary side and secondary side of the transformer were omitted. However, the ratio of turns can be found by dividing the primary voltage by the secondary voltage. In this case 120/24 = 5 winds to 1 wind.

For illustrative purposes assume that the diaphragm valve solenoid = 0.5 amps:

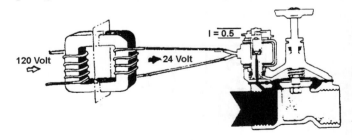

Figure 7.24

1. Solve for watts (power) on the secondary:

$$W = EI = 24 \times 0.50 = 12$$

2. Solve for amps on the primary:

$$\frac{\text{I on secondary}}{\text{I on primary}} = \frac{\text{turns on primary}}{\text{turns on secondary}} =$$

$$\frac{0.50}{X} = \frac{5}{1} = 5X = 0.5; \ X = 0.10$$

3. Solve for watts on the primary:

$$W = 120 \times 0.10 = 12$$

THREE-PHASE POWER

Three-Phase Generator or Alternator

In a three-phase alternator three single-phase windings are combined on a single shaft and rotate in the same magnetic field as is shown. Each end of each winding can be brought out to a slip ring and an external circuit. In a three-phase alternator the voltage in each phase alternates exactly a third of a cycle after the one before it alternates.

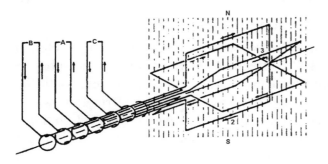

Figure 7.25

Actually, each end of each phase winding is not brought out to a separate slip-ring; rather, the three phases are connected together as illustrated. This makes only three leads necessary for a three-phase winding; each lead is serving two phases. Then each pair of wires acts like a single-phase circuit, substantially independent of the other phases.

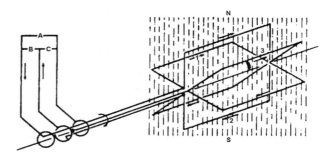

Figure 7.26

The voltage curve shows that when the voltage in Phase 1 is approaching a positive maximum as shown at "start," that in Phase 2 is at a negative maximum and the voltage in Phase 3 is falling.

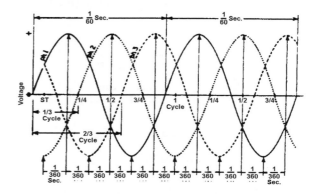

Figure 7.27

Here again the time per cycle is 1/60 sec. in any one winding, but by looking at the bottom of the diagram you see the boosts from the three windings combined are only 1/360 sec apart. Three-phase power is generally produced and transmitted because it is best adapted to motors and involves the least expensive power.

Three-Phase Transformers
Delta Method

A three-phase transformer bank consists of three separate single-phase transformers. The secondaries, one for each phase, may be connected in a variety of ways. The delta scheme shown formerly was the most common; one of the wires is center tapped. This system provides 240-V three-phase power and 120- and 240-V single-phase power. (See Figure 7.28.)

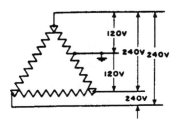

Figure 7.29

Star or Y Method

In power lines today the star or Y scheme is more frequently used; usually there is a fourth wire, the grounded neutral. The diagram shows a three-phase, four-wire Y connection with the center tap grounded. This system provides 208 three-phase and 120 and 208 single-phase.

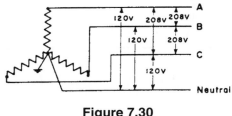

Figure 7.30

In the star or Y system, the three-phase voltage, instead of being 230 V, is usually 208 V. The single-phase current available across any two of the wires (not including the neutral wire) is also 208 V. However, the single-phase voltage between the neutral wire and any of the other three wires is 120 V.

At first glance this may seem all wrong, for if the voltage between wires A and B is 208 V, the voltage between the neutral wire and either A or B might be expected to be one-half of 208, or 104 V, instead of 120 V as previously stated. Remember, however, that in three-phase current, the voltage comes to a peak or maximum at a different time in each phase. At the instant that the voltage in secondary A is 120 V, that in B is 88 V, so that across wires A and B there is a voltage of 120 + 88, or 208 V. The system therefore has the advantage of making it possible to transmit over only four wires (including a grounded neutral) three-phase power at 208 V, single-phase power at 208 V, and single-phase power at 120 V. Quite frequently in a home,

instead of providing the usual 115/230-V three-wire system, three wires of the star-connected system (the neutral and any other two wires) are provided, thus furnishing 120 V for lighting and 208 V (instead of the usual 230 V) for water heaters and similar large loads.

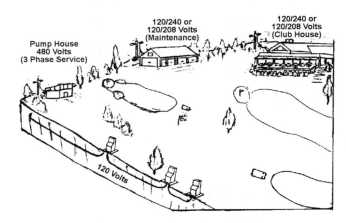

Figure 7.31. Golf course power distribution.

Caution!!! For golf courses equipped with three-phase power and a Y transformer, it is essential that the operator select motors, electrical maintenance equipment, appliances, etc., that correspond to the incoming voltages.

LIGHTNING PROTECTION—GROUNDING

Grounding Procedures

It is a well-known fact that no lightning protection is 100% effective. However, there is no substitute for a good grounding system when field controllers are being used on a golf course irrigation system. *Every controller location must be properly grounded to earth.* Without proper grounding, the lightning and power surge protection devices give little or no protection.

Grounding Process

1. Each field controller location must have a ground rod.
2. The ground rod must be 8 ft in length and have a diameter of 5/8 in. minimum.
3. The ground rod must be fully driven into the ground, as near to the controller as possible.

4. If the soil conditions make it impossible to drive the rod fully into the ground, it should be cut off at ground level and additional pieces should be driven 2 to 3 ft apart. The multiple pieces should be tied together and to the controller with #6 AWG solid copper wire.

5. Wire connections to ground rods must be made by utilizing appropriate copper or brass clamps. These connections should be checked and tightened, if necessary at least every 6 months.

6. In some situations it might be necessary to dig a trench (deep enough to prevent maintenance equipment from damaging the wire) from the ground toward a sprinkler location. Figure 7.34 shows some helpful guidelines for trenching detail. Lay a #6 AWG bare wire in the trench, connect to the ground rod with clamps, and close the trench. This will provide needed moisture for improved grounding.

7. Clay or loam soils have the lowest ground resistance levels. Whenever possible these soils should be selected for a grounding site.

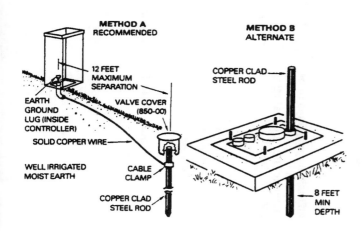

Figure 7.32

8. Chemical treatment of the soil is a good way to reduce the seasonal variations in soil conductivity. Magnesium sulfate, copper sulfate, and ordinary rock salt are suitable materials. Magnesium sulfate is the least corrosive, but rock salt is the cheapest and does the job if applied in a trench dug around the electrode. This method should be considered when deep or multiple grounding rods are not practical. (See Figure 7.33.)

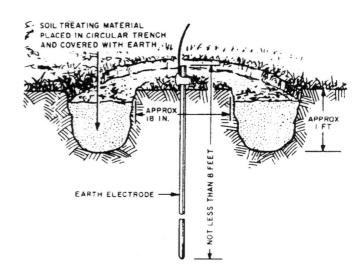

Figure 7.33

9. Separate grounding networks for each controller reduce the chances that a power surge entering one controller will enter another. (See Figure 7.35.)

> *WARNING!!!*
> *Never connect any AC common, modem drain, valve common or pump common to the earth ground. This action would defeat the protection built into the controller and could result in serious damage to equipment and/or injury to personnel in the event of a lightning strike.*

Note: Certain manufacturing companies require an earth ground reading of 10 ohms (maximum) or less. There are several rules to keep in mind to reduce earth ground readings, if necessary.

- Resistance to ground decreases with depth of the electrode in the earth. An 8-ft to 10-ft rod is most effective when it is driven its entire length. The deeper the rod, the better the contact with the deeper moisture levels.
- The length of the rod is what counts; the diameter has little effect in resistance readings.
- Ametek boxes or equivalent placed over the ground rods will retain moisture better than the buried rod. They also will provide a place to test grounding levels and wiring connections.

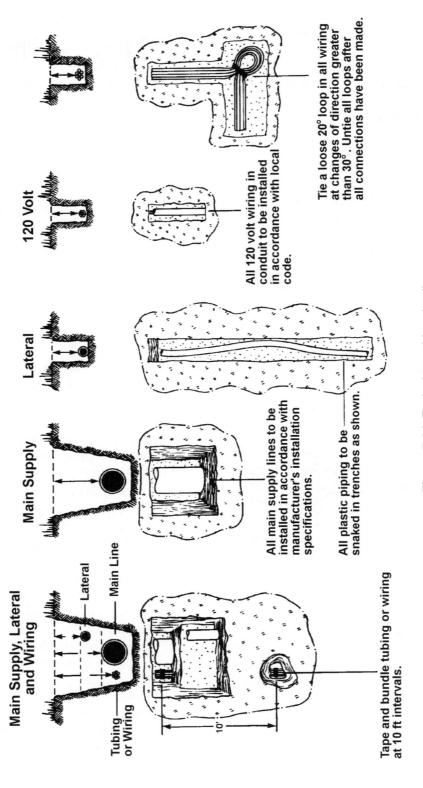

Figure 7.34. Typical trenching detail.

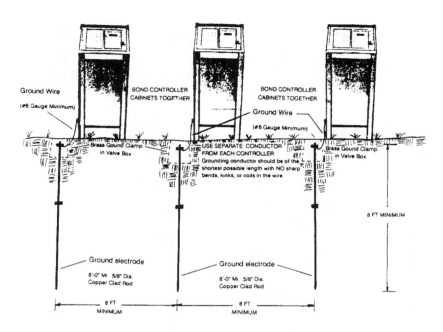

Figure 7.35

- Cadwell connectors actually fuse the ground wire to the ground rod. These should be used whenever possible.
- Two well-spaced rods driven into the earth are in effect two resistances in parallel. However, the rule for two resistances in parallel does not apply exactly. The resultant resistance is not one-half of the individual rod resistance (assuming they are of the same size and depth). Actually, the reduction for two rods of equal resistance is about 40%. If three rods are used the reduction is 60%, and for four, 66%.

Where multiple rods are used, they should be spaced at least two times the length of their immersion apart. This will improve the effectiveness of their grounding ability.

Example

Having two 10-ft rods in parallel with 10-ft spacing, the resistance is lowered to about 40%. If the spacing is increased to 20 ft the resistance is lowered to 50%. Figure 7.36 shows the minimal spacing.

Field Controller Grounding With Solid Copper Sheet

Copper sheet installed as illustrated for grounding field controllers (satellites) offers some favorable attributes as compared to driven ground rods. For example:

- The existing soil type, texture, water holding capacity, depth of the water table, presence of rocks, etc., at the site of grounding are not major factors. The soil as found is removed and replaced with a selected material that has good grounding characteristics.
- Ease of installing the copper plates as compared to driving 8-ft or 10-ft rods. Soils containing rocks or bedrock may make it impossible to force the rod down all the way. This would involve cutting the rod and adding additional shorter rods connected together.
- The moisture gradient and fluctuating water table are not a major consideration as long as the soil around the copper sheets is moistened at all times. A drip irrigation system or equivalent method would be ideal for this purpose.

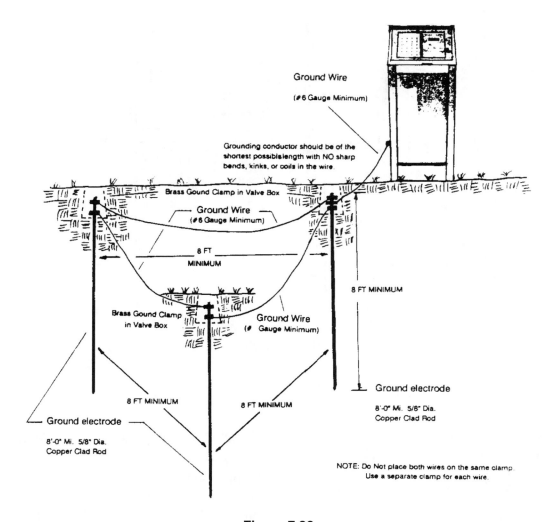

Figure 7.36

- The rod surface contact area for a 3/4-in. × 10-ft rod would equal $2\pi rL = 2 \times 3.14 \times 0.75$ in./2 × 10 ft = 282.6 in.^{2}

    The 18-in. × 24-in. (two sides) sheet surface–soil contact area would equal 18 in.× 24 in. × 2 = 864 in.^{2} (See Figure 7.37.)

WIRE SIZING

Wire Sizing Between the Central Controller and Satellites

There is one basic assumption made and one rule to be followed when sizing wire runs between the *central controller* and the *satellites*.

The assumption is that a minimum of 117 V is furnished to the central controller.

The rule is that 105 V is needed at the satellites for guaranteed proper performance of the controller motors.

These values may vary somewhat with different manufacturers.

Another assumption which must be made is that all wire sizes as shown on the design plan will be what is installed and *all splices* will be correctly made, insulated properly against water seepage or corrosion.

Recommended Splicing Procedures

Following are recommended splicing procedures for low-voltage and 120-V wire connections. (See Figures 7.38 and 7.39.)

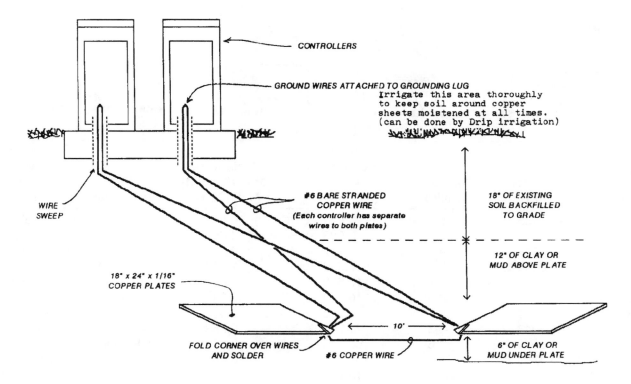

Figure 7.37. Field controller grounding. Not to scale.

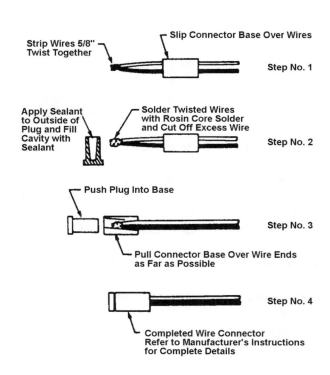

Figure 7.38. Low-voltage wire connection detail (no scale).

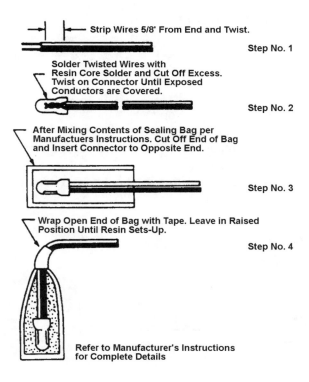

Figure 7.39. Wire connection detail for 120-V wire (no scale).

Table 7.1. Wire Sizes

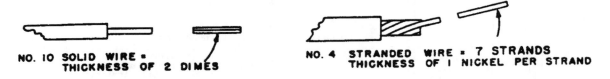

NO. 10 SOLID WIRE = THICKNESS OF 2 DIMES

NO. 4 STRANDED WIRE = 7 STRANDS THICKNESS OF 1 NICKEL PER STRAND

Wire Gauge (AWG)	Diameter (in.)	Diameter (mils)	Cross-Sectional Area (cir. mils)[a]	Resistance (ohms/1000 ft) 77°F (25°C)	Wire Thickness	Wire Type (uf)
18	0.0403	40.3	1,624	6.510	—	Solid
16	0.0508	50.8	2,583	4.090	—	Solid
14	0.0641	64.1	4,107	2.580	1 Penny	Solid
12	0.0808	80.8	6,530	1.620	1 Nickel	Solid
10	0.1019	101.9	10,380	1.020	2 Dimes	Solid
8	0.1285	128.5	16,510	0.641	2 Pennies	Solid
6	0.1620	162.0	26,250	0.403	2 Nickels	Stranded
4	0.2043	204.3	41,740	0.253	4 Dimes	Stranded
2	0.2576	257.6	66,370	0.159	4 Pennies	Stranded
0	0.3249	324.9	105,500	0.100	—	Stranded
00	0.3648	364.8	133,00	0.079	—	Stranded

[a] Note: 1 mil = 0.001 in. diameter; circular mils = mils^2

Circular Mil Method

$$A = \frac{10.8 \times L \times 2 \times I}{E} \text{ or } A = \frac{21.6 \times L \times I}{E}$$

where: E = voltage drop (V)
I = current (amperes)
R = ohms (10.8 for annealed copper wire)
L = length (ft)
A = cross-sectional area of wire (circular mils)

Example

Procedure

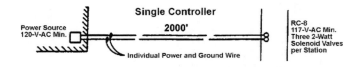

Power Source 120-V-AC Min.

Single Controller
2000'

Individual Power and Ground Wire

RC-8 117-V-AC Min. Three 2-Watt Solenoid Valves per Station

Figure 7.40

Step 1. Calculate controller primary current requirements.

RC-8 Controller only = 0.17 amp
Three solenoid valves per station
= (3 × 0.07) = 0.21 amp
Total = 0.38 amp

Step 2. Calculate maximum allowable voltage drop.

power source (minimum available) = 120 V-AC minimum
desired voltage at controller = 117 V-AC minimum
allowable voltage drop = 3 V-AC

Step 3. Calculate equivalent circuit length

(One) Controller × 2000 ft (equivalent length)

Step 4. Calculate A (circular mils).

$$A = \frac{21.6 \times 2000 \times 0.38}{3}$$
= 5472 CM = No. 12 wire size

The wire sizing method in Table 7.1 was used by many designers for years before the irrigation manufacturers began publishing their own wire sizing procedures. Appendix 6 shows typical examples of current wire sizing techniques.

ACKNOWLEDGMENTS

1. Irrigation Equipment Catalog and Handouts. Buckner Irrigation Company.
2. Irrigation Equipment Catalog. Rain Bird Manufacturing Co.
3. Technical Information Bulletins. Toro Irrigation Company.
4. "Getting Down to Earth." Manual. Biddle Instruments.

Chapter 8

Pump Selection

The following general information defines various pump types, pump classifications, and operating characteristics. A good understanding of this information should assist the superintendent in selecting, installing and operating pumps.

WELL TYPES

Shallow well—generally where the vertical lift from the water source to the pump is 25 ft or less.
Deep well—where vertical lift exceeds 25 ft.

PUMP TYPES

Non-positive—these pumps have an impeller that spins in a casing that does not have close tolerance. Should the valve be closed or the pressure switch fail, the pressure would not build up beyond its normal rating.
Positive displacement—these pumps use gears, pistons, or helical rotors that will discharge water with each rotation or stroke. The pressure will build up until the weakest part of the system fails. These pumps should be protected by a relief valve or equivalent.
Single-stage—one impeller is enclosed in a casing.
Multi-stage—contains multiple impellers enclosed in the casing that produce the same results as pumps in series.

Classification

Non-positive displacement—centrifugal pumps

- shallow or deep well
- straight or multi-phase
- surface mounted or submersible
- electric motor or engine driven

Piston displacement—gear-reciprocating helical rotor

- shallow or deep well
- straight or multi-phase
- surface mounted or submersible
- electric motor or engine driven

Pumps In Series and In Parallel

The analogy between water systems vs. electrical systems is commonly used to clarify or explain one system or the other. Comparing the basic terms shows that a high degree of similarity does exist, such as: generator vs. pump, voltage (electrical pressure) vs. head pressure (psi), amperage (electron flow) vs. gpm flow, ohms (resistance to flow) vs. friction loss (resistance to flow).

Also, both systems comply with Ohm's Laws of series and parallel circuits. For additional discussion refer to page 175 in the section Understanding Electricity.

Pumps in Series

Ohm's Laws

1. Total pressure (psi) = sum of the separate pressures.
2. Total flow (gpm) = same throughout the circuit.
3. Total resistance (R_T)= sum of the separate resistances.

Example

1. Total psi = 20 + 20 + 20 = 60
2. Total gpm = 20

3. $Pump_1$ $R = \dfrac{psi}{gpm} = \dfrac{20}{20} = 1$

4. $R_T = \dfrac{total\ psi}{total\ gpm} = \dfrac{60}{20} =$

   3 times more than single pump

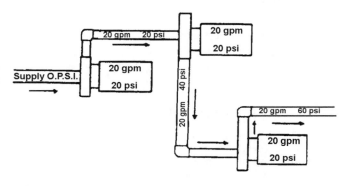

Figure 8.1. Three 20-gpm, 20-psi pumps in series equals 20 gpm at 60 psi.

Pumps in Parallel

Ohm's Laws

1. Total psi = same throughout the circuit
2. Total gpm = sum of the separate gpm

3. $R_T = \dfrac{1}{\dfrac{1}{R_1} + \dfrac{1}{R_2} + \dfrac{1}{R_N}}$

Example

1. Total psi = 20
2. Total gpm = 20 + 20 + 20 = 60

3. R (each pump) $= \dfrac{total\ psi}{total\ gpm}$

   $= \dfrac{20}{60} = 0.333$

4. $RT = \dfrac{1}{\dfrac{1}{3} + \dfrac{1}{3} + \dfrac{1}{3}} = \dfrac{1}{1} = 1.0$

   = 3 pumps in parallel = RT of
   1 single pump or 3 times less
   resistance

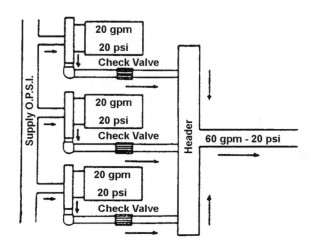

Figure 8.2. Three 20-gpm, 20-psi pumps in parallel equals 60 gpm at 20 psi.

Booster Pumps

The booster pump can be used on domestic lines of inferior pressure or in large golf course irrigation systems where compensations are necessary for pressure losses due to elevation. A booster does as the name implies: it boosts the pressure.

If the pressure at a certain point in the system is *30 psi* at *20 gpm* and the system requires *50 psi* at *20 gpm* at that point, a booster pump that is rated at *20 psi* and *20 gpm* can be installed in the line. (See Figure 8.3.)

Pressure Tanks

Pneumatic pressure tanks are often used where a wide variance of gallonage requirements exists. The pressure tank will relieve the pump from kicking on for a short period of time when a low gallonage demand is made. The tank acts as a pressurized reservoir of water with expanding air forcing water out of the tank to fulfill low and infrequent water demands.

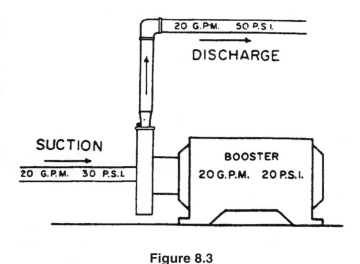

Figure 8.3

Figure 8.4 shows a typical system using a deep well pump with a pneumatic pressure tank. This system has an automatic air replenishing feature which can't normally be used when pumping from a reservoir using horizontal centrifugal pumps.

The water in the pump column (3) drains back into the well after each pumping cycle, and is replaced by air entering through the float vent valve (5). When the pump is started again, some of the air is forced into the tank, replenishing the air-supply there. Excess air is vented to the atmosphere by the float-operated level control valve (8) opening the solenoid valve (12). The float in the control valve (8) will close the solenoid valve when the level is proper, readying the system for the next pumping cycle.

If the pressure tank is too small for low gallonage demands, it will cause frequent and repetitive startup of the pumps. This can also happen if the tank becomes waterlogged or the air is absorbed in the water, causing a loss of the volume of air.

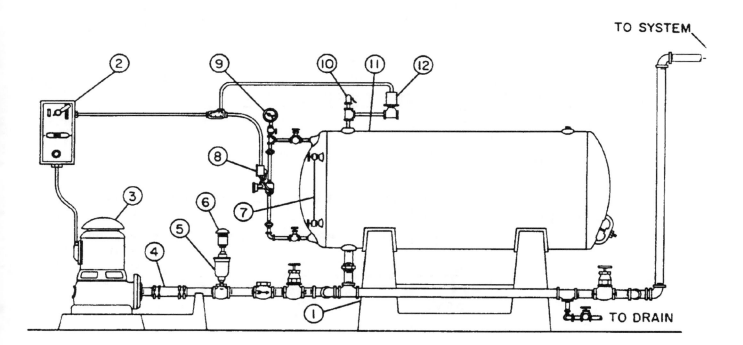

Figure 8.4. 1. Tank saddles; 2. Combination starter; 3. Deep well pump; 4. Rubber hose connection; 5. Float vent valve; 6. Air filter; 7. Water gauge; 8. Automatic dual pressure and level control; 9. Pressure gauge; 10. Pressure relief valve; 11. Pressure tank; 12. Solenoid valve.

An alternative for a pressure tank in a system where the demand varies widely is recirculation using a *dump valve*. Recirculation means to take the water supplied by the pump in excess of the water demanded of the system and dump it back into the supply. The dumping should be done back into the reservoir, because just repiping it back to the inlet can cause severe heating of the water when the demand is small, making the recirculated water volume large. A *pressure relief valve* is used to control the amount of water to be recirculated. Pressure tanks and the related equipment necessary for proper operation can become a maintenance headache, so they should be designed by an expert in the field of pumps and tanks.

Jockey Pumps

At times a small *jockey pump* is used to replenish the tank if low demands exist for a period longer than that for which the tank can provide. (See Figure 8.5.)

An example for the above case would be to have the jockey pump activate, by way of a pressure switch, at 120 psi. It would continue to run until the pressure tank is replenished to 140 psi. If the demand were greater than the jockey pump, the pressure would continue to decrease until it reached the low limit of the pressure switch of the main supply pump. At that point the main pump would activate. If more than one main pump were used, they would be activated in turn as the pressure continued to drop.

Non-Positive Centrifugal Pumps
Principles of Operation

The name of this type of pump comes from the force which is exerted by a body moving in a circular path—centrifugal force. In the pump the liquid is forced to revolve and therefore exerts a centrifugal force on the liquid in the case around the revolving wheel or impeller which is equal to the discharge pressure or head.

The liquid forced out of the casing creates a partial vacuum, permitting atmospheric pressure to force more water into the pump suction, and the operation is continuous.

With no pistons, valves, or close clearances, a centrifugal pump is not positive acting. In other words, a particular impeller is good for only so much

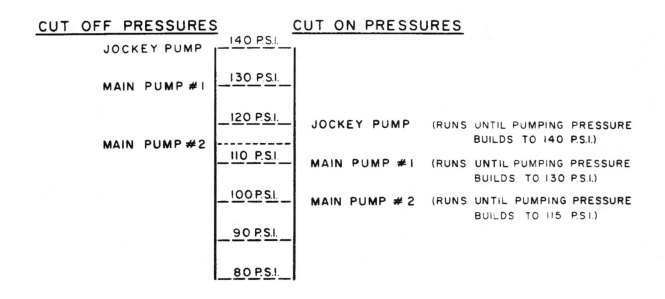

Figure 8.5

pressure or head, and if the actual pressure or head is higher than this, the impeller will merely churn the liquid.

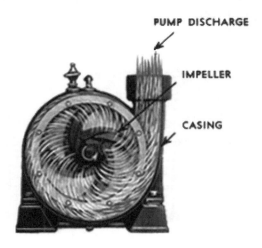

Figure 8.6

Horizontal, Centrifugal, Split-Case Pumps—Single-Stage Case Construction

At the shaft center line, the pump case is split horizontally into upper and lower half casings. This design allows convenient removal of the entire rotating element from the pump casing without disturbing coupling alignment or system piping connections.

The casing half faces are accurately machined and provided with a suitable gasket for sealing.

Figure 8.7

Multi-Stage, Horizontal, Split-Case Pumps

The designation *Type TU* is given to two-stage pumps. The designation *Type TUT* is given to three-, four- and five-stage pumps. Type TU is for capacities up to 4500 gpm, for heads up to 1100 ft. Type TUT is for capacities up to 1400 gpm, for heads up to 1600 ft. Three-, four- and five-stage TUT pumps are built with high loop crossovers similar to Figure 8.8.

	Type TU 2 Stage	Type TUT 3 Stage	4 Stage	5 Stage
Heads	up to 1100 ft	up to 950 ft	up to 1600 ft	up to 1550 ft
Capacities	up to 4500 gpm	up to 1400 gpm	up to 900 gpm	up to 300 gpm
Temperature	up to 300°F	up to 300°F	up to 300°F	up to 300°F
Case Pressures	300–500 psi	350 psi	400–700 psi	700 psi
HP Range	up to 800 bhp	up to 300 bhp	up to 325 bhp	up to 170 bhp
Drives	Electric motor, stationary engine, steam turbine			
Construction Materials	Cast iron, bronze fitted (stainless steel fitted optional).			

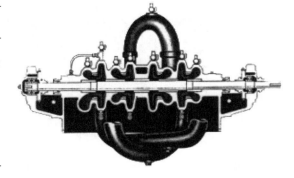

Figure 8.8

Hydroconstant Variable-Speed Centrifugal Pumps—How They Work

Hydroconstant drives control pump or fan speed to hold constant a system condition. Sensing—local or remote—can be used to control:

single point pressure, flow, liquid level, differential pressure, and temperature.

It works on a hydrokinetic principle, using hydraulic oil for torque transmission.

In essence, it works like this:

- An impeller mounted on the input shaft is coupled to a runner mounted on the output shaft by the hydraulic oil shared between them. At maximum oil levels, the output shaft speed is about 97% of the full load input shaft speed. As lesser amounts of oil are shared between the impeller and the runner, the output shaft runs proportionally slower

- In operation, a driver turns the input shaft that's fastened to the impeller. The runner is mounted on the output shaft close to and facing the input impeller. This output runner is enclosed by a cover which is mounted on and turns with the input impeller.

- A nozzle directs hydraulic oil into the cover chamber. A splitter arm between the nozzle and the cover chamber responds to a demand signal to direct either more, less, or a constant amount of oil into the cover chamber.

- To accelerate the output runner and shaft, the splitter maximizes the flow of the oil into the cover chamber. (See Figure 8.9a.)

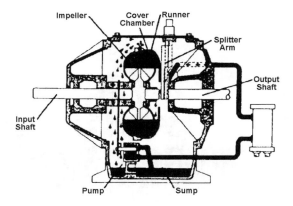

Figure 8.9a. Speed change sequence. Acceleration mode.

- At constant speed, the splitter divides the flow of the hydraulic oil between the cover chamber and the return-to-sump. At constant speed and load, the oil is discharged through the cover's metering ports around its periphery. (See Figure 8.9b.)

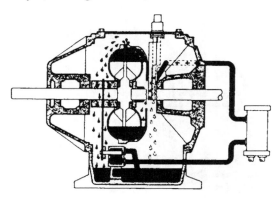

Figure 8.9b. Speed change sequence. Constant speed mode.

- To decelerate the runner, the splitter minimizes the flow of oil into the cover chamber (See Figure 8.9c.)

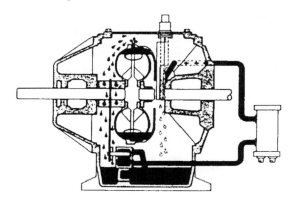

Figure 8.9c. Speed change sequence. Deceleration mode.

The hydraulic oil reaches the cover chambers as follows: In sizes 6A through 17B hydroconstant drives, which drive machines from 3 to 250 hp (60 Hz), the hydraulic oil is circulated from the sump to the splitter by an internal pump. The position of the splitter arm controls the amount of hydraulic oil directed into the cover chamber. It also regulates the speed of the hydroconstant drive's output shaft. The splitter is spring-loaded in one direction. An opposing load control device changes its position.

Several control devices are available. They sense and respond to a deviation from a present value in pressure, temperature, liquid level, flow, or differential pressure.

Through the automatic adjustments of the splitter arm's position, it can move the sensed value toward its present reference value by acceleration or deceleration of the output shaft.

In other words, a Peerless hydroconstant drive provides the most reliable, cost-effective variable-speed control available.

Jet-Centrifugal Shallow-Well Pumps

The nozzle on a jet pump does the same work as the nozzle on a garden hose. Because of its smaller size, it increases the speed of the water (shown here as *drive water*) flowing through it. This action creates a partial vacuum at the nozzle throat. Atmospheric pressure then forces the well water into the vacuum and into the pumping system. In other words, the vacuum created at the nozzle discharge pulls water from the well into the injector. (See Figure 8.10.)

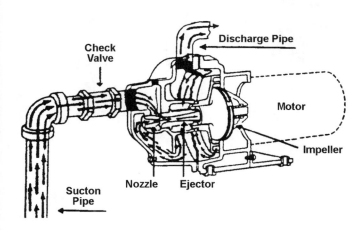

Figure 8.10

The venturi tube, which gradually increases in size, slows down this high-speed water. At the same time it increases its pressure. As a result, a combination of *drive water* and *pumped water* emerges from the venturi at relatively high pressure.

In order to operate the jet, there must be some continuous source of drive water. In conventional jet pumps, this is supplied by the impeller(s) of a conventional centrifugal pump.

Different arrangements of ejector and centrifugal pumps are used in shallow-well and deep-well jets. In shallow-well units, the ejector is mounted in or attached to the pump housing, as shown in Figure 8.11. Connected to the jet is a suction line which extends below the water level down in the well. When the pump starts, the centrifugal impeller forces a stream of water through the ejector. The vacuum created at the nozzle discharge draws water from the well into the pump. Water can be drawn up a maximum of about 25 ft with this arrangement.

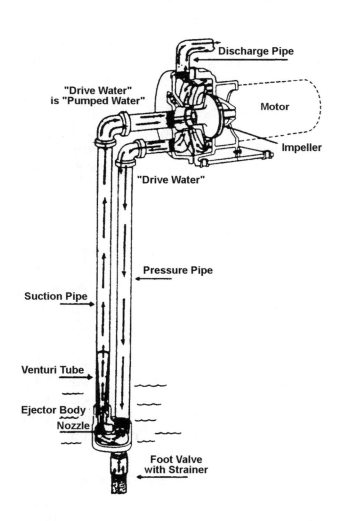

Figure 8.11

How a Deep-Well Jet Works

Deep-well jets differ in that they have the ejector submerged in the water down in the well. In this case the centrifugal pump (on the surface) forces water down what is known as the *pressure pipe,* or *drive pipe.* This pipe is connected to the pressure or nozzle side of the ejector body. As the drive water flows through the ejector, it creates a vacuum, which draws well water into the ejector body. This water, together with the drive water, flows up to the surface through a second pipe, known as the *suction pipe.* Part of the water is recirculated through the drive pipe. The rest is delivered to the pressure tank.

Deep-Well Centrifugal Submersible Pumps

Deep-well submersible pumps have a closely connected motor and pump assembly built as a unit. The motor is specially designed to spend its lifetime in use underwater. The whole assembly is let down into the well on the end of a drop pipe to a position below the water level. In this way the pumping mechanism is always filled with water—primed—and ready to pump. (See Figure 8.12.)

Deep-Well, Submersible, Centrifugal Pumps

Deep-well centrifugal pumps may be of two types: Figure 8.13a shows the submersible unit, in which both motor and pump mechanism are submerged in the well water. In the case of a vertical turbine (Figure 8.13b), the pump unit is connected to a motor at the surface by means of a vertical shaft.

A submersible pump is used for both shallow- and deep-well applications where the physical dimensions of the source of water will accommodate the unit in a submerged position. It is, however, most frequently used as a deep-well type pump.

Operation of Impellers and Diffusers

All of the impellers of the multi-stage submersible pump are mounted on a single shaft, and all rotate at the same speed. Each impeller passes the water

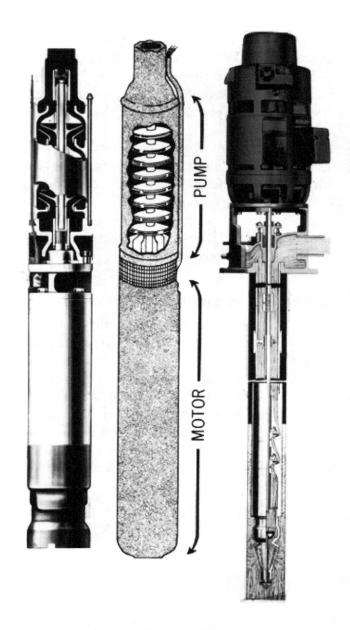

Figure 8.12

to the eye of the next impeller through a diffuser. The impeller increases the velocity of the water and discharges it into the surrounding diffuser. The diffuser is shaped to slow down the flow of the water and convert velocity to pressure. Each impeller and matching diffuser is called a stage. As many stages are used as are necessary to lift the water out of the well at the required system pressure.

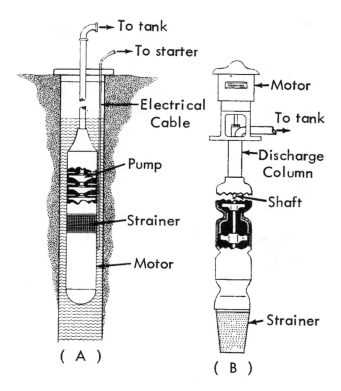

Figure 8.13

Figure 8.14

Portable, Engine-Driven Centrifugal Pumps

For installations where electricity may not be available and is too costly to obtain, or where more than one water source may be used, perhaps an engine-driven centrifugal pump may be used to advantage.

In addition to the diesel and tractor-driven pumps shown below, there is a variety of air- and water-cooled gas engine models available to choose from. (See Figure 8.14.)

Positive Displacement Pumps
Piston Pumps—How They Work

The operating principle of a piston pump is basically the same as the hand-operated piston pump.

As you push down on the handle of a hand-operated piston pump, water is lifted by the plunger, and it flows out of the spout. At the same time, a partial vacuum develops below the plunger. The partial vacuum causes the water in the suction pipe to force open the check valve and fill the cylinder below the plunger. (See Figure 8.15.)

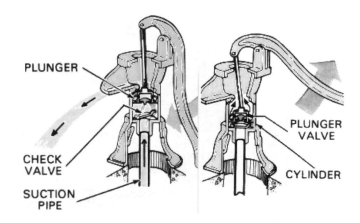

Figure 8.15

As you lift the pump handle the water below the plunger is trapped by the closed check valve in the bottom of the cylinder. At the same time, a valve in the plunger lifts and lets the water pass to the upper side of the plunger. It is then ready for discharge when the pump handle is pushed down and the plunger is raised again. This is called a *single-acting* piston pump. This is the simplest type of shallow-well suction pump.

The shallow-well piston pump uses the double-acting principle. That is, it pulls water from the well on one side of the plunger. Movement of the plunger to the right pulls water from the well into the left chamber and forces water out of the right chamber (b). Plunger movement to left forces water out of the left chamber and pulls water from the well in the right chamber. (See Figure 8.16.)

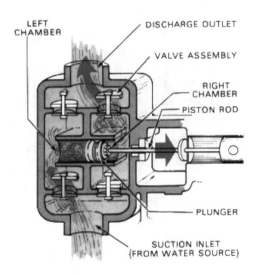

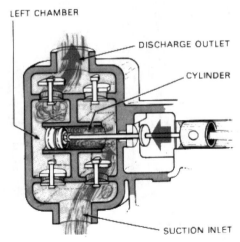

Figure 8.16

Helical Rotor Pumps—How They Work

Shallow-well helical rotor pump. The helical metal rotor C rotates in a molded rubber stator D, trapping water and forcing it toward the discharge end. B is the shaft linkage with the motor, A.

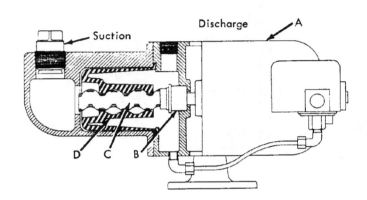

Figure 8.17

Helical rotor pumps are of the positive displacement type. The deep-well units are similar to the shallow-well helical rotor pump. There is a molded rubber stator in which a metal rotor is driven by a shaft and flexible cable. The pump unit is submerged in the well water at the bottom of open-line shaft column. A driver—usually an electric motor—at the surface is connected to the line shaft. Helical rotor pumps are also available with submersible motors. Operation and performance characteristics of the deep well helical rotor pumps are similar to the shallow-well unit. (See Figure 8.18.)

Reciprocating Pumps—How They Work

Deep-well working heads have a power source at the surface connected by rods to a submerged cylinder. Alternate up-and-down movement of the cylinder forces water to the surface. (See Figure 8.19 and 8.20.)

Rotary Pumps—How They Work

A rotary pump is simple in design, has few parts and, like a reciprocating pump, is positive acting. It consists primarily of two cams or gears—spur or herringbone—in mesh. (See Figure 8.21.)

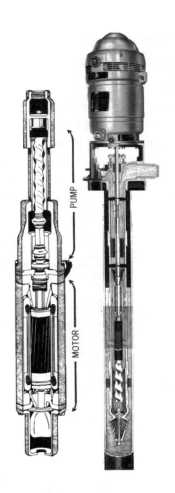

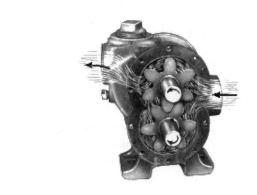

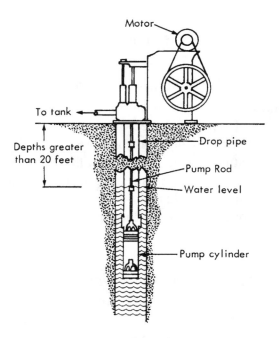

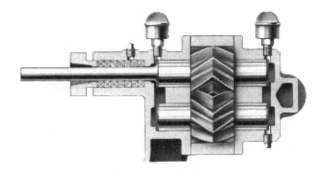

Figure 8.18

Figure 8.20

Figure 8.19

Figure 8.21

THE PHYSICS OF PUMPING

Discharge Side

Let us imagine a pail of water is on the top of a ten-story building, 100 ft above the ground.

Figure 8.22

If it were pushed off, it would fall faster and faster (acceleration), and finally hit the ground. We can calculate how fast it was going when it hit by using the following formula for the law which gives the velocity of a falling body:

$$V^2 = 2gh$$

where: V = velocity in ft/sec of the moving body
g = acceleration of gravity = 32.2 ft/sec
h = head in feet, or distance of fall (100 ft/sec in this example).

Therefore, the pail hit the ground at:

$$V = \sqrt{2gh} = \sqrt{64.4\ \text{ft} \times 100} = \sqrt{6400\ \text{ft}}$$
$$= 80.4\ \text{ft} / \text{sec}$$

Now, if we want to throw the pail back up to the top of the building, it will have to start with an initial velocity equal to the speed it had attained when it hit the ground, that is 80.4 ft/sec.

This is the same law or formula that is used in determining the speed and diameter of the impeller in a pump. Liquid, as it leaves the impeller, in order to be thrown a certain height, must have the same velocity it would have if it fell from that height.

Suction Side of the Pump

How Pumps Lift Water

Atmospheric Pressure

Air has weight: A column of air *one inch square* extended from sea level to the outer limits of the earth's atmosphere will exert a pressure of 14.7 pounds per square inch absolute (psia). This pressure plays an important role in hydraulics, particularly as related to the operation of certain types of pumps.

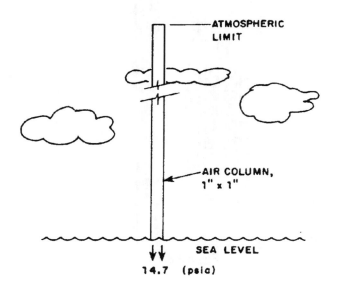

Figure 8.23

Vacuum of Pressure Below Atmospheric

Contrary to popular opinion, pumps do not "lift" water up from the source. Rather, the pump reduces atmospheric pressure on the water in the suction pipe and the atmospheric pressure on the water outside of the suction pipe pushes the water up into the pump. The principle is the same as that of drawing soda through a straw or filling a syringe as shown in Figure 8.24.

Limitations Above Sea Level

The importance of keeping within the suction limitations of the pump cannot be overemphasized. The dynamic suction lift must be calculated carefully due

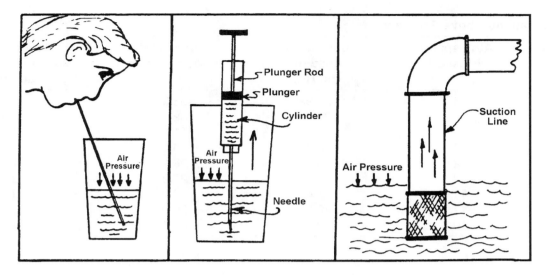

Figure 8.24

to the fact that the suction lift is reduced when the flow capacity increases. Even systems having the water source at an elevation higher than the pump can cause trouble when the friction losses are too high. For any installation, consult the pump manufacturer representative or literature for installation recommendations.

At sea level, the theoretical suction lift for clear, cold water (specific gravity 1.0) is limited to a maximum of 33.95 or 34 ft. (See Table 8.1.)

Factors Limiting the Theoretical Suction Lift

Internal pump losses vary with each pump and with each pump capacity and speed change.

Friction losses include pipe, valves, fittings, and velocity (generally omitted).

Static suction lift (if any).

Vapor pressure can limit the suction lift. As vapor pressure is a function of the temperature, its effect is relatively insignificant for water at temperatures less than 70°F. (See Table 8.2.)

READING GAUGES

Vacuum Gauges

Vacuum gauges are calibrated to read in "inches of mercury." It is, therefore, essential to learn and un-

derstand the relationship of the following values: (1) pounds per square inch, (2) feet of water, and (3) inches of mercury.

> One psi = 2.31 ft of water
> 2.31 ft = 27.72 in.
> Mercury weighs 13.6 times as much as water

Therefore:
> One psi = 27.72/13.6 = 2.03 in. of mercury

Therefore:
> One inch of Mercury = 2.31/2.03 = 1.13 ft of water
> One inch of Mercury = 1.0/2.03 = 0.49 psi

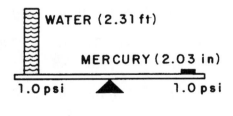

Figure 8.25

Table 8.1. Elevations Above Sea Level

Altitude Above Sea Level (ft)	Atmospheric Pressure (lb/in.²)	Barometer Reading (in. Hg)	Equivalent Head or Water (ft)	Reduction to Maximum Practical Dyn. Suction Lift (ft)
0	14.7	29.929	33.95	0
1000	14.2	28.8	32.7	1.2
2000	13.6	27.7	31.6	2.3
3000	13.1	26.7	30.2	3.7
4000	12.6	25.7	29.1	4.8
5000	12.1	24.7	27.9	6
6000	11.7	23.8	27.0	6.9
7000	11.2	22.9	25.9	8
8000	10.8	22.1	24.1	9

Table 8.2. Water Characteristics

Temperature (°F)	Vapor Pressure (psi Abs)	Vapor Pressure (ft)	Specific Gravity	Approximately Maximum Theoretical Suction Lift (ft)
40	0.1217	0.281	1.0000	33.7
50	0.1781	0.4115	0.9997	33.5
60	0.2563	0.592	0.9990	33.4
70	0.3631	0.815	0.9980	33.1
80	0.5069	1.17	0.9966	32.7
90	0.6982	1.612	0.9950	32.3
100	0.9492	2.191	0.9931	31.4

Vacuum gauge—Reads the total resistance the pump must overcome in raising the water to the pump. The resistance includes: the vertical elevation left and friction losses in the pipe and fittings.

Readings of 20 in. of Hg = 20 × 1.13 = 22.6 ft of water = 22.6 × 0.433 = 9.786 psi

Pressure gauge—Calibrated in psi. Reads the *total resistance* the pump must overcome on the discharge side of the pump. The resistance includes: the vertical elevation lift, friction losses in the pipe valves and fittings, and the sprinkler operating pressure.

Reading of 20 psi 20 × 2.31 = 46.2 ft of head

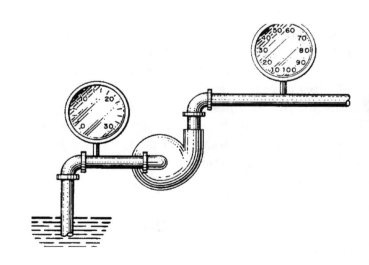

Figure 8.26

Explanation

A reading of 10 in. vacuum shows the pressure inside the pipe is reduced to 20 in. while the atmospheric pressure (sea level) is 30 in. Hg. For each inch of pressure difference, water will be pushed up 1.13 ft. Therefore, $10 \times 1.13 = 11.3$ ft lift (friction and other losses being omitted).

Example: Suction Lift

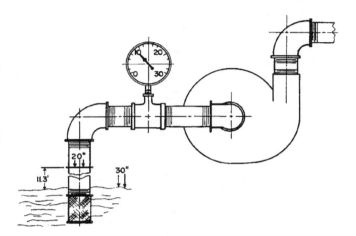

Figure 8.27

Problem: Suction Lift

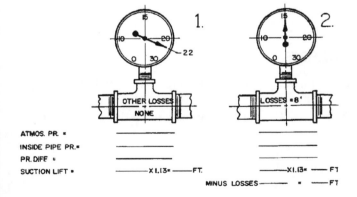

Figure 8.28

Pressure Conversions

The need to convert various pressure scales is essential for understanding and explaining how pumps lift water, also, the meaning of vacuum and pressure gauge readings.

1. Pounds per square inch absolute (psia)

This scale starts at absolute zero pressure or 100% vacuum and goes up to infinity in increments of 1 psi. Atmospheric pressure measures 14.7 psia.

Conversion:

$$psia = psig + 14.7$$

2. Pounds per square inch gauge (psig)

This pressure scale starts at atmospheric pressure with a zero reading and goes up to infinity in increments of 1 psi.

Conversion:

$$psig = psia - 14.7$$

3. Inches of mercury pressure (in. Hg)

This pressure scale corresponds to the absolute scale and goes up to infinity, generally in increments of 1 in.

$$\text{Atmospheric pressure} = 14.7 \times 2.03 = 29.84 \text{ or } 30 \text{ in. Hg.}$$

Conversion:

$$\text{in. Hg pressure} = psia \times 2.03$$

4. Feet of water or water gauge pressure

$$\text{feet of water} = psi \times 2.31$$
$$= \text{in. Hg} \times 1.13$$

5. Inches of mercury vacuum (in. Hg vac.)

This gauge starts with zero at atmospheric pressure and goes down to a reading of 30 in. at absolute zero or complete vacuum.

6. Feet of water lift (see Figure 8.27).

Conversion:

$$\text{feet of water lift} = 30 - (\text{in. Hg vac.}) \times 1.13$$

Note: Vacuum gauges (particularly on tensiometers for soil moisture measurement) are calibrated in units of *centibars,* where 100 centibars measure 100% vacuum, or zero atmospheric pressure.

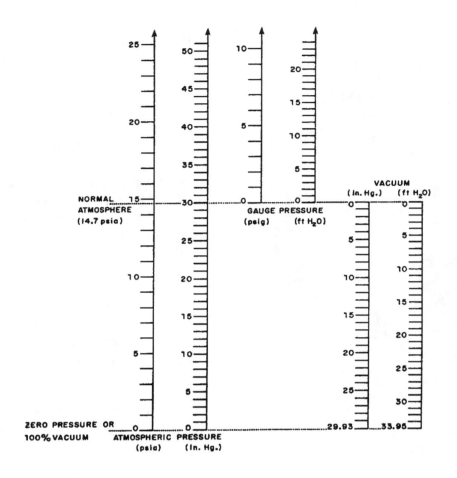

Figure 8.29. Pressure measurement scales and their relationships.

SAMPLE PROBLEMS

1. Pressure gauges are calibrated or read in units of *pounds per sq. in. gauge (psig)*. They are located on the *discharge side* of the pump.
2. Vacuum gauges are calibrated or read in units of *inches of mercury (Hg)*. They are located on the *suction side* of the pump.
3. At atmospheric pressure (14.7 psia) the pressure gauge (psig) reading = <u>0</u> .
4. At atmospheric pressure (14.7 psia) the vacuum gauge reading = <u>0</u> .
5. At complete or 100% vacuum, the vacuum gauge reading = <u>30</u> in. Hg.
6. Mercury weighs <u>13.6</u> times as much as water.
7. A pressure of 1 psi = <u>1.13</u> in. Hg.
8. A pressure of 2.03 in. Hg = <u>2.31</u> ft of water.
9. A pressure of 1 in. Hg = <u>1.13</u> ft of water.
10. in. Hg × <u>1.13</u> ft of water.
11. ft of water, <u>14.7</u> in. Hg.
12. in. Hg pressure = <u>30</u> minus in. Hg vac.
13. psia = <u>14.7</u> plus psig
14. psia = psig minus <u>14.7</u> .

PROBLEMS

Convert the following:

1. 6 in. vac = <u> (30–6/2.03) = 11.8 </u> psia

2. 50.7 psia = <u> 36 </u> psig

3. 10 psig = <u> 24.7 </u> psia

4. 10 psia = <u> 9.54 </u> in. vac

5. 22 in. vac = <u> 8 </u> in. Hg pressure

6. 10 psig = <u> 23.1 </u> ft of H_2O

7. 56 in. Hg pressure = <u> 27.4 </u> psia

8. 22.6 ft H_2O = <u> 20 </u> in. vac

9. Vacuum of 30 in. Hg = <u> 33.9 </u> ft H_2O

10. Vacuum of 6 in. Hg = <u> 24 </u> in. Hg pressure = <u> 11.8 </u> psia = <u> 6.8 </u> ft of H_2O (lift)

NET POSITIVE SUCTION HEAD (NPSH)

In general, the available net positive suction head (expressed in feet) is the difference remaining after subtracting all the losses on the suction side of the pump from the theoretical suction lift at the elevation on installation. (See Table 8.1).

Available NPSH

Available NPSH can be calculated, or on an existing installation it can be determined by field vacuum gauge readings. *Important: The available NPSH must be greater than or equal to the required NPSH.*

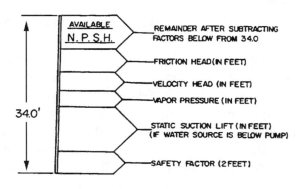

Figure 8.30

Available NPSH Formulas
Planning Installation

Available NPSH = $TSL \pm SE - F_H - V_P - SF$

> *TSL - Theoretical suction lift* (varies with elevation of installation) can be a function of atmosphere pressure or a pressurized tank.
> *SE - Static elevation vertical lift,* water surface (pump down) level to center line of pump. Minus when below pump, plus when above.
> F_H - *Friction head* from pipe, valves, fittings, and velocity.
> V_P - *Vapor pressure*—can omit for water temperature 70°F or less.
> *SF - Safety factor*—allow 2 ft.

Existing Installation

The vacuum gauge reading includes the vertical lift (ft), friction, and other losses.

Example

Find the available NPSH for the following pump installation.

Factors given:

1. installation at sea level
2. water temperature = 60°F
3. static suction lift = 12 ft
4. friction head (pipes and fittings) = 4 ft
5. velocity head = 0.4 ft
6. safety factor = 2 ft

Solution

a. theoretical suction lift at sea level = 34.0 ft
b. negative factors limiting suction lift

1. vapor pressure at 60°F =	0.6 ft
2. static lift =	12.0 ft
3. friction (pipes) =	4.0 ft
4. velocity =	0.4 ft
5. safety factor =	2.0 ft
Minus =	19.0 ft

Available NPSH = (34 ft – 19 ft) = 15 ft

Problem

Available NPSH Installation Elev. 4000 ft

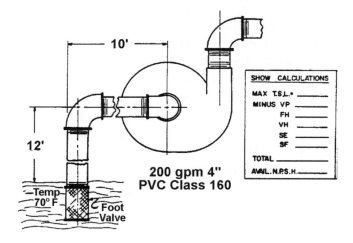

SHOW CALCULATIONS

MAX T.S.L.° _____
MINUS VP _____
FH _____
VH _____
SE _____
SF _____
TOTAL _____
AVAIL. N.P.S.H. _____

200 gpm 4"
PVC Class 160

Figure 8.31

Required NPSH

Required NPSH refers to the internal pump losses and must be given by the pump manufacturer. It varies with each pump and with each pump capacity and speed change. The greater the capacity, the greater the required NPSH.

Example

Find the required NPSH for the following operating conditions:

total head = 30 ft
capacity = 100 gpm
impeller diameter = 7 in.

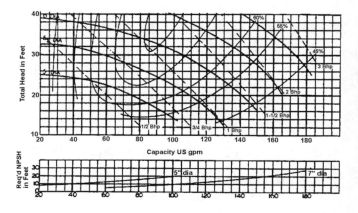

Figure 8.32

Solution

Establish an intersecting point (30 ft, 100 gpm) on the performance curves; project a vertical line down to 7-in.-diameter impeller; read required NPSH on left scale = 10.

Cavitation

The term *cavitation* implies that a cavity or void forms within the pump.

Although much theoretical and experimental work has been done on cavitation, the mechanism of cavitation has not been definitely or completely established. In practice, when a centrifugal pump is installed and operates with less than the required NPSH, the pump tends to discharge the water at a faster rate than it is coming in. If, at some point in the liquid flow, the fluid pressure equals the vapor pressure at a particular temperature, the liquid will vaporize, and a cavity or void will form. As this condition is approached, the sound emitted by the pump changes. At first it sounds as if sand were passing through (with clear water entering). Then the sound or noise may change (as the discharge increases) to give the impression of rocks passing through or a machine gun barrage. The formation and collapse of vapor bubbles can cause dangerous erosion or pitting of some metal parts, often eating holes through the vanes or volute.

Explanation of Cavitation Pictures

On pumps designed with top suction manifolds, it is possible to install transparent plates at the impeller eye and, by means of stroboscopic lights which flash on and off at the same rate as the rotation of the pump, visualize the flow of fluid around the entrance of the impeller vanes. Figure 8.33 shows a picture of such a pump with the end plate removed so that the impeller eye can be seen.

Figure 8.33

Initial stages of cavitation are indicated in Figure 8.34. It is noted that vapor is beginning to form on the entrance edges of the vanes. Little noise or reduction in pump performance is noted in this stage of cavitation. Impeller vanes may show some pitting after extended operation under these conditions, especially if there is any tendency toward corrosion.

Figure 8.34

Figure 8.35 shows the increased formation of vapor which extends over a larger portion of the entrance of the impeller vanes as a result of increasing the suction lift on the pump, which decreases the pressure at this area. Normally, some noise would be evident in the pump, and a reduction in efficiency and head developed by the pump would become apparent at this stage.

Figure 8.35

Figure 8.36 indicates severe cavitation at the impeller vanes as the pressure in the pump is further decreased by further increasing the suction lift. Although the pump will still operate under these conditions, it will be noisy and have a reduced head and efficiency. Vibration would undoubtedly be noticeable, and severe pitting would occur on the impeller if operated for an extended period of time under these conditions.

Figure 8.36

TOTAL DYNAMIC HEAD

Total dynamic head is the expression used for the total resistance against which a pump must operate or overcome to perform satisfactorily. The TDH can be calculated in *ft of water* or in *pounds per square inch.*

Calculating Total Dynamic Head

Total suction head includes the total resistance found on the suction side of the pump.

a. *Static head* is the *vertical* lift or distance from the water line to the center line of the pump.
b. *Friction head* refers to friction losses in the suction pipe valves and fittings.
   Note: (1) Friction head in the suction line should always be held to a minimum, since the atmospheric pressure is the acting force which brings the water to the pump.
   (2) The suction pipe should never be smaller than the size of the connection on the pump.
   (3) When the water source surface is above the center line of the pump, the static suction head is helping

the pump, so it is subtracted from the static discharge head.

Total Discharge Head

a. *Static head* is the *vertical* lift or distance from the center line of the pump to the highest point of discharge or highest sprinkler nozzle.
b. *Friction head* is the friction loss in the discharge pipes, valves, and fittings.
c. *Pressure head* is the sprinkler pressure required to carry the water out into its trajectory. The pressure in pounds per square inch necessary to operate the sprinkler should be converted to feet of head. When a pump discharges into a tank against pressure, the equivalent head in feet is determined from the pressure in the tank (tank pressure × 2.31).
d. *Velocity head* is usually neglected because of its relatively low head. Velocity head must be considered, however, when the discharge pipe has a much smaller pipe size than the suction pipe. The greater the velocity, the more work has to be done by the pump.

Formula used to calculate velocity:

$$1. \ H_v = \frac{v^2}{2g} = \frac{v^2}{64.4}$$

where: H_v = velocity head
V = velocity (in ft/sec)
g = gravity (32.2 ft/sec)

$$2. \ V = \frac{0.408 \times gpm}{D^2}$$

where: V = velocity (in ft/sec)
0.408 = factor
gpm = gallons per minute
D = diameter of pipe in inches

Total Dynamic Head (TDH)—Terms and Their Relationships

Total Dynamic Head = Total Static Head + Total Friction Head = Total Pressure Head

What to Consider in Figuring Dynamic Head

Figure 8.37 shows dimensions to be considered when figuring dynamic head.

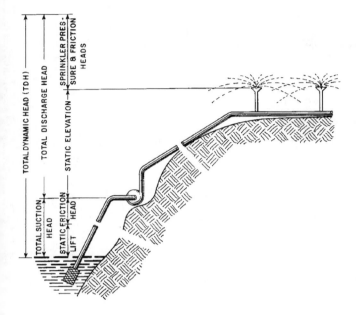

Figure 8.37. Centrifugal pump and sprinkler system.

Definitions

- *Suction lift* or *lift*—the vertical distance from pump to water level.
- *Elevation*—the vertical distance from pump to sprinkler nozzle.
- *Friction head loss*—the head loss due to friction in pipe and fittings.
- *Sprinkler pressure head*—the head which must be developed as pressure to operate sprinklers.
- *Total discharge head*—the head which must be developed by the pump to overcome pipe friction, elevation, and sprinkler pressure.
- *Total (dynamic) head*—The sum total of the suction head and the total discharge head.

SAMPLE PROBLEM

Calculate the Total Dynamic Head

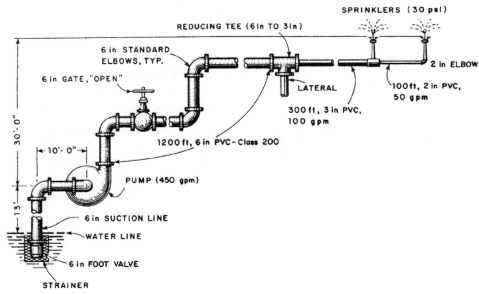

Figure 8.38

Friction Losses 6 in. = 5.1
Suction Side F.L. = 0.57 psi
   Foot valve = _7.7_ ft
   (2) Standard elbows = _32_ ft
   Straight pipe = _23_ ft
   Equivalent ft of steel pipe = _132_ ft
   Friction loss = _1.7_ ft
Discharge Side 6 in. pipe
   Gate valve = _3.5_ ft
   2:1 reduction = _2.8_ ft
   Straight pipe = _1200_ ft
   Equivalent ft steel pipe = _1206_ ft
   Friction Loss = _16_ ft
3 in. pipe V = 4.07 – F.L. = 0.79 psi
   3:2 reduction = _1.4_ ft
   Straight pipe = _300_ ft
   Equivalent ft steel pipe = _301_ ft
   Friction Loss = _5.5_ ft
2 in. pipe V = 4.4 – F.L. = 1.4 psi
   Standard elbow = _6_ ft
   Straight pipe = _100_ ft
   Friction loss = _3.6_ ft

TDH
Suction Side
Static lift = _13_ ft

Friction loss = _2_ ft

$(0.57 \times 1.32 \times 2.31)$
Discharge Side
Static lift = _30_ ft

Friction (6 in.) = _16_ ft

$(0.57 \times 12 \times 2.31)$

Friction (3 in.) = _6_ ft

$(0.79 \times 3 \times 2.31)$

Friction (2 in.) = _4_ ft
$(1.4 \times 1.0 \times 2.31)$
Pressure = _69_ ft
$(30 \text{ psi} \times 2.31)$
TDH = _140_ ft

UNDERSTANDING PUMP CURVES

Introduction

A pump curve is a curved line drawn over a grid of vertical and horizontal lines. The curved line represents the performance of the pump. The vertical and horizontal grid lines represent units of measure of that performance.

Since gravity won't allow water to flow uphill, a pump is used. A pump is a machine used to move a volume of water a given distance which is measured over a period of time and usually expressed in gallons per minute (gpm).

The pump develops energy called *discharge pressure* or *total dynamic head* (TDH). This discharge pressure is expressed in units of measure called *pounds per square inch* (psi) or feet of head.

Examples

1. To establish a pump curve, a gauge, valve, and flowmeter are used on the discharge pipe. First, the pump is run with the valve closed and the gauge is read. This gives the pump's capability at zero flow and maximum head in feet. Next, the valve is open to 5 gpm flow, the gauge is read and recorded on the grid. The process is continued until all the points are marked on the grid.

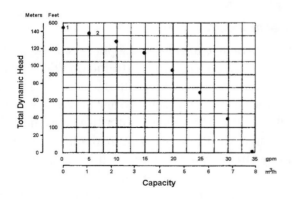

Figure 8.39

2. The points are connected to form a curved line. This line is then called a *head/capacity curve* (H-Q). Head (H) is expressed in feet and capacity (Q) is expressed in gpm.

The pump will always run somewhere on the curve.

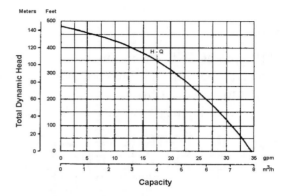

Figure 8.40

3. A different kind of curve you will find used on some grids is called an *efficiency curve*. Efficiency as a percent is shown on the right side of the grid.

Procedure

a. On the grid find 425 gpm at 500 ft TDH. Read vertically up until you touch the efficiency curve line. Go horizontally to the right to find 72% efficiency.

b. Now find 425 gpm at 800 ft. From this point read down vertically to the curve, then across to find the efficiency also at 72%.

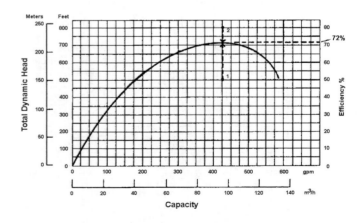

Figure 8.41. Efficiency curve line shown by itself.

175

4. Another curve line shown is called *brake horsepower* (Bhp). Bhp is charted to the right just past the efficiency scale.

Procedure

Find 110 gpm at 71 ft. From this point read vertically down to the Bhp curve line, then horizontally to the right to find 3.1 Bhp.

Also find efficiency = 66%, suction lift = 11 ft.

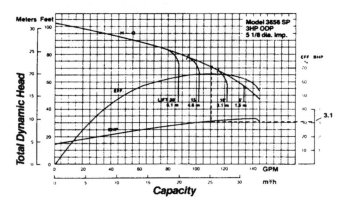

Figure 8.42

5. Pictured here is a single-line curve for a centrifugal pump. Find 70 gpm at 56 ft. Since this point is below the head/capacity curve line the pump will exceed the requirements.

By proper trimming of the impeller, the pump will meet the exact requirements. This will establish a new curve for the pump The curve will run approximately parallel to the curve shown and through the point required, i.e., 70 gpm at 56 ft.

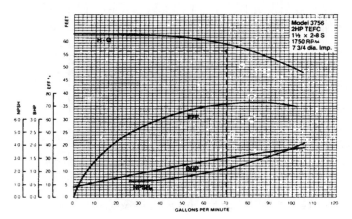

Figure 8.43

Important: If the impeller is not trimmed, the pump will run somewhere on its curve. The head requirement is 56 ft. Follow horizontally to the right until you meet the H-Q curve, then vertically down until the capacity is 80 gpm. Notice that the efficiency, Bhp and required NPSH also change with the capacity of 80 gpm.

If a throttling valve is used on the discharge side of the pump the capacity of 70 gpm can be regulated using the standard impeller. Read vertically up from 70 gpm to meet the curve line, the head will make 58.5 ft.

Remember: Trimming the impeller will make the pump meet the exact design requirements of 70 gpm at 56 ft head.

The head/capacity requirements should always be furnished when the pump is ordered.

Note: Again at 70 gpm at 56 ft find:

efficiency	= 70.5%
brake horsepower	= 1.5
required NPSH	= 2.2

You will have to determine the *available* NPSH, which must be higher than the *required* NPSH for the pumps to work.

Performance Curves and Operating Characteristics

General Information

The performance of centrifugal pumps can be shown by the typical characteristics found on the TDH capacity curve. This curve is plotted from data received by operating the pump with the discharge valve completely closed. There is no water flow and the pressure gauge would show a high reading since the pump is attempting to force out the water. Now, as the valve is opened the pump capacity increases and correspondingly the pressure reading drops. It should be noted that a specific flow capacity is obtained at each different pressure condition encountered. This relationship can be seen on the pump performance curve shown in Figure 8.44.

In this case, the tests were limited to the following operating conditions:

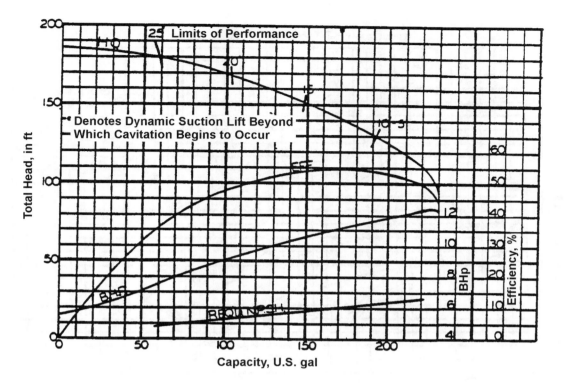

Figure 8.44. Pump performance curve.

1. The liquid being pumped is clear, cold water—specific gravity = 1.0
2. Impeller speed = 3450 RPM
3. Impeller diameter = 7 in.

The curves reveal a clear picture of the operating characteristics of this pump. The highest degree of efficiency of 50% is obtained against a head of approximately 140 ft and capacity of 170 gpm. Note that when the total head decreases, both the total gpm and hp increase.

Multi-Impeller Pumping Curves

Centrifugal pumps are generally made so that one casing or volute can accommodate various impeller diameters. To make a separate curve for each of the several impeller diameters would result in a number of sheets which would be confusing to read.

Therefore, using the actual test data for each size impeller run at different speeds, several points are plotted to form a composite efficiency as well as the horsepower lines on the head-capacity chart. (See Figure 8.45).

Chart Reading Examples

1. With a 7-in.-diameter impeller, an efficiency of 68% is obtained both when pumping 187 gpm with 200 ft head and 222 gpm with 189 ft head. The same efficiency can be obtained with a 6-in.-diameter impeller at 152 gpm with 140 ft head and 189 gpm with 129 ft head.
2. As the total head decreases, both the capacity and horsepower will increase. Consequently, when selecting the horsepower required for a particular application, use the next size larger. When close to the line or when the head may drop for any reason, the larger horsepower may be required. Example: pumping 140 gpm with 178 ft head, a 6.5-in.-diameter impeller requires

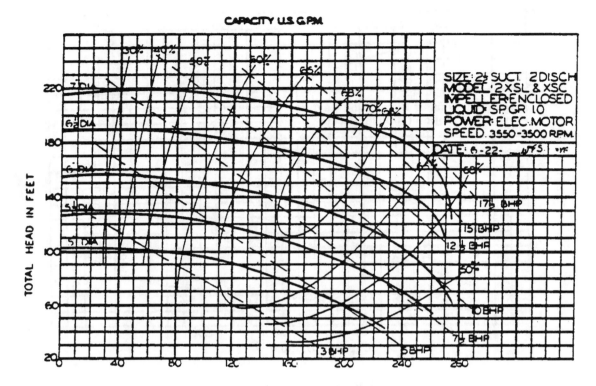

Figure 8.45. Head-Capacity Chart.

7.5 Bhp. Should the head drop to 170 ft, the gpm increases to 170 and the horsepower to 10.

PUMP SELECTION PROCEDURE

The final choice of a single pump or pumping station (more than one pump) should be based on the demands of that system. Pumps that are too big or too small for the job will result in inadequate performance and cost inefficiencies.

Suggested Steps

1. Solve for total gallons per minute. See Table 2.6 on page 50.
2. Solve for total dynamic head (TDH). See page 172.
3. If applicable, solve for net positive suction head (NPSH). See pages 169–170.

Important: Each individual pump must meet the TDH requirement regardless of whether a single-pump or a multi-pump station is used. However, the total GPM output can be divided if two or more pumps *piped in parallel* are used.

4. Refer to the manufacturer's pump catalogs, select the pump model(s) capable of meeting the demands from the "Composite Curve Chart," if provided. See sample below, otherwise refer to the "Pump Performance Curves" for each individual pump model. *Typical pump model shown below.*
5. Generally, the first choice is based on the pump(s) that can do the job having the highest efficiency.

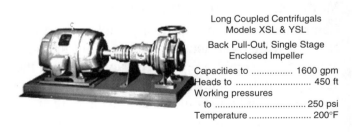

Long Coupled Centrifugals
Models XSL & YSL

Back Pull-Out, Single Stage
Enclosed Impeller

Capacities to .... 1600 gpm
Heads to .... 450 ft
Working pressures
    to .... 250 psi
Temperature .... 200°F

Figure 8.46

SAMPLE PROBLEM

Select a centrifugal pump—electrically driven constant speed—for the demand conditions listed below.

Example 1

System Demands

a. Total Dynamic Head (TDH) = 190 ft
b. Total gallons/min (gpm) = 200 gpm
c. Required NPSH = 10

Procedure

1. From the composite curves chart, select the possible pump models that may meet the design condition requirements.
2. From the performance curve chart for each of the models listed, fill in the blank spaces provided.
3. Carefully analyze and compare the data, then select the pump model which you think best meets the design criteria.

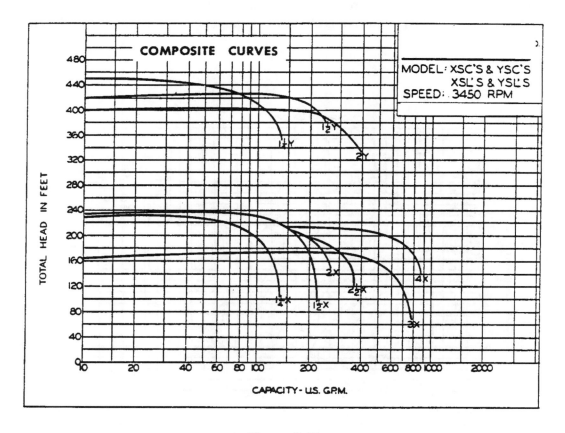

Figure 8.47

Pump Model	Efficiency (%)	Required NPSH	Impeller Size (in.)	Bhp	rpm
1-1/2X	Inadequate—Intersects Above 7-in.-Diameter Impeller				
2X	69	15	7	15	3550
2-1/2X	63	13	7	20	3550
4X*	42	10	7	20	3550
Type M					
Tuas					
3D3S					

* 4X - Too big—designed for about 750 gpm and 180 TDH.

Pump Selection: (Final Choice)

Points to Consider

1. Generally, the first choice is the pump with the highest efficiency.
2. If applicable, allow for future expansion, also for wear and tear with usage.
3. Check the pump power requirements: be sure that the proper phase and voltages are available.
4. Consider advantages and disadvantages of constant speed pumps vs variable speed.
5. Check the pump dealer's and manufacturer's reputation and availability when needed.
6. Where the intersecting point (TDH–gpm) falls above the impeller or horsepower line, select the next higher value.

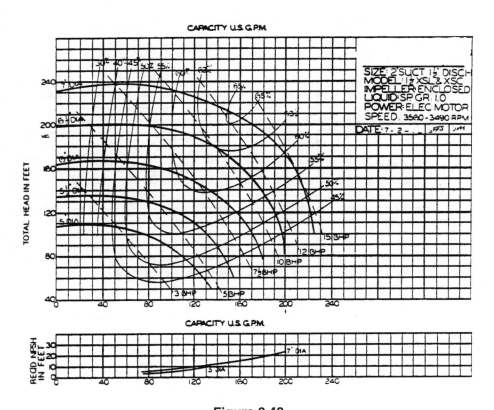

Figure 8.48

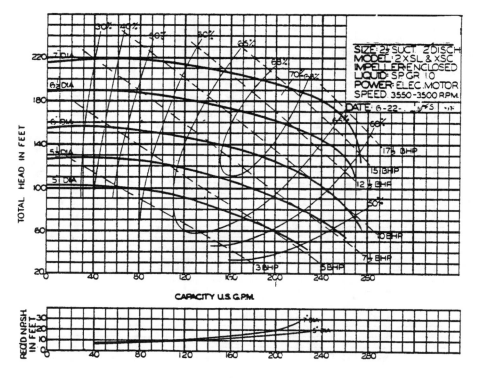

Figure 8.49

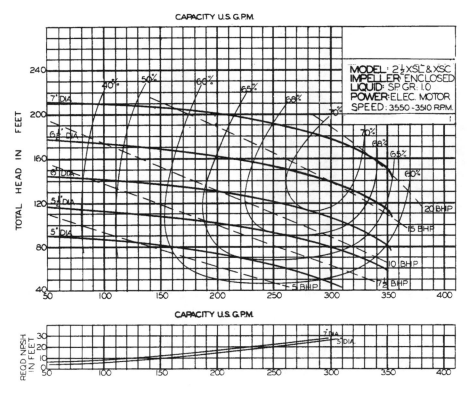

Figure 8.50

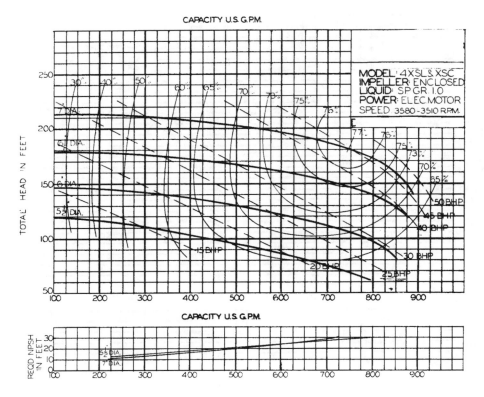

Figure 8.51

HYDRO/constant Size II Type **M**

	3500 RPM	1750 RPM
SPEED	3500 RPM	1750 RPM
CAPACITY	up to 225 gpm	up to 600 gpm
HEAD	up to 440 ft.	up to 320 ft.
HP RANGE	up to 50 hp	up to 75 hp
PUMP TYPE	Type TU	Type TU
	Types PB, A/D also may be used	

Figure 8.52

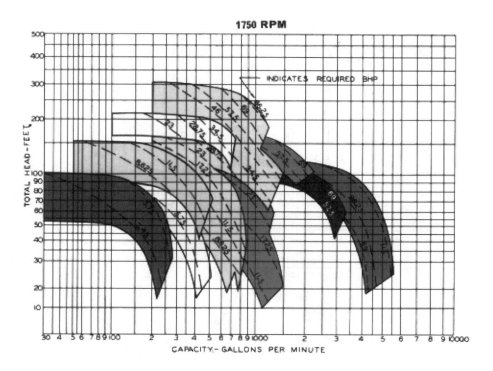

Figure 8.53

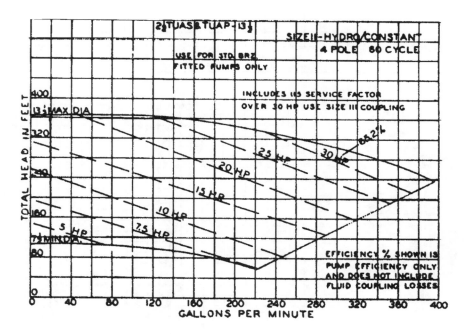

Figure 8.54

Example 2

Select an *internal combustion engine* (gas) to meet the pump demands, i.e., 200 gpm and 190 ft TDH, as given in Example 1.

Procedure

1. Refer to the composite curve chart below. Select Model 3D3s.
2. From the performance curve chart. Fill in the blank spaces.

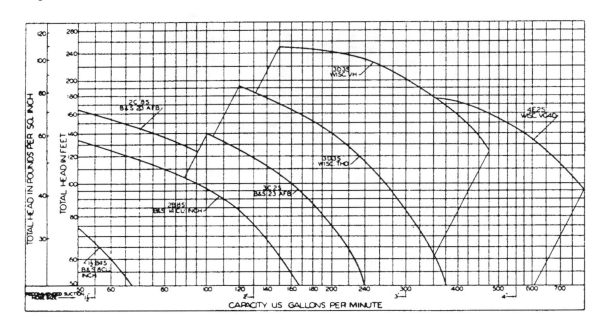

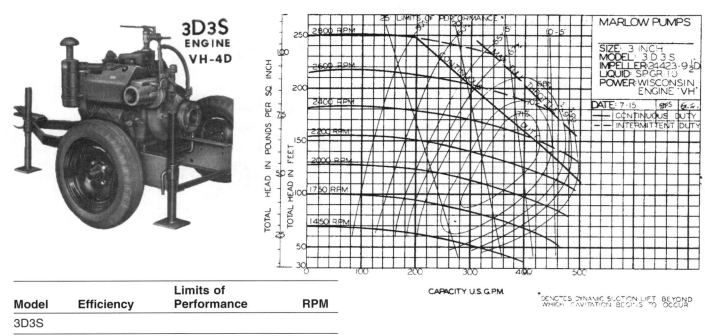

Model	Efficiency	Limits of Performance	RPM
3D3S			

Figure 8.56. Selection chart. Type S pumps with water-cooled, gasoline engines.

STUDENT PROBLEM

Select and record pump model, etc. for the gpm and TDH shown. *Important:* It is essential to calculate the *total dynamic head* (TDH) very carefully to meet the worst condition in order to select the proper pump size. Never select a pump at random or install a pump on hand that may be too large or too small for the job.

Example

Select a pump for the pumping conditions shown below.

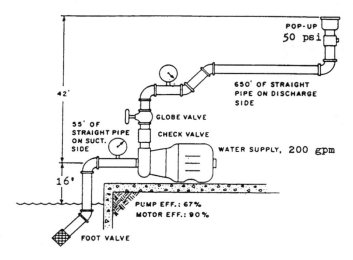

Figure 8.58

Procedure

a. Find the friction loss for each pipe section
b. Calculate friction losses

Use Ring-Tite PVC Class 160
Suction Side
Select Pipe Size _4_
Calculate Friction Loss (ft)
Foot valve = _71_ ft steel pipe
45° elbow = _5_ ft steel pipe
90° elbow = _11_ ft stell pipe
Straight pipe = _55_ ft

Friction Loss = _2.0_ ft (0.76 × 1.4 × 2.31 = 2)

Discharge Side
Select Pipe Size (see above).
Pump to sprinkler—Size _4_ in.
Calculate Frictional Loss (ft)
Check valve = _43_ ft steel pipe
1 globe valve = _110_ ft steel pipe
2 standard elbows = _22_ ft steel pipe
2 45° elbows = _10_ ft steel pipe
Straight pipe = _650_ ft
Total ft of steel pipe = _835_ ft
Friction loss = _15_ ft

Calculate TDH
Suction Side
Lift = _16_ ft
Friction Loss = _2_ ft
Discharge Side
Lift = _42_ ft
Friction Loss = _15_ ft
Pressure = _115_ ft
TDH = _190_ ft

Determine NPSH—Assume installation 2000 ft above sea level. Show calculations.

NPSH =   31.6 ft (lift)
          − 18.0 ft
          13.6 ft

Pump Selection
1. Pump model _____
2. Pump efficiency _____
3. Impeller size _____

SAMPLE PROBLEM

Pump Selection

This problem shows two acceptable methods for calculating friction losses for valves and fittings.

1. Conversion of valves and fittings to feet of straight pipe.
2. Rule of thumb method for figuring friction losses.

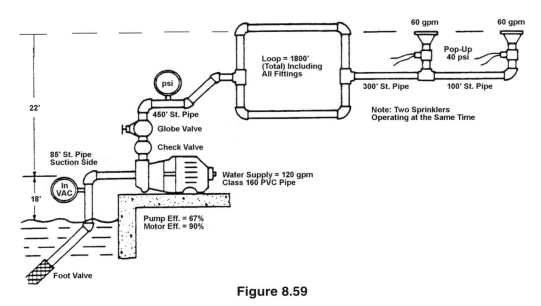

Figure 8.59

Worksheet 1

Suction Side
Select pipe size _3_ in.
Calculate friction loss (ft)

foot valve	= _64_ ft steel pipe
45° elbows	= _4_ ft steel pipe
90° elbows	= _8_ ft steel pipe
steel pipe	= _85_ ft
total	= _161_ ft
friction loss	= _3.7_ ft

Discharge Side
Pump to loop pipe size _3_ in.
Calculate friction loss (ft)

check valve	= _32_ ft steel pipe
globe valve	= _82_ ft steel pipe
2 90° elbows	= _16_ ft steel pipe
2 45° elbows	= _8_ ft steel pipe
steel pipe	= _450_
total steel pipe	= _588_ ft
friction loss	= _13.6_ ft

Loop select pipe size _2.5_ in.
Calculate friction loss _15_
    $(0.72 \text{ psi}/100 \text{ ft}) \times 9 \times 2.31 = 15$
Loop to first head
Select pipe size _3_ in.
Calculate friction loss

90° elbow	= _8_ ft steel pipe
tee	= _5_ ft steel pipe
st. pipe	= _300_ ft steel pipe
total st. pipe	= _313_ ft
friction loss	= _7.2_ ft

First head to second head
Select pipe size _2.5_ in.
Calculate friction loss (ft)

90° elbow	= _8_ ft steel pipe
steel pipe	= _100_ ft
total	= _108_ ft steel pipe
friction loss	= _1.8_ ft

Note: Total friction losses = (3.7 + 13.6 + 15 + 7.2 + 1.8) = 41.3 ft

Rule of Thumb Method

The friction loss for pipelines which include valves and fittings can be determined by the rule of thumb method. Merely add 10% to the gpm carried by each pipe section and select the friction loss value from the appropriate chart or table. The actual length (distance) of pipe (only) is used, and the equivalent feet of straight pipe for valves and/or fittings are omitted.

Worksheet 1a

Problem

Recalculate the above pipeline friction losses using the rule of thumb method. Pipe sizes same as above.

Pipe Section	gpm + 10%	Friction Per 100 ft	Distance (ft)	Total Loss (ft)
Suction Line	132	1.2	85	2.3
Pump to Loop	132	1.2	450	12.5
Loop	66	0.85	900	17.8
Loop to 1st Head	132	1.2	300	8.3
Ist to 2nd Head	6	0.85	100	2.0
Total				42.9[a]

[a] Method 1 vs. Method 2 = A difference of (42.9 ft − 41.3 ft) = 1.6 ft.

Worksheet 2

Calculate TDH

Suction Side

lift	= 8.0 ft
friction loss	= 2.3 ft

Discharge Side

lift	= 22 ft

Friction Loss

to loop	= 12.5 ft
loop	= 17.8 ft
to 1st sprinkler	= 8.3 ft
to 2nd sprinkler	= 2.0 ft
pressure	= 92.4 ft
TDH	= 165 ft

Determine NPSH – TSL at 2000 ft

Assume installation is at
2000 ft above sea level = 31.6 ft

Minus:

	Lift	= 8.0 ft
	F.L.	= 2.3 ft
	S.F.	= 2.0 ft
		= 11.3 ft
Available NPSH		= 19.3 ft

Pump Selection
1. pump model 1.5 XSL
2. pump efficiency 62%
3. impeller size 7 in.
4. horsepower 10

Determine
1. vacuum gage reading 9.1 in. Hg
2. pressure gage reading 67 psi

Determine brake horsepower
Use efficiency given in problem, for pump and motor.

$$\frac{TDH \times gpm}{4000 \times pump + motor\ efficiency}$$

$$Bhp = \frac{165 \times 120}{4000 \times 0.62 \times 0.90} = 8.87$$

STUDENT PROBLEM

Calculate the TDH and Select Pump Model

Calculate the total dynamic head (TDH) to the worst condition, usually to the highest elevation (to Green 1 for this problem). If the pump selected is capable of handling the worst condition, it should have no difficulty elsewhere.

To figure the friction loss for the loop, use the outer distance only. This would be the worst condition, if the laterals were shut down for repairs.

Use the rule of thumb method for calculating friction loss. Review previous problem.

Note: For pump selection the total gpm *does not* include the added 10%.

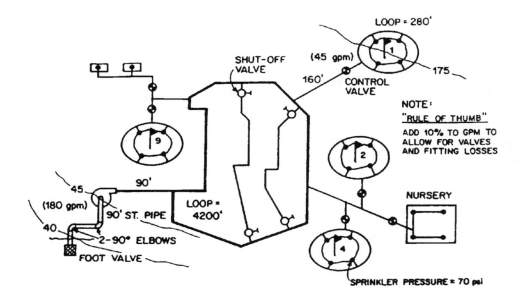

Figure 8.60

1. Solve for pipe sizes and friction losses.
2. Solve for total dynamic head (ft).
3. Select pump model (catalogs to be provided). List all specifications.

	Pipe Size	Friction
a. Suction side	_____	_____
b. Pump to loop	_____	_____
c. Loop	_____	_____
d. Loop to green	_____	_____
e. Loop around green	_____	_____

Total Dynamic Head

Suction side:
Vertical lift = _____ ft
Friction loss = _____ ft
Discharge side:
Vertical lift = _____ ft
Friction loss = _____ ft
Pressure   = _____ ft

Pump Selection:
Model: _____, Efficiency: _____, Bhp: _____, Impeller Size: _____

IMPORTANT PUMPING CHARACTERISTICS

Affinity Laws

The Affinity Laws express the mathematical relationship between several variables involved in pump performance. They apply to all types of centrifugal and axial flow pumps. They are as follows:

Q = Capacity, gpm
H = Total Head, ft
Bhp = Brake Horsepower
N = Pump Speed, rpm
D = Impeller Diameter (in.)

Use Equations 1 through 3 when *speed changes and impeller diameter remains constant.*

1. $$\frac{Q_1}{Q_2} = \frac{N_1}{N_2}$$

2. $$\frac{H_1}{H_2} = \left(\frac{N_1}{N_2}\right)^2$$

3. $$\frac{Bhp_1}{Bhp_2} = \left(\frac{N_1}{N_2}\right)^3$$

Use Equations 4 through 6 when *impeller diameter changes and speed remains constant.*

4. $$\frac{Q_1}{Q_2} = \frac{D_1}{D_2}$$

5. $$\frac{H_1}{H_2} = \left(\frac{D_1}{D_2}\right)^2$$

6. $$\frac{Bhp_1}{Bhp_2} = \left(\frac{D_1}{D_2}\right)^3$$

Example 1

A pump is operating at a speed of 2000 rpm, pumping 500 gpm against a head of 200 ft using a 30-hp engine.

What is the effect when the speed is increased from 2000 rpm to 2400 rpm?

Procedure

Capacity is directly proportional to the change of speed.

$$\frac{Q_1}{Q_2} = \frac{N_1}{N_2} = 500 \times \frac{2400}{2400} = 500 \times 1.2 = 600 \text{ gpm}$$

Head (ft) changes in proportion to the *square* of the speed.

$$\frac{H_1}{H_2} = \frac{(N_1)^2}{(N_2)} = 200 \times \frac{(2400)^2}{(2000)}$$

$$= 200 \times (1.2)^2 = 288 \text{ ft}$$

Horsepower changes in proportion to the *cube* of the speed.

$$\frac{Bhp_1}{Bhp_2} = \frac{(N_1)^3}{(N_2)} = 30 \times \frac{(2400)^3}{(2000)}$$

$$= 30 \times (1.2)^3 = 52 \text{ Bhp}$$

Note: An additional 22 hp is required to obtain an increase of 100 gpm and 88 ft of head. Think carefully before increasing the speed.

Example 2

The same Laws of Affinity hold true when the impeller diameter changes. At a given speed of 2000 rpm, the capacity is 500 gpm, 200 ft head and 30 Bhp. Find the effect when the diameter of the impeller is reduced from 10 in. to 9 in.

Procedure

capacity (Q) = $500 \times 9/10 = 450$ gpm
head (H) = 200×162 ft
horsepower (Bhp) = $30 \times (9)^3/(10) = 22$ (approx.)

Example 3. Impeller Trimming

Pump manufacturing companies can modify (trim) the impeller that will meet the requirements of a specific installation.

Assume the pump requirement was 225 gpm at 160 ft of head (Point 2), as shown on the pump curves below. Notice that this point falls between two existing curves with standard impellers.

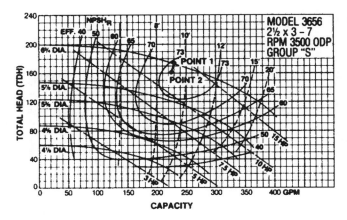

Figure 8.61

To determine the trimmed diameter, draw a line from Point 2 perpendicular to an existing curve line to Point 1. On the graph, locate the impeller diameter (D_1), which is 6-3/4 in. and produces 230 gpm (Q) at 172 ft head (H_1). Then solve for trimmed impeller diameter (D_2).

Procedure

Apply Affinity Law 5 to solve for a new impeller diameter.

Point 1 (Known)   Point 2 (Unknown)

D_1 = 6-3/4 in. D_2 = ?
H_1 = 172 ft H_2 = 160 ft
Q_1 = 230 Q_2 = 225

Rearrange Affinity Law 5 to solve for D_2:

$D_2 = D_1 \times H_2/H_1$
$D_2 = 6.75 \times 160/172 = 6.55$ in. or 6-9/16 in.

To determine whether the new impeller diameter will meet the required capacity rearrange Affinity Law 4 to solve for Q_2.

$$Q_2 = \frac{D_2}{D_1} \times Q_1 = \frac{6.55}{6.75} \times 230 = 223 \text{ gpm}$$

Solving for Power Required

Water horsepower (Whp), sometimes called theoretical horsepower, can be calculated by the following equation:

$$Whp = \frac{gpm \times wt. \times TDH}{33,000} \times SP\text{-}GR$$

where: gpm = gallons per minute pumped
   Wt = weight of the fluid (water weighs 8.33 lb/gal)
   TDH = total dynamic head (ft)
   SP-GR = specific gravity of water is 1
   1 hp = 33,000 ft-lbs of work per minute

The above equation can be reduced as shown below:

$$Whp = \frac{gpm \times 8.33 \times TDH}{33,000} \times 1$$

or simply:

$$Whp = \frac{gpm \times TDH}{3960}$$

Brake horsepower is an expression of the *actual power*. It includes the power (efficiency) loss that will occur within the motor or engine. Expressed as follows:

$$Bhp = \frac{gpm \times TDH}{3960 \times efficiency (\%)}$$

or commonly:

$$Bhp = \frac{gpm \times TDH}{4000 \times efficiency (\%)}$$

Figuring Horsepower and Operating Costs

Example

Select a pump for the following installation requirements: 100 gpm at 375 ft of head and 460 V.

There are two pumps that will make this rating.

1. Model 100H156634—8 Stages 72% efficiency 15 hp
2. Model 150H156634—6 stages 67% efficiency 15 hp

Compare the operating costs of the above pumps:

Procedure

$$Bhp = \frac{gpm \times ft \text{ of head}}{3960 \times efficiency} \times SP\text{-}GR$$

Pump 2:

$$Bhp = \frac{100 \times 375 \times 1}{3960 \times 0.67} = \frac{37500}{2851} = 14.1 \text{ Bhp}$$

Pump 1:

$$Bhp = \frac{100 \times 375 \times 1}{3960 \times 0.72} = \frac{37500}{2851} = 13.1 \text{ Bhp}$$

difference = 1.0 Bhp

Figuring Costs

1 hp = 760 W
Since no motor is 100% efficient, use 1000 W or 1 kW.

If cost for one kWh = \$0.08
Then, 1 Bhp difference × 0.08 = 8¢/hr

If the pump runs for 10 hr/day = 80¢/day
and in 30 days = \$24.00 savings.

HYDRAULIC INSTALLATION GUIDE

Location

Place the pump as near to the water source as possible. Maximum suction lift is dependent upon NPSH required by pump, pipe size, type of valves, elevation above sea level, and type and temperature of liquid being pumped.

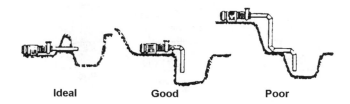

Ideal　　　　**Good**　　　　**Poor**

Figure 8.62

Suction

Suction piping should be of sufficient size to allow a velocity of approximately 3 to 8 fps at design capacity. Use eccentric reducers, straight side up. Keep the number of fittings to a minimum. Avoid using an elbow any closer to the pump than four times the diameter of the suction pipe. This will aid in reducing the turbulence before the flow reaches the impeller eye. A poorly designed suction setup will not allow the pump to operate at full efficiency, resulting in added cost of operation.

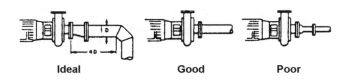

Ideal　　　　**Good**　　　　**Poor**

Figure 8.63

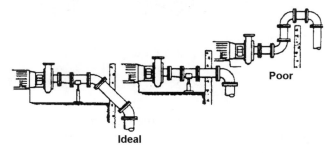

Poor

Ideal

Figure 8.64

The highest point in the suction line should be at the pump. Install the suction pipe on a slope up to the pump. Eliminate high points. Use one pipe size throughout; reduce only at the pump with an eccentric reducer (straight side up). Air pockets caused by improper suction line installation can cause priming and pumping problems.

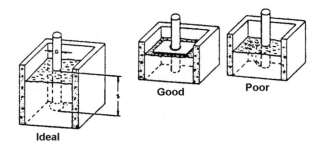

Figure 8.65

Submergence (S) should be at least four times the diameter of the suction pipe (D). If suitable submergence cannot be obtained, a baffle or flotation board around the suction pipe must be provided to prevent vortexes. The diameter of the baffle should be *five times* the pipe diameter. Do not allow vortex action. A bell mouth on an open end suction pipe is ideal. Vortexing can cause reduced head and capacity and may stop pumping action completely. If aeration near the suction cannot be avoided, a baffle must be used to separate air from the water before entering the pump. Failure to eliminate air will result in pump loosing its prime, or at least reduced head and capacity.

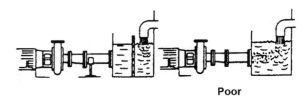

Poor

Figure 8.66

Discharge Fittings

If possible use discharge fittings larger than the size of the pump's discharge; then increase with 8°–12° tapered fittings to the size of pipe required for allowable friction losses.

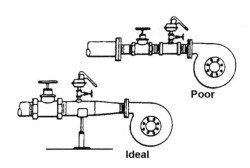

Figure 8.67

Suction and Discharge Piping for Hot Temperature Pumpage

For liquid temperatures above 150°F, expansion joints are recommended on both sides of the pump to reduce the stresses caused by thermal expansion. For pumpage temperatures above 300°F, slow preheat of the pump is recommended to prevent damage to critical parts by thermal shock. Circulate a small amount of hot liquid (actual pumpage preferred) through the pump until temperature is within 100°F.

Piping Loads—Suction and Discharge

Install piping so that no piping loads are imposed on the pump. Piping loads can adversely affect pump component life. Use pipe supports on suction and discharge piping.

Foreign Material

Check the pump, suction fittings, and all piping for bolts, rocks, wire, crating material, or other debris, and remove. Eliminate burrs from cutting welding or tapping pipe which will interfere with water flow. Foreign material left in the suction line can plug or partially plug the impeller, resulting in no flow or reduced capacity and vibration.

A screen or strainer is recommended to keep out any debris larger than what will pass through the pump impeller. The strainer should have a free opening area at least *three times* the area of the suction pipe. Use a sturdy screen to avoid collapse if plugging occurs. Make allowance for some method of cleaning the screen in case it should become plugged.

PUMP SYSTEM CATEGORIES

Currently, there are three major categories of pump systems: (1) bypass, (2) conventional, and (3) variable frequency drive (VFD).

Bypass Pumping Systems

Bypass systems have one or more pumps with fixed-speed motors and a bypass valve that relieves excess flow back to a wet well or to the suction source.

These systems have two major problems, caused by a constant output of their fixed-speed pump(s). First, when the pumps are turned on, they immediately ramp up toward maximum output. This can promote a harmful surge (water hammer) since the bypass valve usually cannot fully divert the excess pressure. Second, most golf course irrigation systems require different levels of performance for various sections and at different times of the day to promote good turf growth and maintenance.

If, for example, a single fixed-speed pump designed for 1000 gpm at 120 psi were used and the sprinkler demand dropped to 600 gpm, which means 400 gpm would be routed through the bypass valve, this would be a waste of energy and money, since flow is at 60% whereas the horsepower is always at 100% of design.

Conventional Pumping Systems

The conventional pump system using the vertical turbine or horizontal suction type pump with constant speed induction motors, will always produce a specific number of gallons per minute (gpm) at a specific number of pounds per square inch.

Conventional systems use a control valve to maintain a constant pressure. This valve is combined with one or more pumps and constant-speed motors which usually are turned on and off by switches that monitor pressure and flow. Any changes in pressure and flow must occur at the control valve.

In a system with a fixed-speed motor(s) the pump has one and only one *optimal* design point, as illustrated in Figure 8.68.

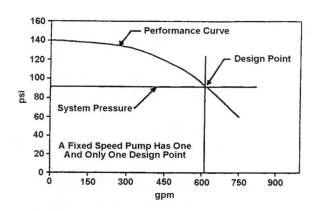

Figure 8.68. Pump curve.

Fixed-Speed Pumping Systems

Components:

1. horizontal centrifugal pumps or vertical turbine pumps
2. CLA-VAL hydraulically operated, pressure-reducing control valve
3. hydropneumatic tank
4. programmable logic controller (PLC)
5. operator terminal interface system (OTIS)
6. pressure transducers

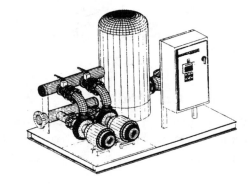

Figure 8.69

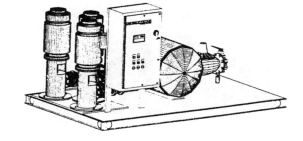

Figure 8.70

Logic Diagram:

1. Irrigation begins.
2. Control valve detects downstream pressure drop and begins to open, releasing stored pressure from hydropneumatic tank.
3. PLC, via the pressure transducer, detects upstream pressure dropping below preset level and starts jockey pump.
4. If, after preset time, the upstream pressure does not return to preset minimum the PLC starts 2nd pump. Pump station continues to start additional pumps as required.
5. Irrigation ends.
6. As flow decreases, PLC sequentially cycles off pumps based on flow signal.
7. Control valve continuously and independently opens and closes as required to maintain constant downstream pressure.

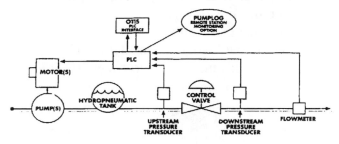

Figure 8.71

Control Valve Operation

When the throttling control opens to a point where more pressure is relieved from the cover chamber than the restriction can supply, cover pressure is reduced and the valve opens.

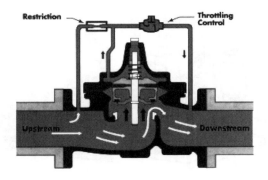

Figure 8.72. Valve open.

When the modulation control closes sufficiently to direct a great enough pressure into the cover chamber to overcome opening forces of line pressure, the main valve closes.

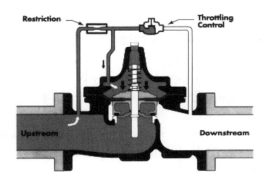

Figure 8.73. Valve closed.

The main valve modulates to any degree of opening in response to changes in the throttling control. At an equilibrium point the main valve opening and closing forces hold the valve in balance. This balance holds the valve partially open, but immediately responds and readjusts its position to compensate for any change in the controlled condition.

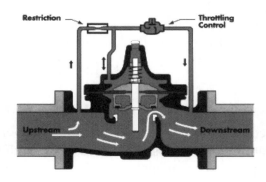

Figure 8.74. Valve throttling.

Constant-Speed Pumping System

In a conventional system, as with the bypass system, the manufacturers add more pumps with constant-speed motors to extend the number of optimum design points. For example, a system with two pumps will have three optimum design points: when one pump is running alone, when the other pump is running alone, and when both are running together.

However, a pumping system cannot operate at peak efficiency when its fixed-speed pumps offer a limited number of optimum design points.

Nevertheless, it is possible to maintain a near constant maximum efficiency. The key to success is to properly plan the irrigation schedule. Throughout the irrigation cycle, the total demand (sprinklers running) must be equivalent to the total pump output based on the optimum design points.

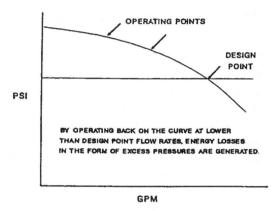

Figure 8.75. Constant-speed pumping system.

Advantages of the Fixed-Speed Pumping Systems

- time-tested with proven technology
- cost-effective
- widely available, and numerous service technicians available

Packaged Pump Systems and Pump Stations

There are many pump manufacturing companies on the market that can configure pump stations to match the unique conditions of your golf course.

Figure 8.76

Before a packaged pumping system can be designed, certain types of information must be obtained:

- Demand requirements, i.e., total gpm
- Lift and discharge head or pressure required.
- Water source—Determine the maximum suction lift. Determine the maximum/minimum variations or effects of storage on pumping demand.
- Type of power available—Determine the most effective form of power (i.e., electrical, mechanical, and standby, if desired) Note whether three-phase power is available. If not, is it obtainable? Pumps over 15 hp generally require three-phase power in 440 or 680 V.

Variable-Frequency Drive Systems (VFDS)

VFD systems also use standard pumps and fixed-speed motors like the other systems. However, solid-state electronics and computer controls give them an almost unlimited number of design points and performance options.

Figure 8.77

Figure 8.78

Variable-Frequency Drive Diagram

The frequency is changed by the sequence of events, as shown in Figure 8.79.

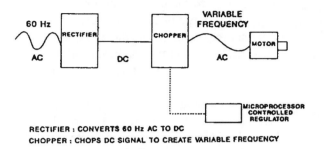

RECTIFIER : CONVERTS 60 Hz AC TO DC
CHOPPER : CHOPS DC SIGNAL TO CREATE VARIABLE FREQUENCY

Figure 8.79. Variable frequency drive (VFD).

Motor Speed

Motor speed is proportional to the input frequency, as shown in Figure 8.80.

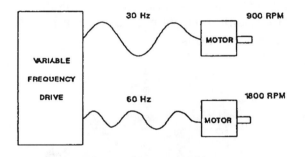

Figure 8.80. Motor speed is proportional to input frequency.

VFD Pump Curves

VFD pumps can handle almost any irrigation requirement by maintaining constant pressure while automatically changing flow. The ability to change speeds gives VFD pumps a broad range of performance curves. (See Figure 8.81.)

Variable-Speed Pumping System (VSPS)—Logic Diagram

The logic diagram shown in Figure 8.82 provides a simplified method for describing the components and principles of variable-speed pump operations.

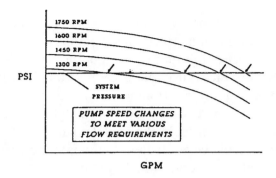

Figure 8.81. Variable-speed pumping system.

1. Pressure transducer sends an electronic signal (proportional pressure) to Smoothflow VI.
2. Smoothflow VI compares actual line pressure with the desired (set point) pressure and then calculates how much to increase or decrease pump speed in order to maintain set point.
3. Smoothflow sends electronic signal to variable-frequency drive (VFD) telling it how fast to drive the motor/pump.
4. VFD varies output frequency to motor so that it runs at the precise speed requested by Smoothflow.
5. OTIS provides local pump station monitoring and control.
6. Optional remote monitoring via PC-compatible computer allows for remote monitoring and data logging.

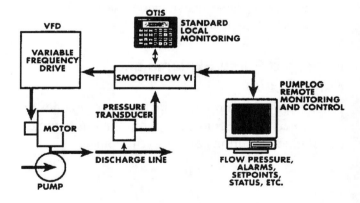

Figure 8.82

Variable-Speed Operation

Preassembled variable-frequency drive pumping systems can be configured to precisely match the unique specifications of your golf course.

Turf pump station systems rely on two basic pump designs. In vertical turbine design, multi-stage impellers below water level eliminate the need for priming. In horizontal centrifugal design, suction pumps draw water directly from the water source, and the system must have valving to hold prime. See Figures 8.83 and 8.84.

A small submersible pressure maintenance (jockey) pump in the 1.5 to 5 hp range, delivering 15 to 45 gpm, is generally included to maintain pressure during non-irrigation times.

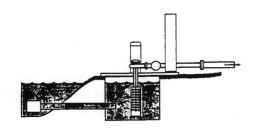

Figure 8.83. Vertical turbine pump design.

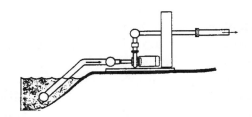

Figure 8.84. Horizontal centrifugal pump design.

Pump Station Selection Guide

The points listed are presented to assist the golf course superintendent and/or golf course committee with the selection process. Should variable-frequency drive (VFD) or constant-speed pumps be used? Vertical turbine or horizontal centrifugal pumps? New installation or upgrade an existing pump system?

You could pay $60,000 or $30,000 for a 100-psi, 1000-gpm pumping station. The low bid may save money up front, but can prove very costly in the longer run.

Points to Consider

Budget—Establish long- and short-term priorities. You get what you pay for, sooner or later.

Gather input data—Check packaged pumping system manufacturers, distributors, and irrigation system designers. Draft an overview of present and future system requirements. Include location, need for filtration, pump system details, utility rates, local codes, and budget.

Contact local, state and federal agencies—Obtain water allotments and/or water usage restrictions, and environmental and energy incentives.

Narrow the pump and power choice—Determine the total dynamic head (TDH) and gallons per minute. Calculate the horsepower requirements. *Note:* A vertical turbine pump costs more up front; however, the efficiency can be 15 to 20% more than that of a centrifugal pump. Its slower speed will also reduce wear and increase pump system life.

Pressure regulation—Traditional valves are diaphragm-throttled. These require maintenance (they may clog) and electronic motor control of surge and pressure. The VFD units, on the other hand, operate without a mechanical control valve. Therefore, effluent and less than pure water are no problem. Continuous automatic monitoring and adjustment of pressure and flow assure maximum energy savings and less stress on the total irrigation system.

Warranties—Check for single source warranty that will cover: delivery, installation, overall system, electronics, pumps, motor, and controller compliance as a unit. Pay particular attention to the reputation of the warranty issuer enforcement clauses in extended warranties.

Factory support and post-installation service—This should include testing for system performance before installation, and field testing assistance after installation. Also, check on the services offered and operator training.

Variable-Speed Operation

This is the normal method of operation for the SyncroFlo variable-speed drive pumping station and all three pressure set point modes:

1. "Normal" set point mode is used to maintain the normal system pressure of the station. Also for "backup" constant-speed operation.

2. "Energy" set point mode is to decrease the system pressure lower flow rates during variable-speed operations to save energy.

3. "Refill" set point mode is used to slowly pressurize or refill the irrigation system during variable-speed operation.

Figure 8.85

If no pumps are running, the jockey pump will start 2 seconds after the system pressure falls 5 psi below normal. The jockey pump is only used to keep the irrigation system pressurized during off hours. Therefore, it will run until the irrigation system is pressurized again or until a main pump starts. If there are no sprinklers running, the system pressure should begin to rise slowly. If the system pressure rises 5 psi above normal and there is no flow, the jockey pump will stop 15 seconds later, assuming that its 1-minute run time has already expired.

If the jockey pump is running, the first main pump will start 2 seconds after the system pressure falls 10 psi below normal or immediately after the system pressure falls 20 psi below normal, which is the low system pressure alarm setting. The inverter soft-starts the first main pump and then runs it at the proper speed to meet system pressure. Once the first main pump is running, the jockey pump will stop 15 seconds later whether or not its 1-minute run time has expired. As the flow rate increases, the inverter will increase the speed of the main pump to maintain system pressure.

If the inverter runs at 100% speed (60 or 50 Hz) for 5 seconds continuously, the inverter will be disengaged from its main pump, and then that pump will immediately be switched across-the-line to run

full speed. Speed is used to determine when to add the next pump, because it is more energy efficient to run a pump at 100% speed across-the-line than it is with the inverter. The inverter soft-starts the second main pump and then runs it at the proper speed to meet system pressure. Or, if the flow rate only requires one main pump to maintain pressure, the inverter will keep the second main pump running very slowly (15 Hz) until it is needed. The station's pressure relief valve will limit the discharge pressure if the flow rate decreases while a main pump is running full speed. If your station has three main pumps, this cycle will repeat when the third main pump is required to run. At full capacity, the variable-speed main pump will be running at 100% speed, the constant-speed main pump(s) will be running, and the relief valve should not have any water flowing through it.

If two or more main pumps are running and the flow rate decreases, the relief valve will open to limit the system pressure and the inverter will slow down the variable-speed main pump as needed. A constant-speed main pump will stop 15 seconds after the flow rate has dropped 20% below the capacity of the other main pump(s), assuming that main pump's run time has expired. Each time a constant-speed main pump stops, the inverter will run the variable-speed main pump at the proper speed to maintain system pressure, and the relief valve will operate as needed.

Packaged Pumping Stations

Prefabricated pumping stations are available which can be designed for practically any pumping requirement(s). These units are prewired, base- or skid-mounted, and completely self-contained for ease of installation. A packaged station may consist of two or three main pumps, a jockey pump, a controller, and other amenities as predetermined. For example, a unit may consist of a lead pump (first stage) that has a variable-speed drive. (See Figure 8.86.)

The first stage is designed to deliver from zero up to a specific gpm at a specific design pressure (based on the demand requirements).

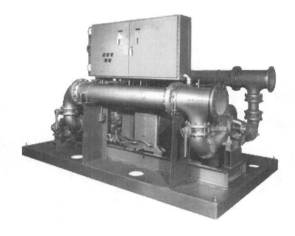

Figure 8.86

The second stage should be a constant-speed drive capable of delivering a specific gpm at the above design pressure, and will operate when the system requirements exceed the capacity of the lead pump. Once the variable-speed motor no longer meets the demand of the system, the second motor (constant-speed) is signaled to start. The variable-speed motor will then automatically adjust to its new required speed.

As the demand from the system decreases, the pumps will shut down in reverse order. When the demand drops to a specific gpm, the jockey pump will continue to operate until the demand reaches zero.

The jockey pump will operate while the main pumps are operating and is capable of maintaining system pressure after the main pumps are turned off.

Figure 8.87

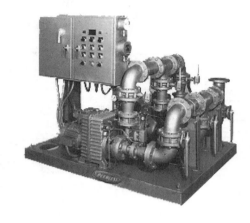

Figure 8.88

PSI - 1800

Figure 8.89

MULTIPLE VERTICAL TURBINE PUMP INSTALLATION

Figure 8.90 shows a type of pump plant installation featuring a number of individual vertical turbine pumps working individually or together to provide constant pressure at various flow rates. An air compressor and pressure tank are used to maintain the desired supply line operating pressure. In-line filtration units are used to clean up the water prior to it entering the irrigation system. This type of pump plant is used in the large irrigation system obtaining its water from a lake, river or reservoir. The filtration is required because these sources are usually contaminated with debris. Typically, this pump plant can be found on a golf course.

This type of pump plant installation uses a pressure tank, approximately three times the system capacity in size, with an air compressor to maintain

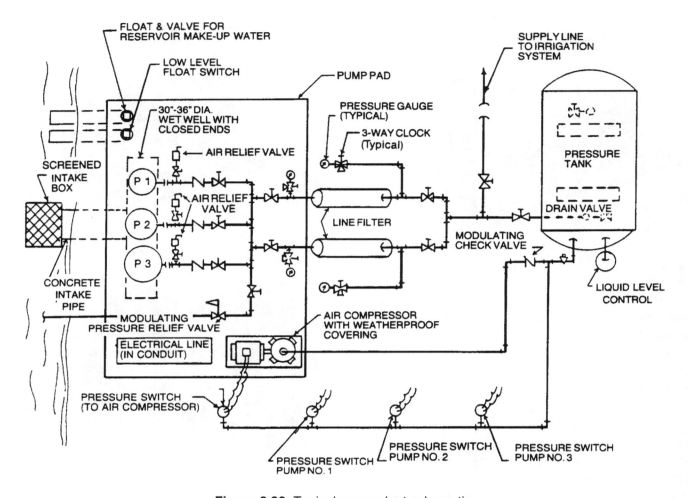

Figure 8.90. Typical pump plant schematic.

the downstream supply line pressure. The air compressor is activated by a liquid level control device at the pressure tank. The air compressor is shut off by a pressure switch at the compressor which is calibrated to maintain the tank at the optimum operating pressure

Each pump is equipped with a mercoid type pressure switch. The pressure switches are set to turn the individual pumps on and off at predetermined pressures. Each pump will provide a different volume of water depending on demand. The pumps will provide these volumes at full desired system pressure.

The filters are installed in such a manner as to remove debris before it reaches the irrigation system supply line. Each filter is equipped with a manual or automatic flush valve, allowing the filter to be backwashed periodically. Pressure gauges are

installed on each filter to indicate when the filter requires cleaning. Cleaning the filters on a timely basis minimizes system pressure losses and increases the life span of the irrigation system components.

The pressure relief valve is adjusted to activate at 10% above system operating pressure and discharges any water back to the water source.

Air release valves are installed on each pump to eliminate any air which might be collected as the pump draws water from the reservoir. Anti-slam check valves protect the pumps from excessive backpressures, thus adding to the longevity and efficiency of the pump plant.

Although larger in size than other types of pump plants, this system can be provided as a prefabricated unit. The pressure tank would be installed separately.

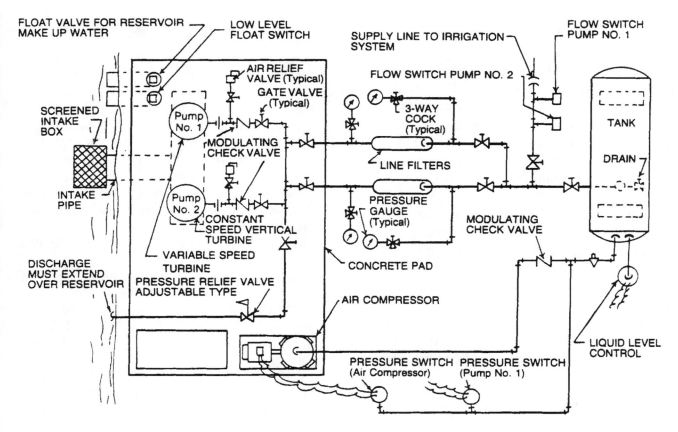

Figure 8.91. Typical pump plant schematic.

VARIABLE-SPEED/CONSTANT-SPEED VERTICAL PUMP INSTALLATION

Figure 8.91 shows a pump plant that utilizes two or more different types of vertical turbine pumps. A variable-speed, constant-pressure vertical turbine is combined with a constant-speed vertical turbine to provide required flow rates at a constant pressure. This type of installation also uses a pressure tank with air compressor to maintain the desired supply line operating pressure. In-line filtration units are used to clean up the water prior to it entering the irrigation system. This type of pump plant is used in the large irrigation system obtaining its water from a lake, river, or reservoir. Filtration is required because these water sources are usually contaminated with debris. Typically this pump plant can be found on a golf course.

This type of pump plant installation uses a pressure tank, approximately three times the system capacity in size, with an air compressor to maintain

the downstream supply line pressure. The air compressor is activated by a liquid level control device at the pressure tank. The air compressor is shut off by a pressure switch at the compressor which is calibrated to maintain the tank at the optimum operating pressure.

The variable-speed pump is equipped with a mercoid type pressure switch. This switch is set to start Pump 1 when a drop in system pressure is sensed. This pump is designed to provide the lesser flow rates demanded. As the flow demand reaches approximately 50% of the maximum system demand, a flow switch starts the constant-speed turbine pump. A separate flow switch is calibrated to shut off the variable-speed turbine pump at zero flow. A mercoid pressure switch is connected to the air compressor to stop it at 10% above system operating pressure.

The filters are installed in such a manner as to remove debris before it reaches the irrigation system supply line. Each filter is equipped with a manual or automatic flush valve, allowing the filters to be backwashed periodically. Pressure gauges are installed on each filter to indicate when the filter requires cleaning. Cleaning the filters on a timely basis minimizes system pressure losses and increases the life span of the irrigation system components.

The pressure relief valve is adjusted to actual at 10% above system operating pressure and discharges any water back to the water source.

Air release valves are installed on each pump to eliminate any air which might be collected as the pump draws water from the reservoir. Anti-slam check valves protect the pumps from excessive backpressures, thus adding to the longevity and efficiency of the pump plant.

Although larger in size than other types of pump plants, this system can be provided as a prefabricated unit. The pressure tank would be installed separately.

ACKNOWLEDGMENTS

1. "Engineering Section." Catalog. Marlow Pumps Manufacturing Company.
2. "Design Information for Large Turf Irrigation Systems." Manual. Toro Manufacturing Company.
3. "Affinity Laws." Booklet. Goulds Pumps, Inc.
4. "Hydro-Constant Pumps" and "Net Positive Suction Head." Technical Data Manual. Peerless Pumps Manufacturing Company.
5. "Planning for an Individual Water System." Booklet. American Association for Vocational Materials.
6. "Introduction to Pump Curves." Booklet. Goulds Pumps Inc.
7. "Variable Speed Pumping Systems." Handout. Flowtronex Pre-Packaged Pumping Systems.
8. "VFD Pump Selection Pump Guide" and "Variable Speed Operation." Technical Data Information Handouts. SyncroFlow Packaged Pumping Stations.
9. "Variable Speed and Constant Speed Pump Installations." Buckner Irrigation Design Manual.
10. "Hydraulic Installation Guide." Cornell Pumps Co.

Chapter 9

Remote Control Valves

CONTROL VALVES

The Table of Contents of a prominent valve manufacturing company currently lists 21 major categories containing an average of 6 different models per category. Obviously, it is beyond the scope of this presentation to consider more than the selected number of valves listed below.

Operating Principles

The globe valve causes high resistance to water flow because of the narrow passageways and sharp turns.

Figure 9.1a

The gate valve, when completely open, provides a straight-through water flow passage with very low resistance.

Figure 9.1b

The ball valve also provides easy passage of water and low resistance.

Figure 9.1c

Check valves are used to limit the water flow to one direction. Check valves can be of the swing variety, Figure 9.2a, which depends on its own weight to close against a backflow.

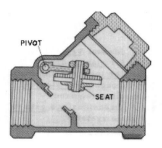

Figure 9.2a

The spring loaded variety, which is closed against backflow by retracting a spring.

Figure 9.2b

The float variety in which the valve is pushed out of the way for regular water flow but pressed into the upstream opening by backflow.

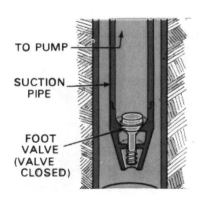

Figure 9.2c

Applications

On the intake side of the suction line, a check valve or *foot valve,* is used to prevent the pump from losing prime.

On the discharge side of the pump, where the pipe slopes upward to minimize hydraulic ram when the pump is shut off to prevent drainage around low heads in a system.

Wafer Swing Check Valve

The short face-to-face length of this valve enables installation in restricted spaces. The spring-assisted closure minimizes the possibility of water hammer. Sizes range from 2 to 24 in.

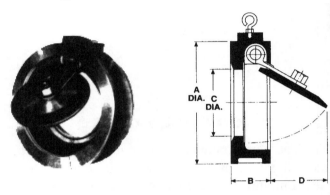

Figure 9.3

Table 9.1

Valve	Dimensions (in inches)				
Size	A	B	C	D	Wt lb
2	4-1/8	1-3/4	1-5/16	1-1/2	4.4
2-1/2	4-7/8	1-7/8	1-11/16	1-3/4	5
3	5-3/8	2	2-1/16	2-3/8	7
4	6-7/8	2-1/4	3	3-1/4	11
5	7-3/4	2-1/2	3-3/4	4-1/4	15.5
6	8-3/4	2-3/4	4-3/4	5-1/16	18
8	11	2-7/8	6-7/16	7-3/16	29
10	13-3/8	3-1/8	7-5/8	8-13/16	49
12	16-1/8	3-1/8	9-1/2	10-1/2	68
14	17-3/4	4-1/4	10-1/2	11-3/4	165
16	20-1/4	4-1/4	12-1/2	14-3/8	200
18	21-3/8	4-1/4	14	16-5/8	220
20	23-7/8	5-1/2	15-1/4	16-7/8	340
24	28-1/4	6	19	20-1/2	551

Atmospheric Vacuum Breakers

The atmospheric vacuum breaker is basically a check valve that generally uses a float seal to seal against backflow. The backflow is directed to atmosphere and vents to atmosphere when there is zero pressure in the line. (See Figures 9.4a and 9.4b.)

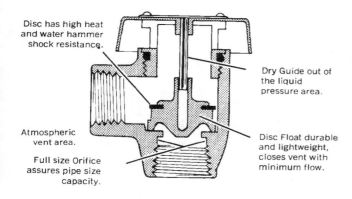

Disc has high heat and water hammer shock resistance.

Dry Guide out of the liquid pressure area.

Atmospheric vent area.

Full size Orifice assures pipe size capacity.

Disc Float durable and lightweight, closes vent with minimum flow.

Figure 9.4a

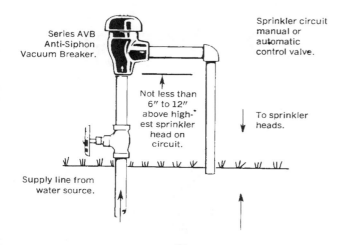

Series AVB Anti-Siphon Vacuum Breaker.

Sprinkler circuit manual or automatic control valve.

Not less than 6″ to 12″ above highest sprinkler head on circuit.

To sprinkler heads.

Supply line from water source.

Figure 9.4b

These units are specifically for the protection of domestic water when irrigation lines use domestic water as a source. In residential systems the vacuum breaker is part of the shutoff valve. The open-end vacuum breakers are installed on the downstream side of the manual or remote control valve. When the control valve is open, the breaker is closed. If the source piping pressure fails for any reason, or if there is a vacuum on the source line while the system is operating, the vacuum breaker opens to stop the siphon action.

These units must be 6 to 12 in. above the highest sprinkler head. Otherwise, water would drain out of the device every time the sprinklers were turned off.

Note: Install in accordance with the local code.

Backflow Prevention Valves

Where municipal or city water is to be used as the source for irrigation, it is mandatory to protect the safe potable water against mixing or backflow of polluted water. The two major considerations in irrigation are (1) the system can be in contact with the soil water and in turn can mix with the water in the system and (2) fertilizers or chemicals may be injected into the system. These conditions are generally classified as toxic hazards, and the appropriate backflow prevention devices must be installed.

The water department and/or the local board of health will specify the type of connections required, and it is necessary to seek their advice and approval.

Constant Pressure

In automatic irrigation systems the backflow prevention devices get more complicated because the device must be under constant pressure. A constant-pressure device may be allowable under constant pressure on the main line before any connection to the irrigation system and between the point of connection and the first remote control valve. The constant-pressure vacuum breaker type is used where it is possible to install it at the highest elevation point because it still depends on breaking the potential siphoning action should any backflow pressure condition exist. (See Figure 9.5.)

Note: No type of atmospheric vacuum breaker can be used in a toxic system if back pressure exists.

Constant Pressure without Back Pressure—Backflow Prevention Devices

There are many types of constant-pressure devices available. In general, they consist of the following basic parts:

- Manual control inlet and outlet valves—used mainly in testing the effectiveness of the unit in the field.
- Double check valves.
- Vacuum breaker—spring-loaded to keep from sticking closed.
- Test cocks—for checking the effectiveness of the device in the field.

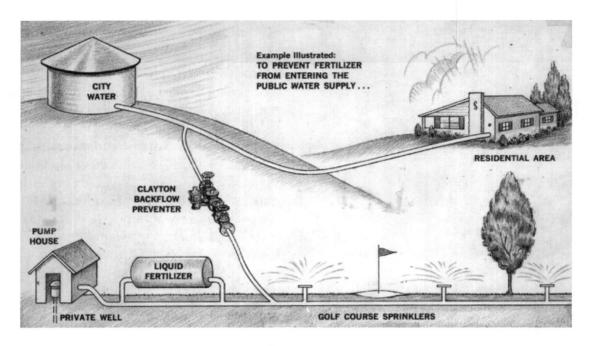

Figure 9.5

Pressure vacuum breakers are spring-loaded anti-siphon devices which require pressure to open and are vented to atmosphere. Although most codes allow one of these units prior to any sprinkler control valves, water can drain back through the device if the pipe is higher than the device. Usually the size is limited to 2 in. IPS (Iron Pipe Size).

Figure 9.7. Double check valve with pressure vacuum breaker.

Under normal operation the line pressure in the system keeps the check valves open and the vacuum breaker closed. If the main line pressure is lost or goes to a vacuum the check valves close and the vacuum breaker opens.

Figure 9.6. Pressure vacuum breaker.

Double-check valve assemblies with gate valves on each end and a pressure type vacuum breaker are sometimes used when the supply line is too large for the pressure vacuum breaker to be in line either singularly or in parallel.

Reduced-Pressure Backflow Preventer

Golf course irrigation systems that are connected to a municipal water system for part or all of their supply should consult the local codes before installing any backflow prevention device.

The *reduced-pressure backflow preventer device* is designed to prevent backflow of water in high hazard applications, such as liquid fertilizer or chemical injection into the irrigation system. (See Figures 9.8 through 9.10.)

Where a pressure condition exists in the irrigation system because of elevation differences or a pump, a reduced-pressure unit must be used. These units have a differential pressure relief valve instead of a vacuum breaker between the two check valves. The relief valve automatically maintains a pressure of 2 to 3 lb less than the inlet pressure between the two check valves. In case of a mainline pressure loss, any backflow water from the irrigation system that gets by the downstream check valve is discharged to atmosphere through the relief valve.

Figure 9.8

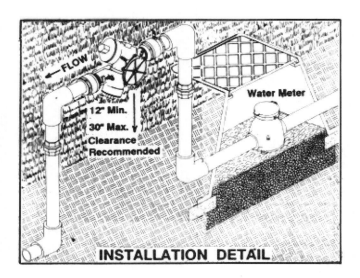

Figure 9.9

Figure 9.10. Reduced-pressure vertical backflow preventer.

Installation

It is recommended that this unit be installed with the check valves in a horizontal position.

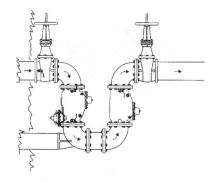

Figure 9.11

Basic Cross Connection and Backflow Prevention Information

The five basic devices that can be used to correct cross connections:

1. An *air-gap* is the physical separation of the potable and non-potable system by an air space. The vertical distance between the supply pipe and the flood level rim should be two times the diameter of the supply pipe, but never less than 1 in. The air gap can be used on a direct or inlet connection and for

all toxic substances. However, this type of protection is usually not practical for irrigation and sprinkler applications.

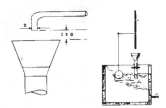

Figure 9.12

2. *Atmospheric vacuum breakers* should be used only on connections to a non-potable system where the vacuum breaker is never subjected to backpressure and is installed on the discharge side of the last control valve. It must be installed above the usage point. It cannot be used under continuous pressure in excess of 12 hr.

   Hose connection vacuum breakers may be used on sill cocks and service sinks.

Figure 9.13

3. *Pressure vacuum breakers* should be used as protection for connections to all types of non-potable systems where the vacuum breakers are not subject to backpressure. These units may be used under continuous supply pressure. They must be installed above the usage point. (See Figure 9.14.)

   Backflow preventers *with* intermediate atmospheric vent may be used as an alternate equal for 1/2-in. and 3/4-in. pressure vacuum breakers and, in addition, provide protection against backpressure.

4. *Double check valve assemblies* should be used as protection for all direct connections through which foreign material might enter the potable system in concentrations which would constitute a nuisance or be aestheti-

Figure 9.14

cally objectionable, such as air, stream, food, or other material which does not constitute a health hazard.

Figure 9.15

5. *Reduced-pressure zone devices* should be used on all direct connections which may be subject to backpressure or back-siphonage, and where there is the possibility of contamination. (See Figures 9.16 and 9.17 and Table 9.2.)

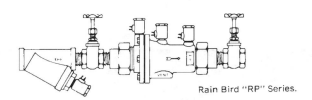

Figure 9.16

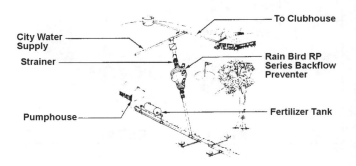

Figure 9.17

Table 9.2. Standards and Testing Laboratories for Backflow Preventer Performance Test and Construction

| Product | Current Standards | | | Testing Labs | | |
	ASSE	AWWA	USC FCCC	Factory Mutual	USC FCCC	NSF
Atmospheric type vacuum breakers	1001					X
Hose connection vacuum breakers	1011					X
Backflow preventer with intermediate atmospheric vent	1021					X
Reduced pressure principle backflow preventer	1013	C-506	Specs Manual for Cross Conn. Controls	X	X	X
Double check valve type backpressure backflow preventer	1015	C-506	Specs Manual for Cross Conn. Controls	X	X	X
Vacuum breakers, pressure types	1020			X	X	X

References: ASSE–American Society of Sanitary Engineers; AWWA–American Water Works Association; USC-FCC–University of Southern California, Foundation for Cross Connection Control Research; NSF–National Sanitation Foundation.

RP-TK Test Kit for RP Backflow Devices
Test Procedure

Reduced-pressure principle device: (See Figure 9.18)

- Connect the No. 2 Test Cock of the device of the "high" hose.
- Connect the No. 3 Test Cock of the device to the "low" hose.
- Close No. 2 shutoff valve of the device.
- Open Test Cocks Nos. 2 and 3.
- Open vent (C) valve.
- Open "high" (A) valve and bleed to atmosphere until all the air is expelled.
- Close the "high" (A) valve. Open the "low" (B) valve and bleed to atmosphere until all air is expelled. Close "low" (B) valve. Close "vent" (C) valve.
- Connect the No. 4 Test Cock of the device to the vent hose.

Test No. 1
Purpose

To test Check Valve No. 1 for tightness against reverse flow.

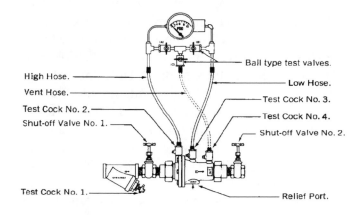

Figure 9.18. Field test equipment required: Reduced-pressure principle backflow preventer test kit. Rain Bird Model RP-TK.

Requirements

Valve must be tight against flow under all pressure differentials. With both the "high" (A) and "low" (B) valves closed, observe the pressure differential gauge. If there is a decrease in the indicated valve the No. 1 Check Valve is reported as "leaking."

Test No. 2

Purpose

To test Check Valve No. 2 for tightness against reverse flow.

Requirements

Valve must be tight against reverse flow under all pressure differentials. Open the "high" (A) and vent (C) valves and keep the "low" (B) valve closed. Open the No. 4 Test Cock. Indicated pressure differential will decrease slightly. If pressure differential continues to decrease, the No. 2 Check Valve is reported as "leaking."

Test No. 3

Purpose

To test operation of pressure differential relief valve.

Requirements

The pressure differential relief valve must operate to maintain the "zone" between the two check valves at least 2 psi less than the supply pressure. Open the "high" (A) valve. Close the vent (C) valve. Open the "low" (B) valve very slowly until the differential gauge needle starts to drop. Hold the valve at this position and observe the gauge reading at the moment the first discharge is noted from the relief valve. Record this as the opening differential pressure of the relief valve. *Note:* It is important to accomplish this bleed very slowly. Close Test Cocks Nos. 2, 3, and 4.

Remove vent hose from Test Cock No. 4. Use vent hose to relieve pressure from test kit.

Remove all test equipment and open No. 2 Shutoff Valve of the device.

Caution! Prevent freezing—drain test kit and hoses prior to placing in case.

Specifications:

- maximum working pressure—500 psi
- maximum working temperature—210°F
- gauge—0.10 psid with ±2% accuracy, full scale.
- hoses—(3) 3 ft with 1/4-in. female threaded couplers.

- adapters—(3) 1/4-in. threaded adapters (3) 1/2-in. bushings, (3) 1/4-in. bushings
- 1—16-in. securing strap
- 1—moisture-resistant instruction guide
- case—shock-resistant molded plastic with special diced foam insert that enables multiple compartment combinations for tools, accessories or similar items

The RP-TK Backflow Preventer Test Kit is a compact portable testing device especially made for testing all reduced-pressure principle backflow prevention devices. The RP-TK is easily connected to any RP device, enabling accurate testing of "zone" differential, fouled check valves, or similar problems that visual inspections cannot locate. The unit is encased in a rugged carrying case for easy handling and accessibility.

Test Unit.

Test Unit & Carrying Case.

Figure 9.19

VALVE CONTROL MODELS AND TYPICAL APPLICATIONS

Normally Closed Electrical System

The Electrical Control System

The controller sends 24V-AC to the control valve, energizing an actuation solenoid. The valve opens, allowing irrigation water to flow to the sprinklers. The valve will remain open until the controller discontinues its 24V-AC output. (See Figure 9.20.)

Note: The electric valves in this series are designed for clean water.

Plastic Diaphragm Valves (Electrical)— How They Work

Closed Position

Irrigation water, metered through the diaphragm orifice, fills the diaphragm chamber, causing in-

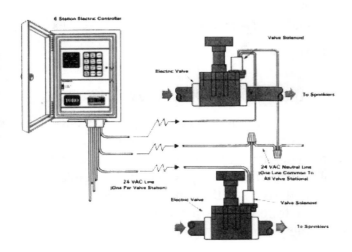

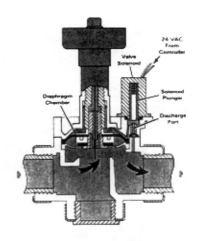

Figure 9.22

Figure 9.20

ternal pressure to build. The pressure within the chamber holds the diaphragm assembly firmly against the valve seat, preventing water flow through the valve. (See Figure 9.21.)

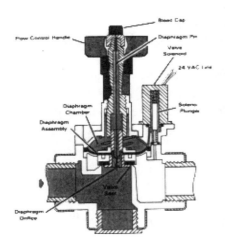

Figure 9.21

Open Position

The valve solenoid, when activated by 24V-AC from the controller, draws the solenoid plunger away from the discharge port, relieving the water pressure from the diaphragm chamber. The irrigation water flows away from the valve seat, allowing water to flow through the valve. (See Figure 9.22.)

Normally Open Hydraulic System
The Hydraulic Control System

Figure 9.23 shows the basic components of a normally open control system. The controller supplies and discharges filtered water to and from the valve diaphragm/piston chamber for operation. The normally open valve diaphragm/piston chamber is isolated from the irrigation water and receives only filtered water from the controller, which enables this valve to be used in dirty water systems.

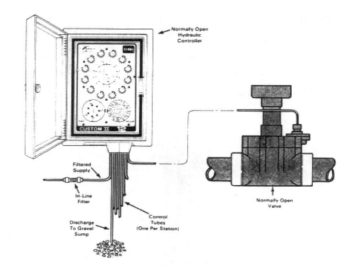

Figure 9.23

Note: In the normally open valve systems the maximum tubing length is 1000 ft.

Plastic Hydraulic Valves— How They Work

Open Position

The selector valve in the controller opens and allows water to discharge through the control tube, relieving water pressure from the diaphragm chamber. Supply water pushes the diaphragm assembly away from the valve seat, allowing water to flow through the valve.

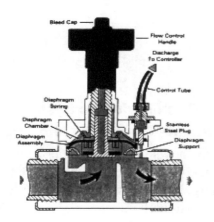

Figure 9.24

Closed Position

Filtered water from the controller fills and pressurizes the diaphragm assembly firmly against the valve seat, restricting the water flow through the valve. The filtered water pressure from the controller must be equal to or greater than the supply water pressure to the valve.

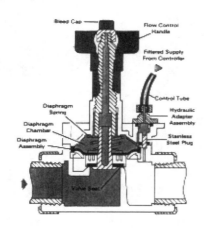

Figure 9.25

PUMP STATION CONTROL VALVING

See Appendix 7.

PUMP CONTROL VALVE SYSTEMS

Vital to all automatic water pumping systems is the need for control. The hydraulic *shock waves* can be potentially harmful to the system unless proper control is available. The type of control depends on the system design and the conditions involved.

The following illustrations observe *some* of the possible pump control valve systems.

Basic Swing Check Valve

The "basic" pump control valve system uses a simple swing check valve to stop flow reversal, which prevents back-spinning the pump and keeps the pipelines full when the pump is off.

Each time the pump starts, the pressure wave created causes the swing check to open suddenly, allowing a damaging surge to enter the system. When the pump stops and flow ceases, the swing check slams shut, causing another surge in the system.

To protect the system against these pressure peaks or surges, a 50-01 pressure relief valve is recommended. This valve senses surges and opens to dissipate the excess pressure to atmosphere.

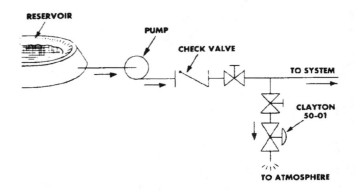

Figure 9.26

81-02 Check Valve

In this system an 81-02 check valve is introduced in lieu of the plain swing check valve. The 81-02 valve is equipped with speed controls. It can be adjusted to allow it to open and close at a regulated rate of speed. This tends to isolate the system from the pump starting and stopping surges.

Again, the 50-01 valve acts to relieve pressure peaks to atmosphere, should a surge occur.

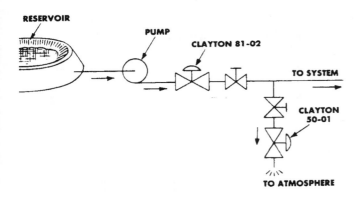

Figure 9.27

91-01 Pressure-Reducing and Check Valve

Further refining of the pump control valve is now made by introducing the 91-01 valve in place of the 81-02.

Along with the check valve with opening speed regulation, we now have the additional function of downstream pressure control. The 91-01 valve will automatically reduce a higher inlet pressure to a steady lower downstream pressure, regardless of varying inlet pressures or downstream demands. Therefore, the system piping is protected from excess pressures.

The 50-01 pressure relief valve functions as a safety factor, relieving pressure peaks to atmosphere whenever surges occur.

An optional feature available on the 91-01 valve provides the practical function of automatically shutting off the pump when the valve closes. This is accomplished with the addition of a *limited switch* assembly. As downstream demands decrease, the valve begins to close. When the valve is closed, the limit switch opens, releasing the pump starter, and allowing the pump to stop.

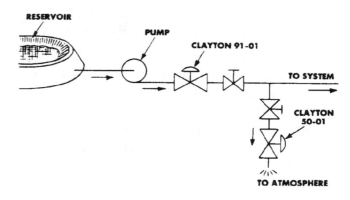

Figure 9.28

92-02 Pressure-Reducing, Pressure-Sustaining and Check Valve

By using the 92-02, the pump control valve is now further improved. The pressure-reducing check feature is the same as that on the 91-01 valve, which provides constant downstream pressure regulation and a regulated check valve operation. In addition, positive pump protection is obtained by including a pressure-sustaining feature. The valve is now able to maintain a constant backpressure on the pump discharge, preventing overpumping and pump cavitation during periods of excessive downstream demands, or low pump suction supply pressure. The sustaining feature also keeps the pump operating within the efficiency range of the pump curve.

As in earlier illustrations, the function of the 50-01 continues to remain the same.

The "limit switch" option is also available for model 92-02.

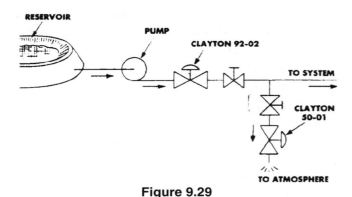

Figure 9.29

AUTOMATIC PUMPING STATION WITH PRESSURE-CONTROL VALVING

The piping diagram in Figure 9.30 illustrates the flexibility and practicality of valves used in various applications. This setup has been installed in a number of golf courses.

The above piping diagram illustrates how automatic valves can be utilized to accomplish the desired results when a water system must meet certain requirements as follows:

- When the maximum pressure in the system must not be exceeded.

- When a pressure must be held against a pump so that during periods of excessive flow a well will not be overpumped.
- When pressure in the system must not be dissipated back into the pump or well areas when all pumps are turned off.

Other advantages with the above setup are:

- When flow ceases, the pressure-reducing valve will close, in turn opening an electrical switch attached to the valve diaphragm assembly, turning off the pump.
- When the pressure in the system drops below the setting of the jockey pump "on" pressure switch, the jockey pump will go on to reestablish the pressure. If the jockey pump cannot keep the pressure from continuing to drop, the main pump "on" switch will turn on, eventually building up the pressure above the "off" set point of the pressure switch used to turn off the jockey pump.
- If a power failure occurs during the pumping cycle and a resulting surge is encountered in the downstream piping, this surge can be relieved by a 50°F pressure relief valve.
- During periods of prolonged low flow rates in the normal pumping cycle, the 50-01B pressure relief valve will open and relieve enough water for the pumps to remain cool until normal flow rates are again resumed and/or the 92L valve with the limit switch assembly has called for the pump to be turned off.

Note: Figure 9.30 illustrates the flexibility and practicality of the valves in various applications. This setup has been installed in a number of golf courses.

ACKNOWLEDGMENTS

1. "Pumping Station Control Valving." Technical Bulletins and Handouts. CLA-VAL Manufacturing Company
2. "Normally Open and Normally Closed Electric and Hydraulic Systems." Technical Data Bulletin. Toro Irrigation Company.
3. "Basic Cross Connection and Backflow Prevention." Sprinkler Irrigation Equipment Manual. Rain Bird Irrigation Company.
4. "Planning for an Individual Water System." Booklet Association for Vocational Instructional Material.

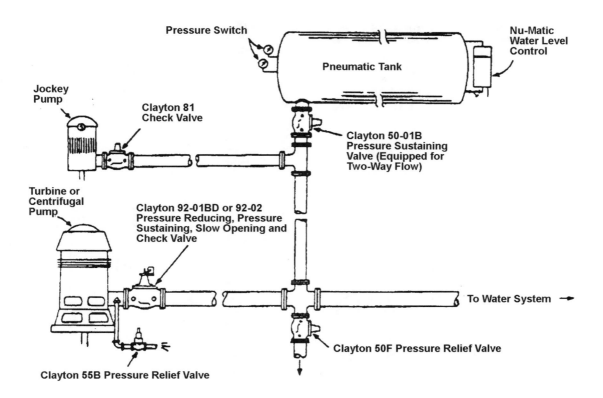

Figure 9.30

Chapter 10

Automatic Controllers

AUTOMATIC CONTROL SYSTEM SELECTION

Automatic controllers for golf course irrigation vary in type, complexity, versatility, degree of sophistication, etc. Basically, all controllers have one major objective, which is to sequentially turn on/off one or more sprinklers for a preset time period.

The final choice of the automatic controller for a particular installation is based on how much versatility the buyer is willing to pay for.

In this discussion and presentation of the available types of automatic controllers it must be recognized that they are representative of the present state of the art.

Since the invention of microchip technology, there has been a phenomenal growth and development that is mind-boggling. The present computerized controllers will undoubtedly be outdated in the near future.

The commercial literature used in this section was selected from material on-hand and considered as typical of what is available on the market today. The automatic controller selected will undoubtedly correspond to the sprinkler model (manufacturing company).

In no way should the use of this material be interpreted as being an endorsement of any product. The primary purpose is to provide visual aids to those students and/or individuals not acquainted with golf course irrigation system design and automatic controllers.

CONTROLLER SYSTEMS

The following illustrations are intended to point out some relationships between pipe layout and sizes, type of controller, and amount of built-in versatility.

Central Control System

Assume that the layout of the golf course was such that a designer decided to divide the fairways into five equal zones (15 sprinklers per zone) as shown. The water source and pump are located at the pond.

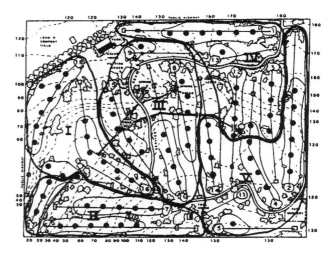

Figure 10.1

217

Irrigation System Design

A maximum of *one* sprinkler is to operate in each of the five zones at the same time. The automatic controller will be located in the pumphouse by the pond.

Other Conditions

1. Fully automatic—single-row, valve-in-head (VIH)
2. Total number of fairway sprinklers = 75 (15 per zone)
3. Irrigation cycle—water all heads every night
4. Irrigation time—20 minutes per station (as calculated)
5. Sprinkler discharge rate—50 gpm, total = 250 gpm
6. Total stations = 15 (5 heads per station)

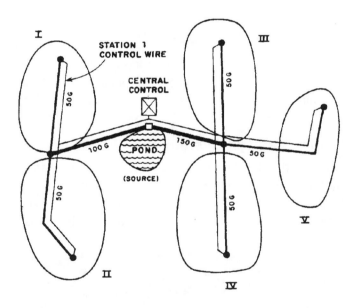

Figure 10.2

Considerations

Advantages:

- Savings on pipe sizes, fittings, and installation costs. Notice the relatively short main pipelines which are designed to carry 150 and 100 gpm. The laterals within each zone are required to carry 50 gpm.

Disadvantages:

- Lack of versatility—The operator cannot adjust for the different watering needs that may exist in each zone. These variations have been stressed throughout. They are: soil types and conditions, type of turf grass, root depth, thatch, slopes, traffic, etc.
- It is questionable whether five remote control valves can be operated per station.
- Cost of wire for 15 stations. Also, large wire size needed for electrical load and distances involved.
- Difficulty of checking whether the sprinklers are on in each zone, even with a walkie-talkie.
- Difficulty in finding control wire breaks.

Field (Zone) Controller System

This system design is based on running a total of five sprinklers within a zone. The controllers will be located in the field—one controller per zone as shown.

Operate one sprinkler head per station.

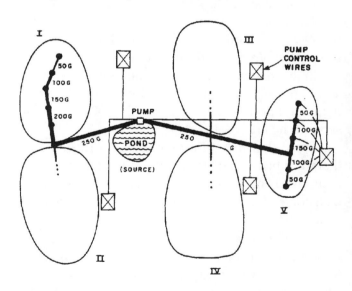

Figure 10.3

Considerations

Advantages:

- Provides considerable versatility. Each head can be individually controlled to adjust to the

different watering needs that may exist in a particular location.
- Shorter lengths of control wiring.
- Observation of the area may be possible from the field controller.

Disadvantages:

- Cost of increased pipe sizes, fittings, and installation.
- No central control.
- Controller wires must be run back to the pump.

Central/Satellite Control System

System design is based on the above conditions (II) with a *central controller* located at the pump or at another convenient location and *field controllers (satellites)*—one for each zone.

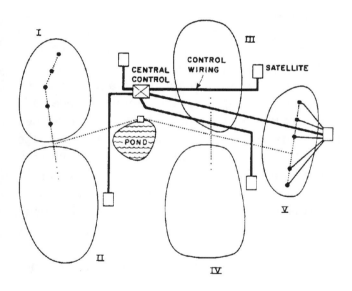

Figure 10.4

Considerations

Advantages:

- Provides considerable versatility. Further, irrigation literature lists the various built-in and optional features for the central satellite control system. These can be selected to suit the user's wishes.

Disadvantages:

- Each additional field controller and optional feature, such as syringing, adds to the cost of installation.
- As the complexity increases, the chances for breakdowns and failures increase as well.

Important: It is essential for the golf course superintendent to know exactly what the system is designed to do—its versatility as well as its limitations.

Further, he should have access to a complete set of plans showing the precise locations of pipelines, valves, control wiring, and heads, as well as the watering program and other instructions.

TYPES OF CONTROLLERS

There are many different types of controllers having many features on the irrigation market. Currently, there is no set of standards or means for rating automatic controllers.

To assist in the introduction of controller systems to students and others with limited background in this subject the following presentation is based on an arbitrary *level of control* as published by a prominent irrigation company. The level of control is rated from 1 to 5. A controller with a rating of 1 has the least features and versatility and a controller with a rating of 5 has the most.

Also included are other comparable controllers on the market that can match or possibly exceed the level of control as given.

It should be understood that the limited number of controllers included in this section were selected from a complete line of available controllers taken from literature on-hand.

Field Controllers

The literature shows that some of the field controllers are designed to stand alone and are operable from the field with their own program, whereas others can be connected to a central controller and operate as part of a central/satellite system.

The time-tested solid-state controllers are used almost exclusively. They are user friendly, easy to understand and program, accurate, and reliable. They provide many additional features, including a wide array of flexible watering options, dual programming, 14-day watering cycle and up to 23 start times.

The size of the field controller is a very important decision for the golf course superintendent and/or the system designer to make. The final choice should be delayed until all the pertinent factors are considered, such as: the type of system to be purchased, topographical features, line of sight, type of system (single-row or multi-row), number of heads to run from one station, and integration of a remote control.

Current controller sizes range from 8 to 51 stations. The area covered can be as small as one fairway, including greens and tees, or a larger zone area. Naturally, the 51-station controller will cover more area than the smaller size.

Types:

1. solid-state
2. electronic
3. electromechanical or "mechatronic"
4. combination of solid-state central and satellite or electromechanical central and solid-state satellite

Automatic Controller Levels

Level Number 1 Features
(See Appendix 8.)

Notice the features listed under the Vari-Time 4000. The central has up to 6 control modules. Each module controls up to twenty 11-station satellites for a total of 220 stations.

This system has a solid-state central and can control either electric or hydraulic satellites.

Level Number 2 Features

The Vari-Time II is an advanced solid-state central with additional features as listed.

This central controller has up to 6 modules which are capable of controlling up to 72 11-station satellites for a total of 792 stations.

Downside of Level 1 and Level 2
(See Appendix 8.)

a. One-way communication from the central controller to the satellite. Instructions are downloaded and the satellites will utilize the information to run the program accordingly. However, there is no feed back to the central if a problem occurs.

b. To turn on a station or syringe from the field the operator must go to the appropriate satellite. Furthermore, it may be necessary to run through all the stations to get to the one that is wanted.

c. In case of rain it is necessary to go to the central controller to shut down the system.

Level Number 3 Features
(See Appendix 8.)

The OSMAC irrigation system is designed for either new or retrofitting existing installations.

This system consists of a computerized central controller, solid-state satellites and a hand-held remote control radio.

Basically, this is a one-way communication system from the computer to the satellites. However, a degree of two-way communication is accomplished by integrating a portable radio. Furthermore, the portable controller can command the system to run any station or group of stations either sequentially or simultaneously from anywhere on the golf course.

This feature makes the *line of sight* no longer an issue. Also, notice the list of features offered by this controller system.

Note: There are a number of comparable hand-held remote control radios.

Level Number 4 Features (See Appendix 8.)

This system employs true two-way data communications allowing the central to transmit program data and receive status data from the satellites.

Level Number 5 Features
(See Appendix 8.)

The NW8000 central allows for two-way communications between the central controller and field satellites. This communication allows you to monitor activity in the field and supply the user with valuable information.

Satellite Sensors (Optional): Allows the user to monitor activities at the satellite, such as current watering status, temperature, flow, pressure, rainfall, triac voltage (can notify user via alarm before failure occurs), amperage draw and alarm status.

COPS 4—COMPUTER ORIENTED PROGRAMMING SYSTEM

This system is comprised of a computer interface, software, and field controllers. The buyer is free to choose a personal computer from the best local source. In order to provide varying needs and future growth of a golf course, this system has true two-way communications. (See Appendix 9.)

Two-Way Communications

A review of the central and field control literature (on-hand) indicates that currently there are a number of different methods available for two-way communications between central and satellite controllers. Each type of system has some merits as well as some downsides. Naturally, the greater the number of features and options the higher the price tag. On the upside, more versatility and sophistication should result in better water distribution practices, water conservation and overall efficiency, which can possibly offset the initial added cost.

The type of controller system ultimately selected by the golf course will depend on many factors, such as: new construction or retrofit, the desires of the management, the budget, etc.

Types of Two-Way Communications
Telephone Communications
(See Appendix 9.)

Conventional telephone lines can be used to communicate with the central controller from remote sites across the state or across the country. The modem interface links the PC central to a system of field controllers or one individual field controller.

Every field controller (satellite) is assigned a telephone number and a dedicated modem which can be dialed from the central controller. Once the communication link between the satellite and central has been established, information can be downloaded.

Hardwire Communications
(See Appendix 9.)

The communication cable is connected between the interface or interface/modem and all field controllers and weather stations by using direct burial wire. The wire is dedicated to transmit only a communication signal. The wire in the cable shall be 18-gauge stranded wire encased in high-density polyethylene with a minimum wall thickness of 0.015 in.

Hardwire systems are mainly used for new installations using the trenches for pipeline which would keep the cost of placing the wire at a minimum.

Radio Communication (See Appendix 9.)

Radio communication systems are relatively new to the irrigation field. They are predominantly used in the retro-fit market or where no trenching is being done. The field units will fit in or onto the existing controllers regardless of make or model. The receiver will work on Rain Bird, Toro, Hardie, Nelson, Weathermatic and all other controllers from 2 to 24 stations.

True two-way communication is achieved by installing transmitters in the satellites, which provides the ability to send information back to the receiver in the central controller. Paging also operates under a similar technology. A radio transmitter and interface are located at the central controller and receivers are located in the field satellites. Signals are sent to the satellites with instructions for operations.

HYDRAULICALLY OPERATED CONTROLLERS

Central control systems can employ mechanical, solid-state, computer, or a variation of these methods.

Hydraulic systems enjoy wide acceptance in areas prone to a high incidence of electrical storms. Lightning can severely damage the controller system. Proper grounding can reduce or eliminate its effects.

In systems using hydraulically operated valves, tubing is used as the communication link between the valve and controller.

Tubing is usually polyvinylchloride (PVC) or polyethylene plastic (poly). Poly tubing has the great benefit of not bursting when water freezes in the pipeline. Its ability to expand when necessary and return to its original shape when the water inside thaws makes its use in freezing climates highly desirable. Its softness, however, makes it vulnerable to damage by burrowing rodents.

For golf courses PVC is generally used because of its resistance to damage from rodents.

Compression type couplings and tees are used for PVC tubing because of its hardness. PVC tubing of 1/4 in. to 5/16 in. i.d. is usually used.

Fittings for polyethylene tubing are of the insert type, and an exterior collar is usually necessary to prevent splitting of the tubing on the insert fitting.

Tubing size is generally limited to 1/4 in. o.d. and 1/8 in. i.d.

Maximum Height Differentials Under Which Hydraulic Systems Operate

In a normally open, hydraulic system, the valves should never be more than 25 ft higher than the controller, nor more than 75 ft lower than the controller.

In a normally closed, hydraulic system, the valves should never be more than 25 ft higher than the controller nor more than 70 ft lower than the controller.

Components of a Hydraulic Normally Open Valve-in-Head System

In normally open valve-in-head systems, the supply water pressure from the controller must be equal to or greater than the irrigation water pressure. (See Figure 10.5.)

OPERATION OF VALVE-IN-HEAD SPRINKLERS

Normally Open Type

Closed Position

Filtered supply water, transferred under pressure through a control tube from the satellite controller, fills and pressurizes the piston chamber, holding the piston assembly against the valve seat, restricting water flow. (See Figure 10.6.)

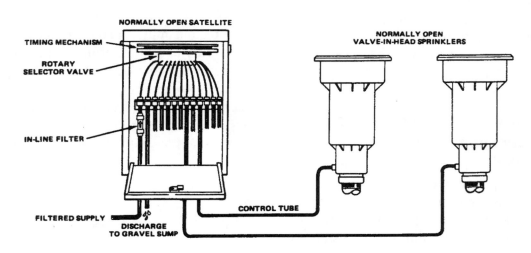

Figure 10.5

Open Position

The valve opens when the controller allows supply water to discharge from the control tube, relieving pressure from the piston chamber. Irrigation water lifts the piston assembly from the valve seat for sprinkler operation.

Normally Closed Type

Closed Position

Irrigation water is metered into the piston chamber through an orifice in the center of the piston assembly. From the piston chamber, it is channeled through a short control tube to the closed pilot valve. Pressure builds within the piston chamber and holds the piston assembly against the valve seat. (See Figure 10.7.)

Open Position

The pilot valve actuator, when activated by the controller, opens the pilot valve, relieving pressure in the short control tube and piston chamber. Irrigation water lifts the piston assembly from the valve seat for sprinkler operation.

ACKNOWLEDGMENTS

1. "Irrigation System Design." Manual, also Catalogs and Handouts. Rain Bird Manufacturing Company.
2. "Design Information for Large Turf Irrigation Systems." Toro Manufacturing Company.
3. "Irrigation System Design." Manual, also Catalogs and Handouts. Buckner Manufacturing Company.
4. "Sprinkler Equipment." Catalog. Hunter Sprinkler Manufacturing Company.
5. "Sprinkler Equipment." Catalog. Weather-Matic Sprinkler Manufacturing Company.

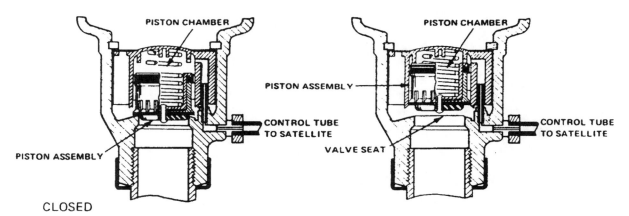

Figure 10.6. Normally open.

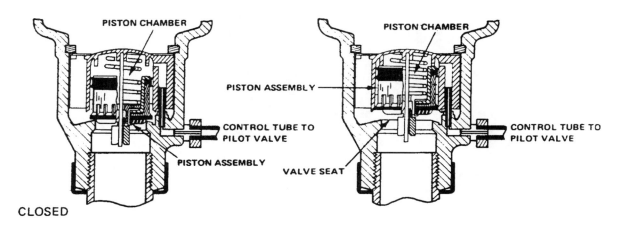

Figure 10.7. Normally closed.

Chapter 11

Scheduling Irrigation

EXPLANATION OF SOIL-WATER-PLANT TERMS AND THEIR RELATIONSHIPS

Soil Composition

Soil is a porous mixture of inorganic or mineral particles, organic matter, air, and water. The term "soil," for our purpose, may be best defined as "the upper layer of the earth that may be dug, plowed, etc." There are many types of soil, the three main divisions being sands and gravels, silt, and clay. Each has definite drainage characteristics and may vary in thickness from area to area. For this reason, and also to ascertain the water table level, a soil survey is usually taken at the start of any underdrain project.

Soil Texture

Soil texture is the basic particle size of the mineral part of the soil, and serves as the basis for soil classification.

Soil Classification

Soil Particles	Diameter Limits (mm)[a]
very coarse sand	2.0–1.0
coarse sand	1.0–0.5
medium sand	0.5–0.25
fine sand	0.25–0.10
very fine sand	0.10–0.05
silt	0.05–0.002
clay	below 0.002

[a] 1 in. = 25.4 mm

Note: Small soil particles have more surface area per net volume than large particles. Soil consisting of small particles has more pore space than soil of large particles.

Most crops in deep, uniform soils use moisture more slowly from the lower root zone than from the upper soil, as shown in Figure 11.1. The top quarter is the first to be exhausted of available moisture. The plant then has to draw its moisture from the lower three-quarters of root depth. Well-drained soils permit and encourage deeper root systems with superior drought resistance.

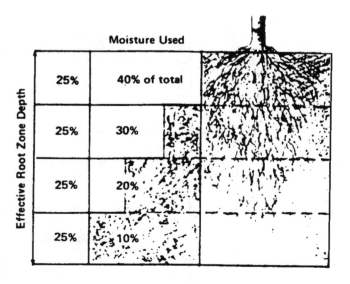

Figure 11.1

225

Soil-Water-Plant Relationships

Plants require soil, water, and air to survive. Assuming that a typical volume of soil consists of about 50% mineral and organic matter and 50% pore space, plants would soon perish if the pore spaces were filled with water (saturation), due to a lack of air.

Figure 11.2

As water drains by gravity from the pore spaces it is replaced by air. At field capacity, about 50% of the soil moisture is replaced, as shown in the diagram. This condition is optimal for plant growth.

Plants experience little difficulty in extracting water from the soil particles. However, non-stress conditions of moisture may produce lush or succulent types of plants that are less than ideal for golf courses.

Wilting Point

As the moisture is depleted by transpiration from the plant and evaporation from the soil, the tension forces increase as the moisture film around the particles becomes thinner. The plants are subjected to increasing stress conditions. At a soil tension of approximately 15 atmospheres, many plants cannot extract moisture from the particles (extreme stress condition) and may wilt and die.

Soil Moisture Classification
Gravitational or Free Water

Gravitational or free water is the water in the range between saturation and field capacity and will be

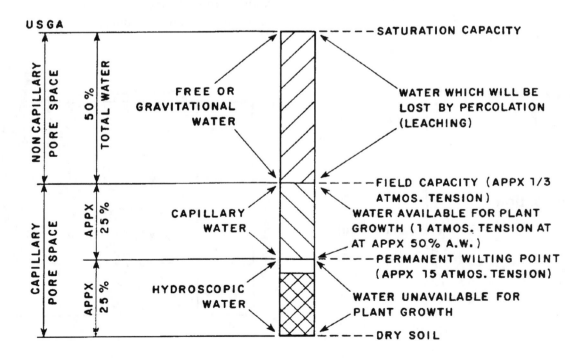

Figure 11.3

removed 24 to 48 hr following a saturating rain. This water is not available for plant use. (See Figure 11.3.)

Capillary or Available Water

Capillary water is the water in the range between field capacity and the wilting point. At field capacity the soil moisture is readily available and the plants are not stressed. However, as the moisture is used or depleted, the stress to plants increases until the available water drops to the wilting point.

Hydroscopic or Unavailable Water

Hydroscopic water is the water in the range between the wilting point and air or oven-dry soil. This *thin* film of moisture is held by high tension forces and is unavailable for plant use.

Non-Capillary Pore Space

This is the space from which gravitational water (between field capacity and saturation) is removed and replaced by air.

Capillary Pore Space

This is the space from which capillary and hydroscopic water (between oven-dryness and field capacity) will be removed by drying.

Typical Water-Holding Characteristics of Different Textured Soils

The average number of inches of "available water" for each type of soil corresponds to the values shown in the table below. However, it must be pointed out that the values will vary with compaction, organic content, etc. (See Figure 11.4.)

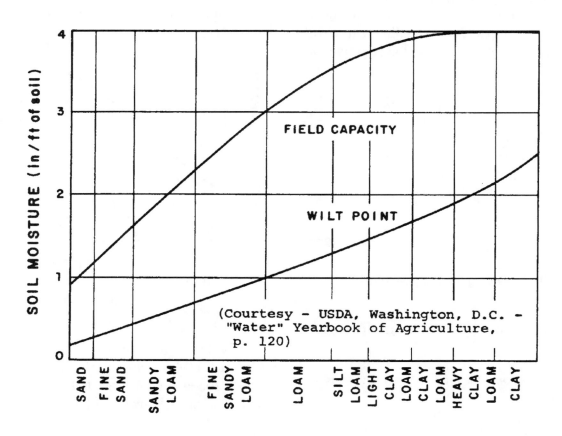

Figure 11.4

FINDING EFFECTIVE ROOT ZONE DEPTH FOR THE TYPE OF TURFGRASS TO BE IRRIGATED

Studies have established that the depth of plant roots is related to the height of the cut. The shorter the cut, as on greens, the shorter the depth. Ground water available to plant depends on root depth which, in turn, affects the irrigation interval. For these reasons it is necessary to determine the effective depth of turfgrass roots.

Method for Finding the Effective Root Zone

During a careful profile study of the soil, the depth at which most of the plant's roots are concentrated can be observed.

Caution: Where there is a concentration of roots above the true soil level, this indicates a thatch problem. The superintendent must get rid of the thatch with a continuing program of vertical mowing, sweeping, and nutrient and water control in order to establish and develop a root system in the true soil. This must be accomplished before the effective root zone depth can be determined.

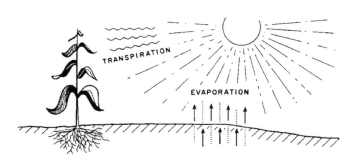

Figure 11.5. Finding the daily water usage

Note: Water in the soil is depleted in two major ways:

1. *By transpiration*—the process by which plants consume water.
2. *By evaporation*—where water is lost to the atmosphere directly from the soil surface.

The *evapotranspiration rate* is influenced by four principal climatological factors that may change daily. These are:

1. average temperature for the day
2. sunshine duration during the day
3. average relative humidity for the day
4. average wind speed for the day

Numerous experiments have established that of the four, only *temperature* and *sunshine duration* are significant. Therefore, relative humidity and wind speed are of lesser importance.

Method for Finding the Daily Evapotranspiration Rate for (e.g.) Massachusetts Soils

Since the average monthly temperature and the percentage of annual daytime hours in a month are constant for a given latitude, the average daily water consumption rates can be computed.

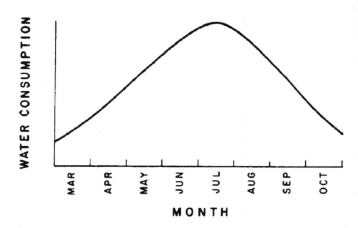

Figure 11.6

Table 11.1 selected from Leaflet No. 246, "Irrigation for Massachusetts Farms," can serve as a guide for estimating the daily evapotranspiration rate. This table, however, is based on average weekly temperature.

Table 11.1. Estimated Evapotranspiration Rates (inches of water)[a]

Mean Weekly Temperature (°F)	Evapotranspiration Rate	
	Daily	Weekly
52	0.072	0.50
54	0.090	0.63
56	0.100	0.70
58	0.109	0.76
60	0.120	0.84
62	0.131	0.92
Ave. 64	0.141	0.98
66	0.150	1.05
68	0.160	1.12
70	0.170	1.19
72	0.180	1.26
74	0.192	1.34
Max. 76	0.203	1.42

[a] Based on meteorological observations at the University of Massachusetts, Amherst.

Other tables used: Irrigation system designers are primarily concerned with the daily maximum evapotranspiration rate because the system must be designed to meet this peak condition.

These rates can be found in numerous irrigation books, bulletins, and guides. One example: From "Sprinkler Irrigation Handbook"—published by Rainy Sprinkler Sales, Peoria, Illinois.

The of peak moisture use was checked with information obtained from various agricultural colleges, experimental stations and government agencies, and was found to compare with the average of all recent data.

RAINFALL AND EVAPOTRANSPIRATION DATA

United States and Canada

Six major factors affect the need to water. These are: the length of the growing season, rainfall or precipitation rate and amount, evaporation rate, transpiration rate, type of soil, and type of grass. The collective action of all these factors determines the water deficit or water surplus. Turfgrass managers are concerned primarily with the water deficit—the difference between what nature furnishes to the gross

and the amount of water needed for satisfactory growth. The water deficit must be supplied by irrigation if growth is to be maintained.

A number of investigators concerned with calculating theoretical and actual water needs have shown that after adjusting for day length, a close relationship between mean temperature and potential evapotranspiration exists. Evapotranspiration or consumptive use represent all water lost from plants through transpiration and by evaporation from soil and plant surfaces. The leaf surface controls the plant's ability to absorb solar energy. Hence, a coefficient may be calculated for specific types of plants. When this adjustment, as calculated for turfgrasses, is applied to potential evapotranspiration for a given climate zone, the result (expressed in inches of water) represents the total water needs. This is the key to successful water management, including sprinkler design.

The figures shown in Table 11.2 are averages, and should be used only as a guide in designing a system that meets minimum watering needs. Soil conditions, water salinity, and other unique characteristics need to be considered. Therefore, the tabulated figures can not be considered as absolute. Use them only as the basis or starting point for your planning.

1. Figures shown in Table 11.2 represent inches of moisture.
2. Rainfall is based on a 30-year average (1930–1961).
3. Potential evaporation calculated from modified Blaney-Criddle formula.
4. Consumptive coefficients for lawn grass as developed by the Soil Conservation Service in 1960—reported in a paper entitled, "A Method for Estimating Irrigation Water Requirements of Lawns," presented by T.H. Quackenbush and J.T. Phelan; 1963 Meeting of ASA at Denver, Colorado.

   Note: A complete booklet may be obtained by writing to: The Toro Company, Minneapolis, MN 55420. The booklet contains weather data tables for all the States and Provinces.

MASSACHUSETTS WATER USAGE RATES

Water lost from turfgrass by evapotranspiration is much lower in cool climate regions as compared to the hot and desert regions. Therefore, it is essential to use the established water usage rate for your local area.

Example

Find the peak water usage deficit (in./wk) for Central Massachusetts.

From Tables 11.2 find:

a. month of July
b. RF   = 3.60 in.
c. PVT = 6.38 in.
d. DIF = –2.78 in.
e. in./wk = –2.78/4 = 0.70 in.

Note: For Massachusetts, it is common practice to base the irrigation system design on a water usage rate of 1 in./wk.

FINDING THE IRRIGATION INTERVAL

The irrigation interval is found by determining the number of days it takes to reduce the soil moisture from field capacity to the irrigation point.

As the irrigation interval (days) increases, the area to be irrigated decreases. That is, a two-day interval allows watering one-half the area per day (or night); a three-day interval, one-third of the area, etc. Also, as the irrigation time per day (or night) increases, the number of sprinklers required to operate at one time decreases.

A longer irrigation interval, then, makes it possible to complete an irrigation cycle over a longer period of time without damage to the turfgrass with fewer sprinklers, less gallonage, smaller pipe and pump sizes, and lower installation costs.

Solving for the Irrigation Interval (Days)

Formula:

$$II = \frac{FC \text{ (in. water)} - IP \text{ (in. water)}}{\text{daily ET rate}}$$

(use the maximum for your area)

where: II = irrigation interval
FC = field capacity
IP = irrigation point
ET = evapotranspiration

Example

Find the irrigation interval for a golf course consisting of partly sandy soil (water holding capacity = 2.0 in.) and sandy loam soil (water holding capacity

Table 11.2. Massachusetts Water Usage Rates

	Jan.	Feb.	Mar.	Apr.	May	June	July	Aug.	Sept.	Oct.	Nov.	Dec.	Total
Western (Pittsfield)													
RF	3.39	2.69	3.56	3.90	4.00	3.96	4.41	3.73	4.43	3.25	4.06	3.41	44.79
ET	0.00	0.00	0.46	1.51	3.33	4.78	5.77	4.93	3.05	1.61	0.58	0.00	26.02
Diff.	3.39	2.69	3.10	2.39	0.67	–0.82	–1.36	–1.20	1.38	1.64	3.48	3.41	18.77
Central (Springfield)													
RF	3.86	3.10	4.09	3.84	3.58	3.71	3.60	3.79	3.95	3.23	4.14	3.60	44.49
ET	0.00	0.00	0.67	1.79	3.66	5.30	6.38	5.50	3.55	1.89	0.76	0.00	29.50
Diff.	3.86	3.10	3.42	2.05	–0.08	–1.59	–2.78	–1.71	0.40	1.34	3.38	3.60	14.99
Coastal (Boston)													
RF	4.04	3.37	4.19	3.86	3.23	3.17	2.85	3.85	3.64	3.33	4.11	3.73	43.37
ET	0.00	0.00	0.73	1.71	3.37	4.95	6.18	5.48	3.58	2.05	0.87	0.30	29.22
Diff.	4.04	3.37	3.46	2.15	–0.14	–1.78	–3.33	–1.63	0.06	1.28	3.24	3.43	14.15

Code: RF = rainfall; ET = evapotranspiration; Diff. = + or – difference between RF and ET

= 4.0 in.). The evapotranspiration rate (ET) = 0.20 in. per day.

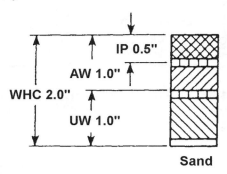

Sand

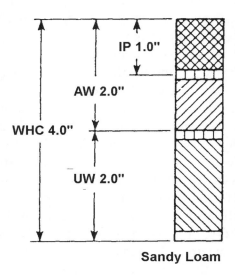

Sandy Loam

Figure 11.7

Figure 11.7 shows that the irrigation interval based on the sandy loam soil would be 5 days. The water level in the sandy soil would be reduced, during this period, down to the permanent wilting point, causing damage to that area of the course.

On the other hand, the irrigation interval based on the sandy soil is 2.5 days. Therefore, the water lost in the sandy loam soil will be replaced before its wilting point is reached.

SOLVE FOR SYSTEM PROGRAMMING AND SCHEDULING

Use of Terms
Programming

The number of sprinklers that will run from a single station of a controller.

Scheduling

The station operating time and sequence.

Programming Irrigation

Once the irrigation system and choice of controller(s) will be wired to a station, programming irrigation means the operating sequence and length of watering time.

Fairways

Every sprinkler on an irrigation system must be assigned to a station (zone) on a controller in regard to its starting and stopping times. This can vary with a number of considerations, such as:

1. electric or hydraulic controller system
2. block or valve-in-head method
3. pipe layout, pipe sizes, and their water carrying capacities
4. land slopes and topographical features
5. constant- or variable-speed pumps

Keeping the above considerations in mind, the designer and/or superintendent must program the sprinkler assignments based on their knowledge and experiences related to the area to be irrigated.

Generally, the number of sprinklers per zone is limited to two or three; however, under certain conditions as many as five heads may be assigned to a station. This may depend on topographical factors or the desires of the golf course committee.

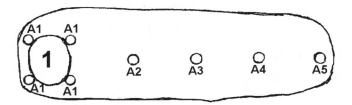

Figure 11.8a. Fairway 1: Single-row, valve-in-head system. Green: All heads run at the same time.

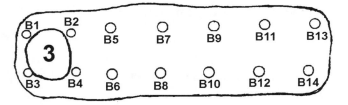

Figure 11.8b. Fairway 2: Two-row, two heads per valve, sequentially run. Green: Front two heads, then back two heads run at the same time.

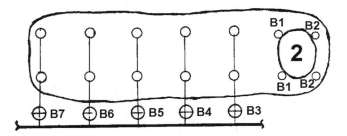

Figure 11.8c. Fairway 3: Two-row valve-in-head system. Green: One head (only) runs at the same time.

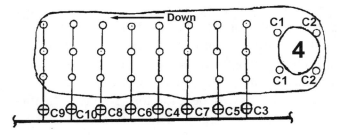

Figure 11.8d. Fairway 4: Multi-row, three heads per valve, down slope, run staggered. Green: Front two heads then back two heads run at the same time.

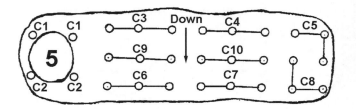

Figure 11.8e. Multi-row, three heads per station, across slope, run as shown. Green: Upper two then lower two run at the same time.

Greens

There are many ways to program greens. The final choice is usually determined by the designer and/or the golf course superintendent.

The factors to consider include: the pipe layout and pipe sizes. Although the center-sod-cup method is still used in a few installations, it is not generally recommended. Running all the sprinklers around the green at the same time and for the same period of time may cause some serious watering problems. Greens with varying slopes and unique construction features can result in over- or underwatering some areas. The alternatives are first to operate the two approach heads and then the back two heads. The greatest flexibility is obtained by the valve-in-head method, where each head is programmed to operate independently for a predetermined time period.

Tees

Tees can be programmed to operate individually or as a cluster, usually when they are in line of sight. The number running at the same time may vary when the designer programs the system to balance the total demand to the total gpm output of the pump.

Figures 11.8a through 11.8e show *some* of the configurations for programming an irrigation system. Systems having one sprinkler assigned to one station allow the operator to interchange the control wires to attain various configurations.

Programming—Valve-in-Head Irrigation System

On golf courses it would be foolish to design a system which would have the tees, fairways, roughs, approach, greens, bunkers, hills, and valleys all controlled by one controller station. The superintendent would have the designer's head for a tee marker, as these areas all represent different watering requirements and thus require different zones of control. Flexibility and versatility must be built into a system design. It can't be added on to the system once it's installed. This means more controllers with fewer sprinklers per station.

Design Advantages of a Valve-in-Head System

The valve-in-head concept allows the designer unlimited versatility in selection of areas to be watered at the same time for the same length of time. Usually it is recommended that no more than five sprinklers be run on any one station of the controller, since control capabilities become awkward after that point.

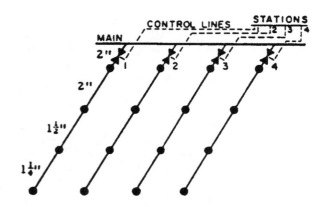

Figure 11.9b. Block design.

In areas where the soil is extremely dense, so application has to be limited as much as possible, the *skip method* can be used. Every third head would run so there is no overlapping of operating sprinklers. Figure 11.10, based on a single-row fairway system, shows how this is accomplished.

High areas can be run together, roughs can be separate from the fairway areas, backup heads around greens can be separately controlled and, in general, all varied use areas and topographically inconsistent areas may be separated so that one station will control one type of area. Figure 11.11 is a topographical sketch of a fairway showing how sprinklers can be programmed so the areas they cover are similar in regard to water demands of the turf. Similarly numbered heads would run on the same station.

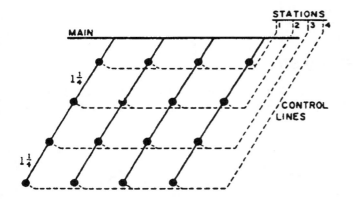

Figure 11.9a. Valve-in-head design.

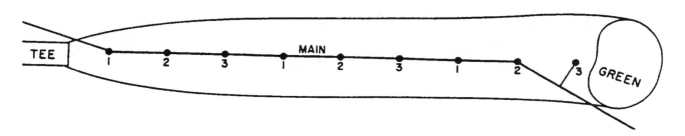

Figure 11.10

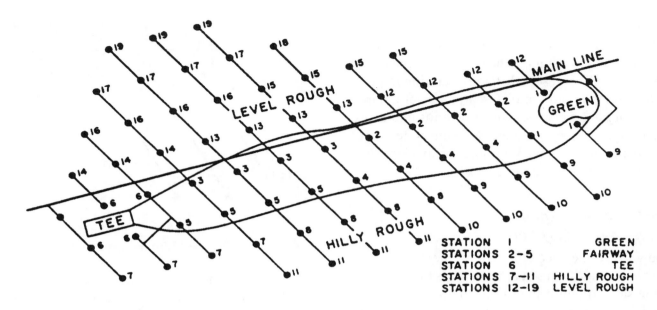

STATION 1 — **GREEN**
STATIONS 2-5 — **FAIRWAY**
STATION 6 — **TEE**
STATIONS 7-11 — **HILLY ROUGH**
STATIONS 12-19 — **LEVEL ROUGH**

Figure 11.11

STUDENT PROBLEM

Prepare a program for the area (Figure 11.12). Assume you are using a 20-station controller having four to five valve-in-head sprinklers per station. On the sketch, show the similarly numbered heads that will operate per station.

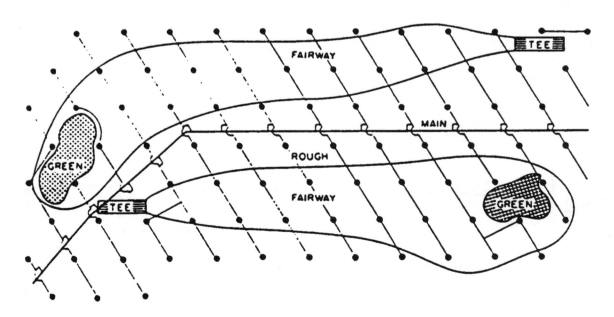

Figure 11.12

Sample Program and Watering Schedule

The golf course, as shown in Figure 11.13, was divided into four areas (zones). It is designed to operate only one controller in each of the four areas, one head per station. By strategically locating the controllers in each area, the operator can see the sprinklers in action.

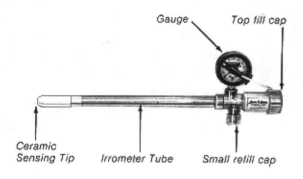

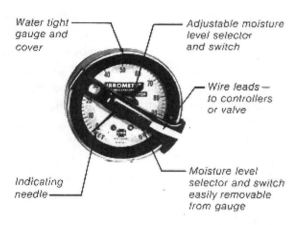

Figure 11.13. Automatic soil moisture control.

Scheduling Irrigation

There are various types of soil moisture measuring devices, water usage gauges, weather stations, etc., on the market which can be utilized, with caution, to establish the irrigation schedule.

Method No. 1—Tensiometer (Irrometer)

It is possible to use some type of moisture meter as an indicator for scheduling irrigation in some locations.

Automatic control with moisture sensors can be effective if used under the right conditions and in the proper manner. However, in large areas like golf courses where watering needs may differ not only from fairway to fairway but within the fairway itself, it is difficult to select a "typical" site to install sensors that are representative of each area. They may be used effectively as indicators by wiring them to a signal light. This could serve as a guide to the operator as to whether or not irrigation is needed.

Operation

The irrometer tube is filled with water through the fill cap. The ceramic tip is placed in the soil at the desired sensing depth. The moisture level selector is manually positioned at the optimal soil moisture level. As the soil moisture is depleted, water is pulled from the irrometer through the ceramic tip, creating a vacuum, which registers on the indicating needle. When the indicating needle contacts the moisture level selector a switch is closed, allowing the next scheduled watering cycle to operate. The controller should be scheduled to water daily or as frequently as possible.

The first limitation of the tensiometer is that though it can measure up to one atmosphere (30 inches of mercury) of tension, in actual practice it is only effective up to about 0.8 atmospheres. Now if we take 1/3 atmosphere as the field capacity and 15 atmospheres of tension as the permanent wilting point, it can easily be seen that the instrument is effective in only a small portion of the range of available moisture. However, in actual practice, the situation is not as bad as it might look on the surface, because relatively small percentages of the available moisture in the soil are held at the higher tensions.

A tension of 30 in. vacuum is generally regarded as the lower limit of "available water" to plants. Therefore, the tensiometer is suited to this range of soil moisture measurement.

The vacuum (tension) gauge is generally calibrated in centibars instead of inches of mercury. One atmosphere or 14.7 psi is equal to 100 centibars.

where: 0 centibars = 14.7 psia = 0 in. Hg vac.
100 centibars = 0 psia = 30 in. Hg vac.

A gauge reading of 0 equals saturation. A reading of 100 equals approximately the lower limit of available water.

The range of the tensiometer is suited to turfgrass irrigation. Areas such as golf courses that would commonly have compacted soil conditions and a reading of *70 to 80 centibars* would indicate a dry condition.

Installation

It is recommended that a minimum of two irrometer instruments be used in conjunction with each controller. However, as many irrometer locations as necessary can be used in parallel. The irrometer locations are then wired in parallel to the controller. This provides for shallow sensing to control the application of water according to the daily use of the plant and deeper irrometers to control tap root moisture, as it is gradually depleted over a period of time.

Figure 11.14. Typical installation.

Method No. 2—Budget Method

Following is a review of terms and basic factors concerning soil-water-plant relationships.

Saturation

Often a volume of soil is compared to an ordinary sponge. Water applied to an ordinary sponge will be absorbed until all the pore spaces are filled (saturated). Adding water beyond this point will only result in surface runoff or will be lost by percolation.

Field Capacity

For a period of about 24 hr after a soil has been saturated, water will continue to be lost from the pore spaces by the pull of gravity, and is replaced by air. The amount of water held by the soil against the pull of gravity is expressed as the *field capacity* of the soil, commonly expressed as inches of water per foot of depth.

The water holding capacity varies for each soil type.

Available Water

Of the total water held by a soil, only a certain amount is available for plant use. As the *available water* (AW) is depleted, the soil particles hold onto the remaining water with greater and greater tension until the plants can no longer compete. At this point, the plants permanently wilt and die.

The available water naturally varies with the depth of the crop root zone. For example, a crop having a root zone of 6 in. will have approximately one half of the AW shown in the above table.

Irrigation Point

Commonly when about one half of the available water is depleted, irrigation is started and the soil is filled back to field capacity.

Evapotranspiration Rate (ET)

Water is depleted from a soil in two major ways: by evaporation and by transpiration. The ET rate is influenced by four principle climatological factors that may change daily: (1) net radiation, (2) average temperature, (3) relative humidity, and (4) wind speed.

The daily ET rate also varies with the crop grown.

Actually, a soil can be considered a reservoir for plant use. Water is added by rainfall and lost by evaporation/transpiration. Therefore, keeping a budget of water added and water lost could serve as a guide or indicator as to if and when supplemental irrigation is required.

Method No. 2—Weather Station Budget Method

The water usage rate (in./wk) for a given geographical location is generally based on official weather station records over an extended period of years. Review pages 229–230.

For example, if the water usage rate was found to be 1 in./wk, this is the maximum average weekly value. Obviously, the actual weekly usage will vary with the daily local weather conditions: temperature, hours of sunlight, wind speed, humidity, and rainfall.

Budget Method

The state-of-the-art computer controllers can be scheduled by sensors to a weather station that measures rainfall as well as continually monitoring one or more of the weather conditions listed above. The sensors are calibrated to determine the evapotranspiration (water usage) rate.

The computer maintains a running record of the amount of available water in the soil by the *budget method*. Inches of rainfall collected by the rain gauge are added, whereas the water usage is subtracted. When the calculated water level drops to the predetermined irrigation point the computer will automatically start the irrigation system.

ET Prediction Model Method

Evapotranspiration (ET) prediction model software, a microenvironmental monitoring station, and a computer are three key elements that have raised the precision and accuracy of golf course water management to the highest level in irrigation history.

ET is the amount of water transpired from the plant and evaporated from the soil. The daily ET value is the key factor predicted by the software,

with a modifiedPenman model being the most widely applicable across North America.

The success of ET prediction modeling in performing its ET-sensitized irrigation scheduling is dependent on its ability to acquire microclimatic data from the golf course. The microenvironmental station provides the necessary climatic parameters. Net radiation, temperatur, wind speed, and relative humidity, plus rainfall and wind direction, are constantly monitored, stored, and periodically uploaded to the central computer.

The microenvironmental station can be hard-wired with cable to the central computer from as far as 20,000 ft. It should be positioned on the golf course to gather the most representative microclimate data.

Features Available

Computer:

- Data collection, logging and analysis.
- Constant communication with weather sensors.
- Seven days of storage of hourly data.
- Rugged, lightweight metal construction.
- Fully gasketed housing.
- Water-resistant sensor connectors and cables.
- Powered by rechargeable, sealed gel-cell batteries.
- Simple-to-service sensors and internal components.
- Batteries charged by external low-voltage (16-V) transformer or optional solar power.
- Sensor monitoring of six microenvironmental parameters:
  – Net radiation.
  – Air temperature.
  – Relative humidity.
  – Wind speed.
  – Wind direction.
  – Rainfall.
- Self-diagnostic test mechanisms:
  – Internal moisture sensor.
  – Battery voltage level.
  – Test port for local sensor check.

ET Prediction Software:

- Calculates ET values, with modified Penman Model preferred.
- Stores daily and historic ET data.
- Monitors and displays current microclimate conditions.

Method No. 3—Operator's Experience, Common Sense, and Good Judgment

Unfortunately, there are no standard rules or established procedures that are generally applicable for solving the irrigation schedule. Each installation has a unique set of design conditions which make scheduling a very complex problem.

After a system is installed, it usually takes up to three years or more to finalize the schedule. Some of the variables that must be considered are: (1) soil types, (2) type of turfgrass, (3) cultural system used, (4) local daily weather conditions, and (5) design and versatility of the irrigation system.

Soil Types

Soil types can vary from clay to sandy areas even within a fairway itself. It is not possible to check every section of the golf course. Furthermore, the water holding capacity of a soil will vary with the degree of compaction, organic matter, slopes, etc.

From a practical standpoint, the amount of soil moisture present is determined by making periodic observations of the wet to dry to wet cycle. *A study should be done of soil cores taken during and after a saturating rain and continued until the soil appears dry and the plants begin to wilt; after some practice one will soon be able to determine the soil moisture level by examination.*

The intake rate can vary considerably with the presence of thatch condition, degree of compaction, type of soil, slope, etc.

Type of Grass Affects Water Use

Watering practices must also be adjusted to the requirements of the predominating grass on the fairway. Some kinds of grasses will thrive at lower levels of available soil moisture than others.

The fescues and Kentucky bluegrass, for example, can grow on drier soils than the bent grasses, and do not require as frequent watering.

Further, the cool weather grasses, such as the fescues and bluegrasses, grow slowly during periods of high temperature and cannot be forced into rapid growth by watering. The principle effect of frequent heavy watering of these grasses during hot weather is to stimulate rapid weed development.

The Weather Conditions in the Area

Refer to the discussions on evapotranspiration and irrigation interval.

Design of the Irrigation System

The design of your system may play a major role in scheduling, operating time and sequence. It can vary with the type; manual, semi-automatic, automatic, or computerized and the built-in versatility.

Operating Time

If a certain type of turfgrass has a requirement of 1 in. of water per week under a given set of weather conditions, that requirement can't be met by dumping 1 in. of water on that turf at one time. If this were done, the turf manager's job would be greatly simplified—mainly because he wouldn't have any turf left to manage in a short period of time.

The method of providing the turf's water requirement is what is important. The water has to get down to the roots to stimulate growth, not flow across the top of the plants. Soils which resist water penetration complicate the problem, as do physical restrictions of the irrigating equipment available. Most large turf sprinklers apply water at a rate of 0.3 in./hr on up to 0.7 in./hr. Even the low figure is too high for the infiltration rates of many soils. Applying water at that rate for more than a few minutes will create runoff, so the low areas become lakes and the high areas become barren mounds.

15 min of continual watering on low-infiltration rate soils may result in:

5 min of efficient watering
5 min of 50% efficient watering
5 min of runoff

To alleviate this problem, water should be applied only as long as the ground will absorb it. The irrigation should then be suspended until the ground will accept more water. This may mean applying water for five minutes, waiting an hour, applying another five minutes, waiting another hour and applying another five minutes of water. This is much more efficient than applying water for 15 minutes, resulting in most of the last 10 minutes of water application running off onto an area not being covered by that particular sprinkler.

Summary

The superintendent must know exactly how the irrigation system is designed—its versatility, water application rate, limitations, etc. The system must be scheduled to apply the correct amount of water to meet the needs found. For these reasons, it is common practice in the design and sales of automatic irrigation systems to merely provide their customers with a watering tool. The tool is designed to provide a specified number of inches of water or precipitation per week when used as directed.

The superintendent is the soil and turfgrass management expert. Therefore it is his duty to apply this tool to the fullest advantage. Nothing can substitute for the operator's *experience, common sense,* and *good judgment* in planning a watering schedule.

WATERING SCHEDULE

Figure 11.15 shows placement of pipelines, pumps, valves, and controllers for a single-row automatic system on a 9-hole golf course.

Fairway Schedule

50 min. operation every other day gives 1 in. of precipitation per week.

Tee and Green Schedule

20 min. operation every day gives 1 in. of precipitation per week.

Operational Data

There are two fairway controllers and one tee and greens controller in each of the four areas (I, II, III, IV).

Operate only one controller in each of the four areas at one time.

Example

Set up one fairway controller from each area to operate Monday, Wednesday, and Friday.

Set the other fairway controller from each area to operate Tuesday, Thursday, and Saturday.

Set the tee and greens to operate every day after the fairway controllers have finished their schedule.

Watering Schedule (Partial shown in Figure 11.16)

Fairway Schedule

50 min operation every other day gives 1 in. of precipitation per week.

Tee and Green Schedule

20 min operation every day gives 1 in. of precipitation per week.

Notes: Field controllers #1 and #13 shall have a rain sensor, temperature sensor, and pressure sensor installed at that location. These controllers shall also have a flow sensor installed near the pump location and wired back to the controller with two #14/1 wires.

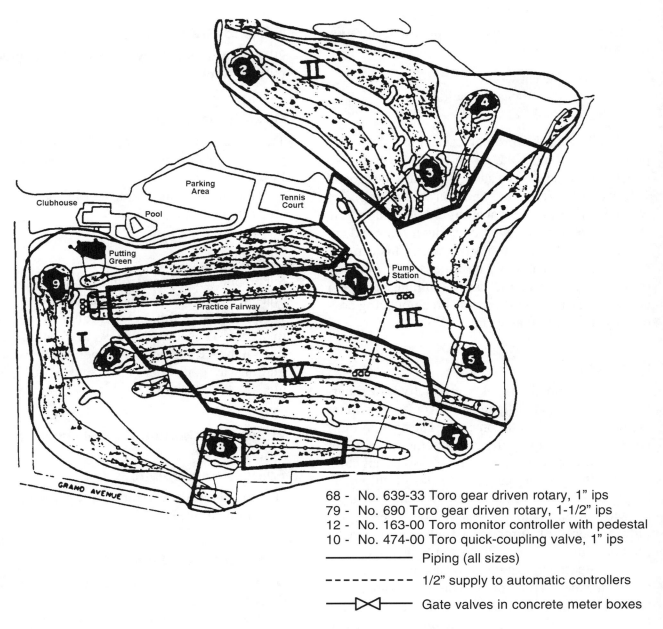

68 - No. 639-33 Toro gear driven rotary, 1" ips
79 - No. 690 Toro gear driven rotary, 1-1/2" ips
12 - No. 163-00 Toro monitor controller with pedestal
10 - No. 474-00 Toro quick-coupling valve, 1" ips

————————— Piping (all sizes)

----------- 1/2" supply to automatic controllers

——⋈—— Gate valves in concrete meter boxes

Figure 11.15. 9-hole golf course. Single-row automatic system.

LEGEND:

- ●   TORO 754-02-54 Check-O-Matic Double Row Fairway Sprinklers

- ○   TORO 690 Series Check-O-Matic Single Row Rough Sprinkler
  (Existing Sprinklers To Be Removed From Existing System)

- ●   TORO 658-06-56 Valve-In-Head 2-Speed Green Sprinkler

- ●   TORO 655-06-56 Valve-In-Head Part Circle Green Sprinkler

-     TORO 640 Series Valve-In-Head Green Syringe Sprinkler
  (Not Shown On Plan. Estimated 5-6 Sprinklers Per Green)

- ●   TORO 754-06-53 Valve-In-Head Full Circle Tee Sprinkler

- ●   TORO 655-06-55 Valve-In-Head Part Circle Tee Sprinkler

- ●   TORO 640 Series Check-O-Matic Tennis Court Sprinkler

- ⊕   TORO 216 Series Pressure Regulating Electric Control Valve

- ▫   TORO 286-66-01 Actuator For Syringe Sprinklers

- ☐   TORO 132-76-08 Network 8000 Field Controller

- ◖   TORO 132-94-00 Network 8000 Central Computer

- ○   TORO 474-00 1" Quick Coupler Valve

- ⊛   Brass IPS Ball Valve (Green & Tee Shut-Off Valves)

- ⋈   Cast Iron Main Line Isolation Gate Valve
  (Left In Open Position)

- ⊖   Cast Iron Main Line Isolation Gate Valve
  (Left In Closed Position)

- ▼   CRISPIN IC10 Air Relief Valve

- ☢   Pump House

- —   Class 200 PVC Piping

- ---   120 VAC Controller U.F. Power Wiring

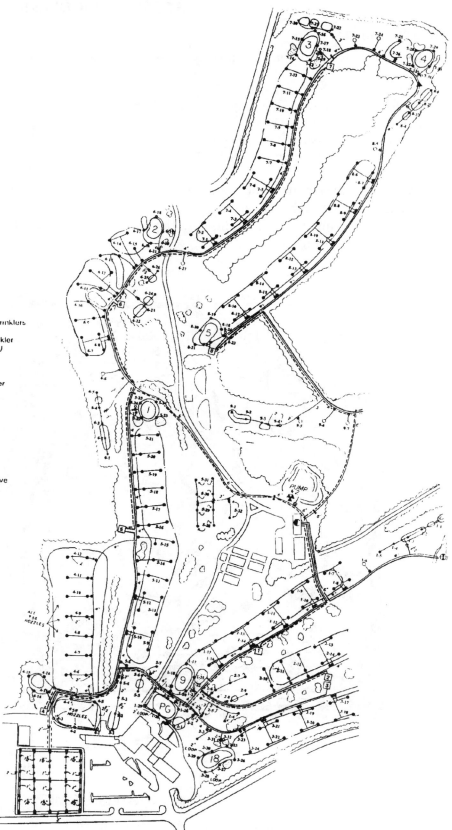

STUDENT PROBLEM

Programming

On the map shown, program the fairways (only) based on the design conditions listed.

total number of fairway heads = 75
number of heads to operate at the same time = 5
number of areas (zones) = 5
number of heads per zone = 15
number of heads running per zone = 1

Program the system to sequentially run 8 heads (one per station) the first night (1, 2, etc.), and the *remaining* 7 heads the next night (1a, 2a, etc.).

Note: Start at the heads at the highest elevation in each area. The irrigation schedule can be derived as shown on pages 48 through 50.

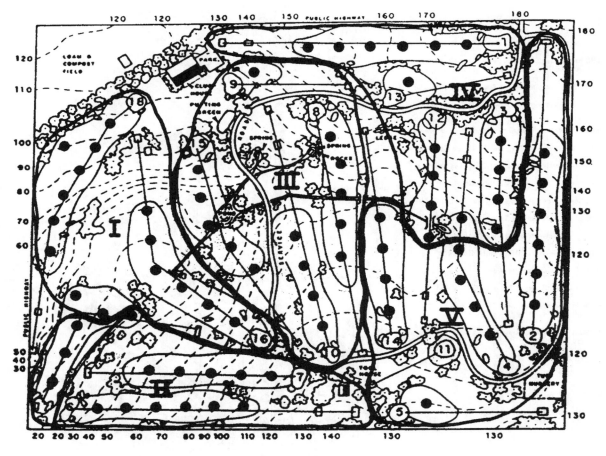

Figure 11.16

ACKNOWLEDGMENT

Sprinkler Irrigation Catalog. Rain Bird Sprinkler Manufacturing Corp.

Chapter 12

Guidelines for Basic Surveying of Tile Drainage Systems

THE TRIPOD LEVEL

There are many different makes of level upon the market, with prices ranging from $25 to $400 depending upon the class of work for which the level is to be used. A good level can be obtained for about $150.00. Great care must be exercised at all times in handling the level, otherwise it will become out of adjustment. The level consists of a stand or tripod upon which is mounted a telescope which has a level tube attached. Level screws are placed on the sides or underneath the telescope, and these, when properly operated, will level the instrument.

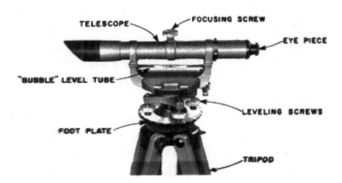

Figure 12.1. Dumpy level.

The Level

The level is used to obtain rod readings of different places upon the earth's surface. Practice setting up the instrument so that the operation will be quickly and properly done on the job. Important points to consider are:

- Make sure the level is in adjustment.
- Make sure the tripod legs are firmly planted in the ground. (Watch this, especially in wet or frozen ground.)
- Make sure the plate on top of the tripod is nearly level before attempting to set the leveling screws.
- Make sure the telescope glasses are clean.
- Never leave the instrument.
- Allow no one to touch the instrument except yourself, and then do not hang things on the instrument or bump the tripod legs with your feet as you move about. Even the slightest jar can upset adjustment.
- Wipe off the instrument thoroughly after it has been exposed to wet weather.
- Carry the instrument out or into buildings under your arm so that the head is forward. When outdoors, carry it over the shoulder head to rear with the tripod legs forward. Carry only the instrument and see that the rest of your surveying party has equipment equally distributed.

Setting Up the Tripod Level

1. Spread the legs of the tripod three or four feet apart and push them firmly into the ground, adjusting them so that the leveling head is approximately level and the telescope is approximately at eye height.
2. Tighten the tripod screws.
3. Loosen the telescope clamp and swing the telescope directly over a pair of leveling screws.

4. If need be, adjust the leveling screws so that they are snug against the leveling head. Then adjust the leveling screws by tightening one and loosening the other simultaneously to center the bubble in the bubble tube.

Figure 12.2a. Bubble off center.

Figure 12.2b. Bubble centered.

To do this, move your thumbs toward each other to move the bubble in one direction and away from each other to move the bubble in the other direction. The bubble follows the direction taken by the left thumb.

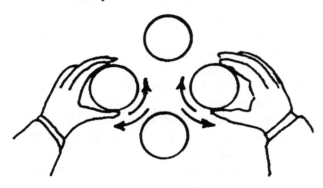

Figure 12.3

5. Turn the telescope over the other pair of leveling screws, and again center the bubble. If the leveling screws become tight, loosen one screw of the other pair that was adjusted previously.

6. The adjustment made by each pair of leveling screws affects the adjustment of the other pair; therefore, turn the telescope back and forth over the pairs of leveling screws until the bubble stays level or nearly level throughout a complete rotation of 360°. The leveling screws should be drawn up snug, but again, leveling accuracy will

not be improved by making the screws extremely tight. *Be sure leveling screws are firm upon the plate—not tight.*

7. Sight through the telescope, preferably against a clear sky, and observe whether or not a series of fine lines are visible. These lines are called cross-hairs, and should split the circle of vision from top to bottom and from side to side at the center. Sometimes there are two additional horizontal cross-hairs, one above and one below the center horizontal cross-hair. These are used for stadia surveys.

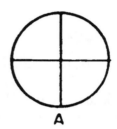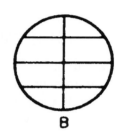

A   **B**

Figure 12.4. A. Ordinary cross-hairs. B. Cross- and stadia-hairs.

The center cross-hairs are used, the vertical one assisting the levelman in determining whether or not the leveling rod is being held vertical. All readings are taken at the point where the center horizontal cross-hairs cut the rod. If the cross-hairs are not visible, an adjusting screw is attached to the eyepiece of the telescope, which, when operated properly either in or out, will bring the cross-hairs into focus. Always get the best focus possible.

8. Point the telescope toward the rodman and aim it as you would a rifle by sighting over the top. Now sight through the telescope and find the rod, using the focusing screw on the side or top of the telescope to assist. Bring into the clearest vision possible and then make your reading. First, get the footmark, and then tenths; then, very, very carefully, count hundredths. Check your reading and record.

Checking the Level

Any piece of equipment can perform the job for which it is intended if it is in proper adjustment.

This is especially true of surveying instruments. Adjusting screws and certain other movable parts makes it possible to adjust the instrument. Adjusting screws usually are made of bronze and are easily damaged or jammed. Adjustments are relatively simple if certain procedures are followed and care is exercised. Generally, adjustments should be made only by or under the supervision of someone trained in the proper techniques of the job.

The next best thing to knowing how to adjust a level is knowing when it is in need of adjustment and where this service can be secured.

The following procedures for checking a level are the ones used for checking a dumpy level, but are applicable to checking any type of level. There are three items that should be checked periodically to ensure accuracy and precision of the instrument:

1. Checking to see that the axis of the bubble tube is perpendicular to the vertical axis of the instrument.
2. Checking to see that the axis of the bubble tube is parallel to the line of sight.
3. Checking to determine that the horizontal cross-hair is truly horizontal when the instrument is level.

Axis of the Bubble Tube—Perpendicular to the Vertical Axis

After the instrument is leveled as outlined above, turn the telescope directly over two of the leveling screws. Bring the bubble to the exact center of the tube by careful adjustment of these two leveling screws. Then rotate the instrument 180° so that the telescope is again over the same two leveling screws. If the bubble moves off center, the bubble tube needs adjustment. This adjustment can be made by raising or lowering one end of the bubble tube with the adjusting screw, bringing the bubble halfway back to center. Relevel the instrument and repeat this procedure if necessary. Caution: adjusting screws are brass; apply only enough pressure to hold the level vial firmly in position. (See Figure 12.5.)

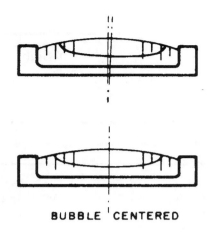

BUBBLE CENTERED

Figure 13.5

Axis of the Bubble Tube—Parallel to the Line of Sight

This check can best be done outdoors. A simplified version of the two-peg check is recommended. Select two points of about equal ground elevation about 300 ft apart. Call the points B and C. Next, chain and subdivide distance BC and set point A midway between B and C (Figure 12.6). Set up the level at point A, with one pair of leveling screws on line BC, and carefully level the instrument.

Drive solid stakes at B and C to the same elevation. Use a level rod with target for setting the top of stake B level with the top of stake C. The two stakes, B and C, will be the same elevation even though the instrument is in error because the distances AB and AC are equal.

Now move the level to point B (Figure 12.7), setting the instrument between B and C and very near B, with the eyepiece about 2–3 in. from a level rod held vertical on stake B. Level the instrument, keeping one pair of leveling screws on line BC. Look through the objective end of the telescope and set the target on the level rod so the target line bisects the small circle of the eyepiece. After the target has been set at B, move the rod to point C, and sight at the rod with the level. If the horizontal cross-hair bisects the target, the instrument is in perfect adjustment. For use in conservation work, an error, C_1C_2, of plus or minus 0.03 ft is allowable. If the error is greater than this amount, the instrument

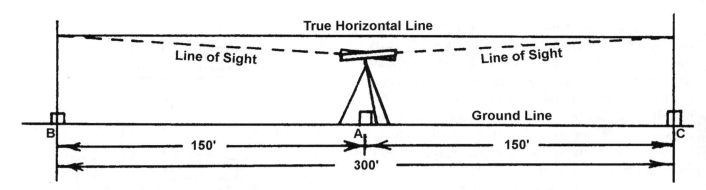

Figure 12.6. Measure AB equal to AC, then B and C can be set at the same elevation regardless of the accuracy of the instrument.

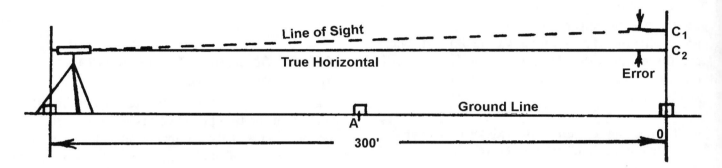

Figure 12.7. The error in the instrument in the distance BC.

should be adjusted by someone experienced in the adjustment and repair of surveying equipment.

Checking the Horizontal Cross-Hair

After leveling the level it is recommended to periodically check the telescope cross-hair, which must be perfectly horizontal for taking accurate rod readings.

Set horizontal cross-hair on target or other well-defined point.

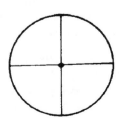

Figure 12.8

Swing telescope slowly from side to side; if cross-hair remains on point, adjustment is perfect.

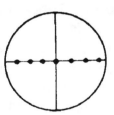

Figure 12.9

If cross-hair moves away from point, adjustment is needed. This can be easily done while the level is still set up and leveled at stake B and sighted at the target on the rod held on stake C. Be sure that the center of the horizontal cross-hair bisects the target; then rotate the level back and forth slightly from side to side. If the ends of the cross-hair cut the target in the same place as the center of the cross-hair did, it is in adjustment.

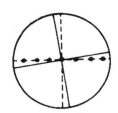

Figure 12.10

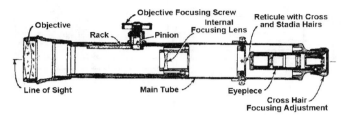

Figure 12.11. Cross section of internal focusing telescope.

If the ends of the cross-hair do not remain on the spot as did the center, the cross-hair should be adjusted. Usually, if the cross-hair needs adjustment, it is best to have it done by someone experienced in the adjustment and repair of surveying equipment.

The Leveling Rod

The leveling rod is generally made of wood, and has feet and divisions of the foot marked in red and black, respectively. The foot divisions are either in inches and fractions of an inch or in tenths and hundredths of a foot. Of the many kinds of leveling rods there are two classes, namely, the self-reading and the target rods. Sometimes there is a combination of these two features. The self-reading rod permits the levelman to read results directly from the level instrument while the target rod is read only by the rodman. The self-reading rod is more desirable, especially if the rodman is a different person on each job; generally the levelman keeps all notes taken.

Rod Reading

Leveling rods are so divided as to permit reading to 1/100 ft. The full foot numbers are in red and 1/10 foot numbers in black. Each 1/10 foot is divided into 10 divisions by means of black lines and white spaces. The width of each line or space is 1/100 foot.

Each 1/100 foot is read at the top edge of the corresponding division mark.

Many times when the rod reading is taken close to the instrument, the rod foot divisions are out of the path of the telescope. The levelman will need to know the proper foot reading as well as that in fractions; so whenever such a condition presents itself, he will ask the rodman to raise the rod until he (the levelman) catches the red foot mark. The rod should be raised up in a straight line above the place where the reading is to be taken. As soon as the foot mark is seen, the rodman should lower the rod to the original point for a final check. Practically all surveys depend upon the rodman for speed. Any delays on his part will tie up perhaps several men in the surveying party.

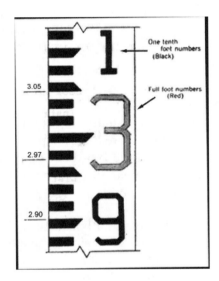

Figure 12.12

Holding

The leveling rod should be held vertical or plumb, with the proper figures toward the instrument. Be sure the rod is not upside down and that it rests upon a solid or firm foundation. Grass, soft earth, etc., will allow the rod to sink. If necessary, push the rod firmly down upon the earth to secure a good base. By using both hands, one on either side of the rod— never on the front face because the view of the levelman will be cut off at times—and then gently testing for plumb by removing one hand and then the other to see if the rod will fall, the rodman can very quickly bring the rod to a vertical plumb position and hold that while the levelman makes his readings. Watch the levelman, and when he is finished he will give a signal by waving both hands. Should he desire a turning point he will swing one arm or hand in a circle above his head; or should the rod be out of plumb, a one-hand motion indicating which way to swing the rod will be given.

FIELD NOTES

The field notes of the surveyor must contain a complete record of all measurements made during the survey. When necessary, sketches, diagrams, and narrations should be made to clarify the notes. The best field survey is of little value if the notes are not complete and clear. The field notes are the only record that is left after the survey party departs the field survey site. Notes must be kept on the standard survey forms and not on scraps of paper for later transcription.

Field Notebook

The survey notes are usually kept in a field notebook. These notebooks are of two types: the permanent bound book and the looseleaf. The following information must appear in each book no matter which type is used. (See Figure 12.13.)

- Instruction for return of book, if lost.
- Index of field notes contained in the book.
- List of party personnel and their duties.
- List of instruments used, to include serial numbers, calibration data, and dates used.
- A generalized sketch and description of the project.

PRINCIPLES OF LEVELING

Leveling Terms

Benchmark (BM) = Reference Point

a. Temporary reference points may be assigned to a onetime job. Oftentimes the benchmark and starting point are the same.

b. Where it may be necessary to recheck or redo the leveling job in the future, a permanent object such as a boulder, iron pipe, concrete step,

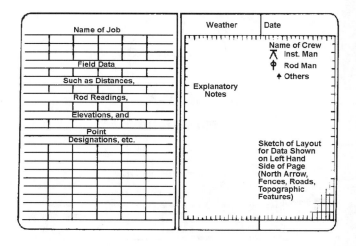

Figure 12.13

manhole cover, etc., can be used. Here the benchmark and starting point are not necessarily the same, as shown in Figure 12.14.

Figure 12.14

c. Bench mark elevation—An arbitrary elevation (usually 100.00 ft) is assigned to the benchmark to avoid the use of negative numbers, which would be the case if zero (0+00) were assigned. (See Figures 12.15a and 12.15b.)

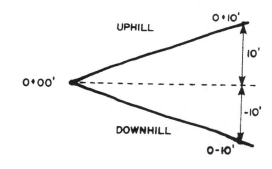

Figure 12.15a

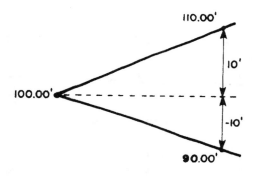

Figure 12.15b

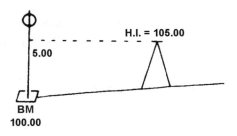

Figure 12.17

In practice, the assigned elevation is important for downhill leveling. The assigned elevation should be chosen on the known or estimated vertical drop in feet. For example, 100.00 ft for drops less than 100.00 ft, 200.00 ft for drops less than 200.00 ft, etc.

Principles of Leveling

1. Either a permanent or temporary benchmark is selected and an elevation is assigned. (Assume 100.00.)

2. Backsight (BS). After the initial setup of the instrument, as well as each subsequent setup, the first reading is always a backsight.

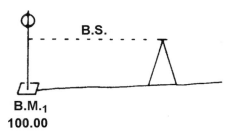

Figure 12.16

3. Height of the instrument (HI). The height of the instrument is determined by adding the backsight reading to the known elevation at a given station. (See Figure 12.17.)

Rule: elevation + BS = HI

4. Foresight height of the instrument is determined; all readings are foresights.

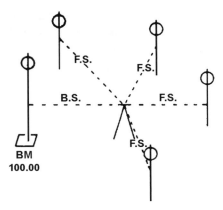

Figure 12.18

5. Elevation. The elevation at the base of the rod at the point of the foresight reading is determined by subtracting the foresight from the height of the instrument.

rule: HI – FS = elevation

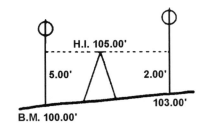

Figure 12.19

Station Identification

In leveling jobs every rod reading taken and noted is called a station. Each station is identified by name or by a convenient code. The following are some of the commonly accepted methods used to identify stations.

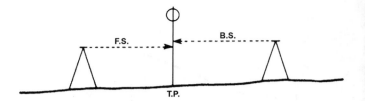

Figure 12.20

   (a) Benchmarks (BM_1–BM_2–BM_n...)

      Benchmarks can be assigned to beginning and ending points as well as other specific reference points along the way.

   (b) Starting Point (SP)

      A starting point is generally assigned when it differs from the beginning benchmark.

   (c) Turning Point (TP)

      This is where a foresight reading and a backsight reading are taken at a specific rod setting, as shown in Figure 13.20.

   (d) Alphabetical Letters (A–B–C...)

      Letters can be assigned to identify stations where foresight readings only are taken. (See Figure 12.21.)

   (e) Measured Distances (usually feet)

      Stations are identified in number of feet from the beginning of the line. This procedure is commonly used for profile studies. (See Figure 12.22.)

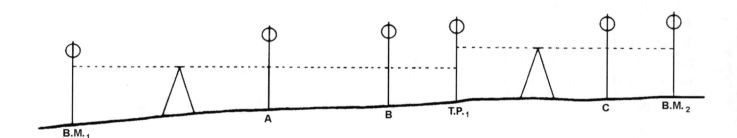

Figure 12.21

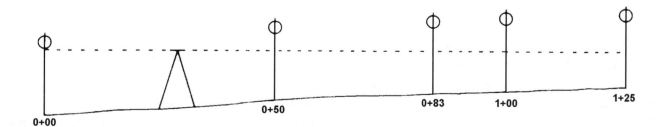

Figure 12.22

SAMPLE PROBLEM

Recording the Job in the Field Notebook

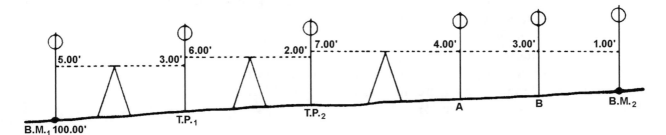

Differential Elevation (Tee and Green)

Station	BS	HI	FS	Elevation
B.M.$_1$	5.00	105.00		100.00
T.P.$_1$	6.00	108.00	3.00	102.00
T.P.$_2$	7.00	113.00	2.00	106.00
A			4.00	109.00
B			3.00	110.00
B.M.$_2$			1.00	112.00

Math Check: 18.00 − 6.00 = 12.00

Mathematical Check

1. Add backsight column = 18
2. Add foresights at *turning points* and last foresight only = 6
3. Subtract the smaller number from the larger.
   Difference = elevation difference = 12

Summary

Leveling Procedure

   1. Assign benchmark elevation.
   2. Set up instrument.
   3. Determine height of instrument.

         rule: elevation + BS = HI

   4. Determine elevation of station.

         rule: HI − FS = elevation

STUDENT PROBLEM

Differential Elevation

1. Study Problems 1, 2, 3, and 4 to gain an understanding of determining elevation difference procedure.
2. On each of the *sketches* identify the *turning points* and *stations.*
3. Make a mathematical check:
   a. Add the BS Column
   b. Add the FS Column *only* at the turning points and *last* FS reading
   c. Subtract the difference—compare the actual elevation difference.
4. Check allowable errors (Problem 4)

$$AE = 0.030 \times \sqrt{\frac{\text{total distance}}{100}}$$

Problem 1

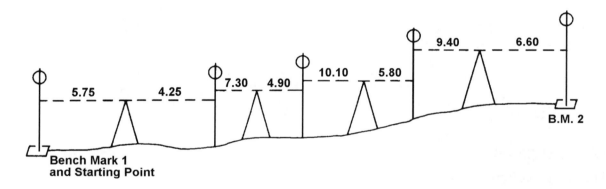

Station (St.)	Backsight (BS)	Height of Instrument (HI)	Foresight (FS)	Elevation (Elev.)	Distance (Dist.)
BM + SP	5.75	105.75		100.00	
TP 1	7.30	108.80	4.25	101.50	
TP 2	10.10	114.00	4.90	103.90	
TP 3	9.40	117.60	5.80	108.20	
BM 2			6.60	111.00	

Problem 2

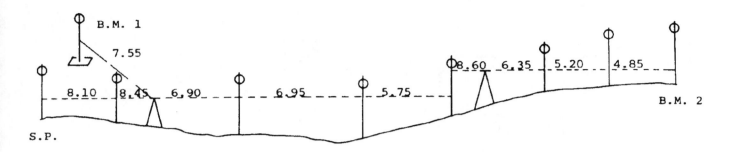

St.	BS	HI	FS	Elev.	Dist.
BM 1	7.55	107.55		100.00	
SP			8.10	99.45	
A			8.45	99.10	
B			6.90	100.65	
C			6.95	100.60	
TP 1	8.60	110.40	5.75	101.80	
D			6.35	104.05	
E			5.20	105.20	
BM 2			4.85	105.55	

Problem 3

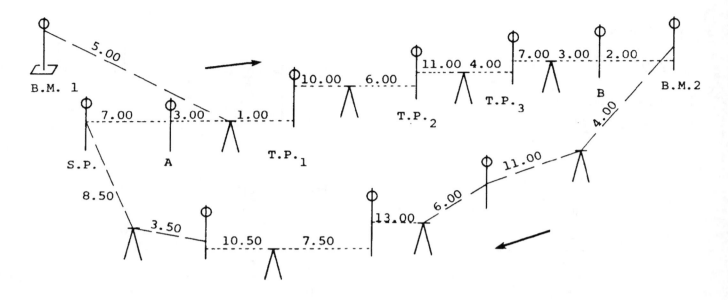

St.	BS	HI	FS	Elev.	Dist.
BM 1	5.00	105.00		100.00	
SP			7.00	98.00	
A			3.00	102.00	
TP 1	10.00	114.00	1.00	104.00	
TP 2	11.00	119.00	6.00	108.00	
TP 3	7.00	122.00	4.00	115.00	
B			3.00	119.00	
BM 2	4.00	124.00	7.00	120.00	
TP 4	6.00	119.00	11.00	113.00	
TP 5	7.50	113.50	13.00	106.00	
TP 6	3.50	106.50	10.50	103.00	
SP			8.50	98.00	

Problem 4

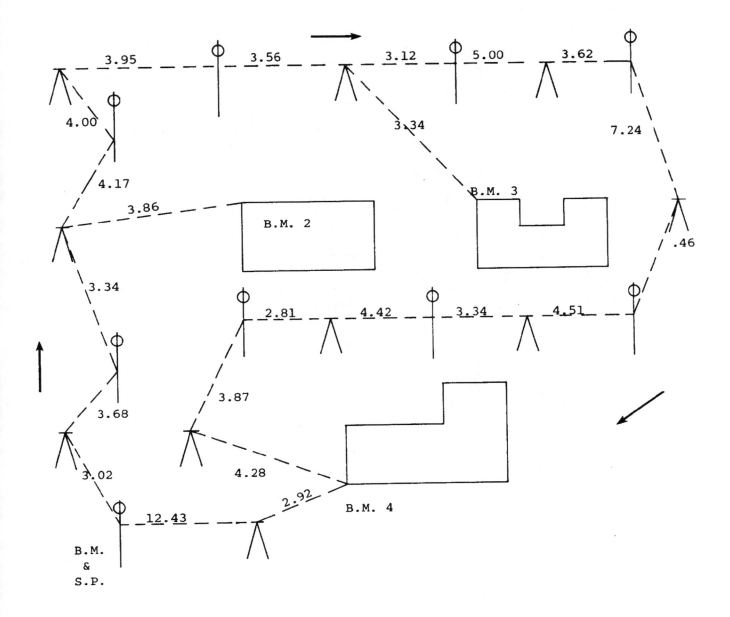

St.	BS	HI	FS	Elev.	Dist.
BM 1	3.02	103.02		100.00	250
TP 1	3.34	102.68	3.68	99.34	270
BM 2			3.86	98.82	330
TP 2	4.00	102.51	4.17	98.51	300
TP 3	3.56	102.12	3.95	98.56	320
BM 3			3.34	98.78	370
TP 4	5.00	104.00	3.12	99.00	320
TP 5	7.24	107.62	3.62	100.38	280
TP 6	4.51	111.67	0.46	107.16	310
TP 7	4.42	112.75	3.34	108.33	260
TP 8	8.87	113.81	2.81	109.94	
BM 4	2.92	112.45	4.28	109.53	
BM 1			12.43	100.02	

Profile Leveling

Profile leveling is the process of determining the elevation of a series of points at measured intervals along a line. Profile leveling is the same as differential leveling except that the foresights are taken at measured points. In this case a number of foresights may be taken between turning points and they are subtracted from the HI as long as the instrument remains in one position. A turning point is established each time the instrument is moved, in the same manner as in differential leveling.

A profile survey is particularly useful in drainage, terracing, and construction of waterways. In these cases the course to be surveyed must first be taped and staked at regular intervals and at intermediate points where the surface of the ground changes abruptly. It is assumed, in the course of a profile survey, that the ground surface is represented by a straight line between the points where elevation is determined. It is therefore the responsibility of the men taping the course to select the intermediate points to give the proper representation of the ground surface.

SAMPLE PROBLEM

Assume a trench is to be dug along a straight line between the two range poles.

    a. Plot the profile of the hillside to scale on graph paper.
    b. Determine the percentage of slope.

Procedure

1. Make a careful visual inspection of the hillside, and mentally note any changes in slope.
2. Stakes are placed at each point where a variation of slope occurs—wider apart on gentle slopes and closer together on steep slopes or drastic changes. Good judgment dictates the number and spacing of stakes.
3. Accurately measure the horizontal distances between stakes.

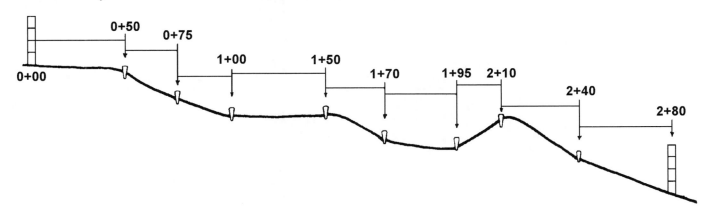

4. Record the data in the field notebook as shown.

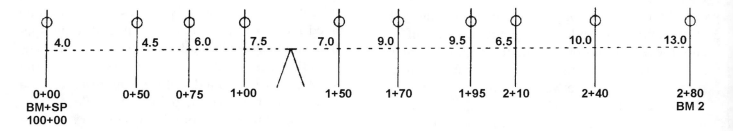

5. Calculate the percentage of slope

$$\frac{\text{elevation difference (ft)}}{\text{horizontal distance (ft)}} \times 100$$

$$= \frac{9}{280} \times 100 = 3.2\%$$

6. Plot profile to scale on graph paper.

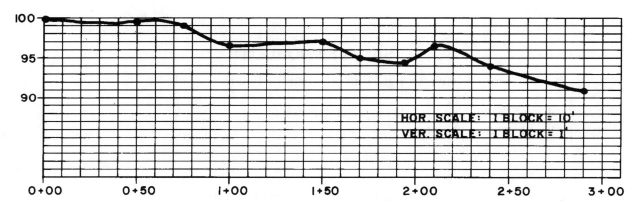

PRACTICE PROBLEM

Record the field data from the following sketch.

1. Determine the elevation at each stake.
2. Make a mathematical check.
3. Calculate the percentage of slope.
4. Plot the profile to scale on the graph paper provided.
   Note: It is considered a better technique to set the turning point (TP₁) off of the course.

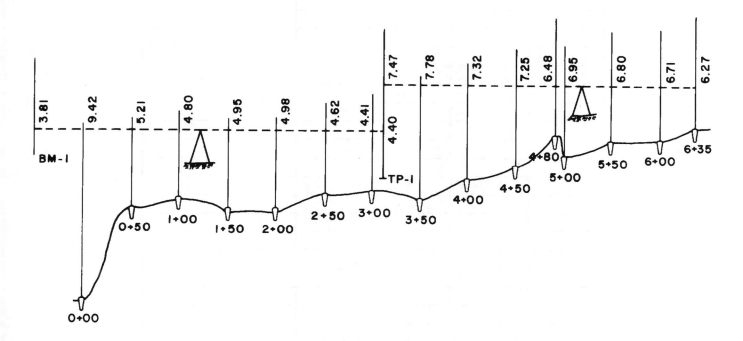

Record Field Data

Profile Leveling Tile Drainage					

Plot Profile

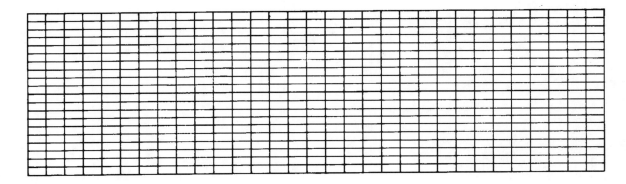

Calculate: Percentage slope

BATTER BOARDS

To assist workmen in obtaining the proper grades and depths between stations while digging trenches for drain lines, it is necessary to use what are called batter boards. The side stakes are usually two-by-fours, driven into the ground very firmly at each station; sometimes every 50 ft, depending upon accuracy required. Cross-pieces are generally of one-by-fours or one-by-sixes, and are set so that the top edge of each of these cross-pieces will be exactly of the same grade or slope as the intended ditch bottom. Cord is then stretched from board to board to aid in obtaining ditch bottom between stations.

The procedure for setting batter boards to grade can be best explained by the following sequence of illustrations. The illustrations are based on the same design conditions, except for changes in profile. (See Figures 12.23a and 12.23b.)

The leveling rod can be used to measure the height of the batter boards, which are nailed or clamped in place.

Carpenter's levels may be used to check or assist in keeping the batter boards level.

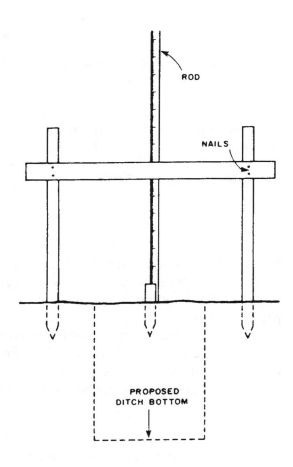

Figure 12.23b

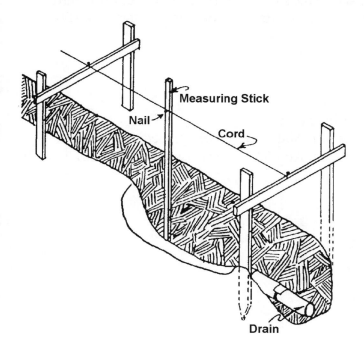

Figure 12.23a

SAMPLE PROBLEM

Assume a line of batter boards are to be installed to grade for a tile drainage ditch.

Design Conditions

Total length of line	=	500.00 ft
Elevation difference	=	5.00 ft
Batter board spacing	=	100.00 ft
Depth of ditch	=	3.00 ft
Measuring stick	=	6.00 ft

Suggested Steps

1. Place stakes 100 ft apart (horizontal distance).
2. Determine the elevation at each stake, using the level.
3. Plot the profile accurately on graph paper.
4. Calculate percent of slope.
5. Determine the elevation of the ditch bottom at each stake.
6. Determine the elevation of the batter board at each stake.
7. Determine depth of cut at each stake.
8. Determine height of the batter board at each stake.

Procedure

1. Determine the starting point. *Note:* Since the profile shows the slope (in this case) to be uniform through-out, any one of the stakes (stations) could have been used as the starting point. Here, start at 0 + 00.
2. Determine the tile line grade (% slope).

$$\% \text{ slope} = \frac{5.0}{500.0} \times 100 = 1.0\%$$

or 1.0 ft rise/100.0 ft horizontal distance.
3. Determine the elevation of the ditch bottom at the starting point.
4. Determine the elevation of the ditch bottom at each station. Add 1.0 ft to each preceding elevation.
5. Determine the elevation of the batter board at each station. Add 6.0 ft (measuring stick length) to ditch elevations.
6. Determine the depth of cut.

   cut = ground elevation − ditch elevation

7. Determine the height of the batter board.

   height = batter board elevation − ground elevation

Illustration 1

Uniform grade throughout (ideal case).

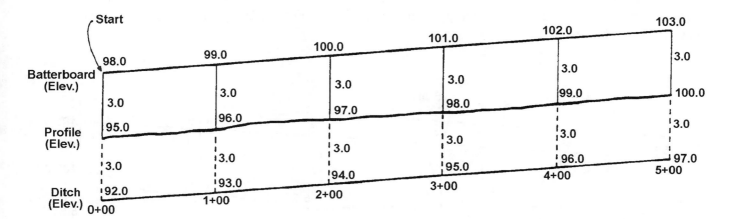

Figure 12.24

Illustration 2

Grade not uniform throughout. (All other conditions same, as above.)
   Note: On gentle slopes, the height of the batter boards may vary somewhat.

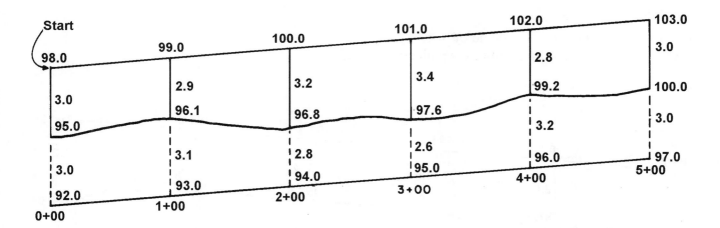

Figure 12.25

Illustration 3

High ground elevations at beginning and end.

Note: The starting point was changed to 2 + 00 ft in order to maintain a 3.0-ft cut for most of the length of line. Dashed lines (- - - -) show height of the batter boards if 0 + 00 ft was used as the starting point.

Important: From the starting point add 1% going uphill and subtract 1% going downhill to find the elevation of the ditch bottom.

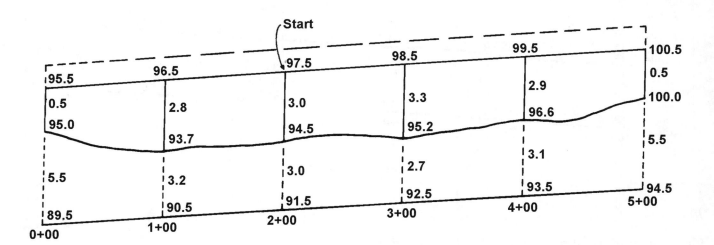

Figure 12.26

Illustration 4

Low elevations at beginning and end of line.

Note: Change the starting point as explained previously.

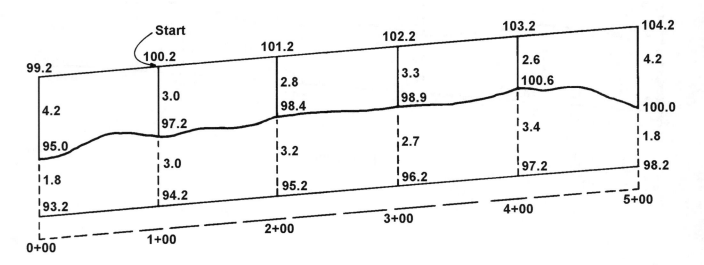

Figure 12.27

PLOTTING CONTOUR LINES

A contour map is a useful tool in planning land improvements. A map can be developed by taking the elevation of a series of measured points and interpolating elevations between the measured points. One way to make a field survey is to lay out a grid across the field at 100-ft intervals and then determine the elevations at the intersection of the grid points. The assumption is made that the ground surface is approximated by a straight line connecting the grid points. If this is not true, then intermediate points should be taken between the regular grid points. After the field survey is completed, the elevations can be plotted on graph paper, as shown.

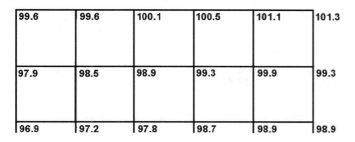

99.6	99.6	100.1	100.5	101.1	101.3
97.9	98.5	98.9	99.3	99.9	99.3
96.9	97.2	97.8	98.7	98.9	98.9

Figure 12.28

The contour map can be developed by knowing a few characteristics of contour lines. The important ones are: (1) contour lines connect points of equal elevation, so they can never cross, (2) contour lines close someplace on the surface of the earth, so they are continuous, (3) contour lines point up into draws, (4) contour lines point down on ridges, and (5) a depression or a knob will show as a closed circle, while other lines will start at the edge of a map and continue across it.

Procedure

1. Determine the highest contour elevation that can be plotted.

   Note: Find the highest elevation shown = 100.8 ft. Therefore, highest contour line = 100.0 ft.

2. Roughing in: The contour lines can be placed roughly in position—as illustrated (x - - - - x) on the map—to get a general idea where they run. *Note:* The 100.0 ft contour line runs between elevations that are higher and lower than 100.0 ft. (See Figure 12.29.)

3. Accurate placement of the contour lines is accomplished by interpolation.

Interpolation

1. At the point where the contour line being plotted crosses a grid line, determine the number of tenths above and below the contour line. (See Figure 12.12.)

2. Determine the total number of tenths and set up a ratio based on the scale of the map. In this case, 1 block = 100.00 ft.

Example

Starting from the left side:

1. $\begin{array}{c} 100.8 \\ 100.0 \\ 99.2 \end{array} \Big] \begin{array}{c} 8 \\ 8 \end{array} = 16 \quad \therefore \quad \dfrac{8}{16} = \dfrac{x}{100} = 50$

   Plot 50/100 below 100.8 or 50/100 above 92.2

2. $\begin{array}{c} 100.6 \\ 100.0 \\ 99.8 \end{array} \Big] \begin{array}{c} 6 \\ 2 \end{array} = 8 \quad \therefore \quad \dfrac{6}{8} = \dfrac{x}{100} = 75$

   Plot 75/100 below 100.6 or 25/100 above 99.8

3. $\begin{array}{c} 100.2 \\ 100.0 \\ 99.2 \end{array} \Big] \begin{array}{c} 2 \\ 8 \end{array} = 10 \quad \therefore \quad \dfrac{2}{10} = \dfrac{x}{100} = 20$

   Plot 20/100 below 100.2 or 80/100 above 99.2

4. $\begin{array}{c} 100.2 \\ 100.0 \\ 99.3 \end{array} \Big] \begin{array}{c} 2 \\ 7 \end{array} = 9 \quad \therefore \quad \dfrac{2}{9} = \dfrac{x}{100} = 22$

   Plot 22/100 away from 100.2 or 78/100 from 99.3

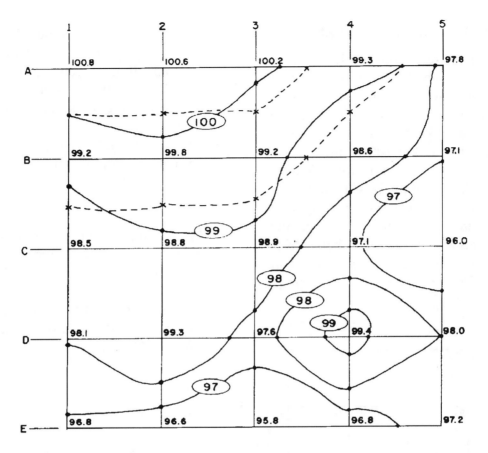

Figure 12.29. Scale: 1 block = 100.00 ft. x - - - - - x = Rough Placement. •———• = Accurate Placement.

Connect the points (free-hand) and identify the contour elevation.

Note: With a little practice most of the calculations can be done mentally, which will reduce the plotting time considerably.

TOPOGRAPHICAL SURVEY

Practice Problem

Assume that the field data of a given area was plotted to scale as shown in Figure 12.30.

Procedure

1. Starting at the highest even-foot elevation, in this case 101.0 ft, plot the contour lines using drops of 1.0-ft intervals as illustrated.

2. Make a rough estimate of the location of the contour line by placing a dot or x along the path.

3. Then, place an accurate contour line using the calculation procedure.

STAKING OUT A BUILDING

When excavation for the building starts, the corner stakes (A, B, C, D) will be lost. To preserve their location and to establish the elevation of the building foundation, it is common practice to set up batter boards about four feet outside the building lines.

The outline of a simple building and construction of a typical batter board are shown below. The letter N indicates a nail driven into the top of the boards. Strings are stretched from nail to nail and should cross precisely over the stakes, as illustrated at point C.

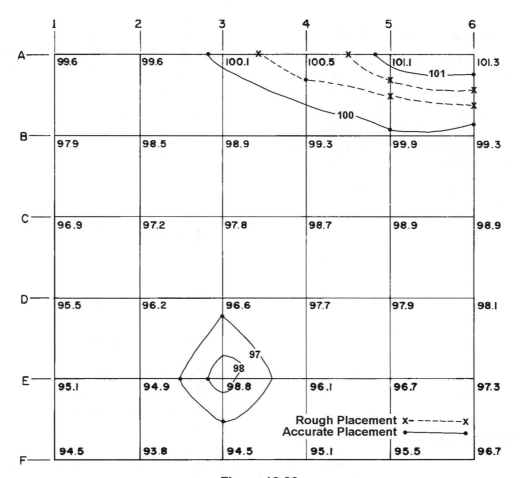

Figure 12.30

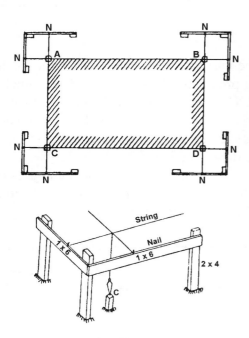

Figure 12.31

Setting Corner Stakes, Using Level

Assuming that the front of the building is AB; that the location of A is known; that the direction of B is known; and that all the angles are equal to 90°, center and level your instrument over A and then sight on B. Using a tape, measure the frontage of the building from A and set point B. With the instrument still at A, turn a right angle and set point C at the proper distance from A. The instrument is next set over point B and a sight is taken on A. By setting off an angle of 90° to the left, the direction to D may be determined, and D may be set at its required distance from B by taping. Check the length of CD with a string or tape to ensure that its length is the same as AB. It is excellent practice to measure both diagonals (AD and BC). If you have laid out the building correctly they will be the same length.

Setting Corner Stakes, Using Steel Tape

If a right angle is desired, as for a corner of a building, the multiples of 3-4-5 can be used. Steel tapes will be injured if bent at sharp angles, but by looping this can be avoided. (See Figure 12.32.) To accurately stake out the right triangle requires three men. The first man holds 0 and 80 ft, the second man holds 15 ft and 25 ft, and the third man holds 45 ft and 55 ft. Now carefully stretch the tape until the three sides intersect.

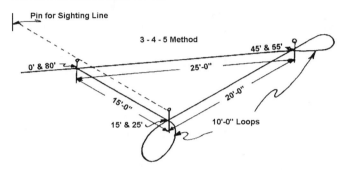

Figure 12.32

Setting Up Batter Boards

When the corner stakes are in place, the surveyor's level is set at a convenient location, preferably near the center of the building. The top edges of the horizontal batter boards are all set at the same elevation, preferably an even number of feet above the top of the foundation. By the leveling processes previously discussed, the appropriate rod reading for the top of the foundation may be figured. If this rod reading is decreased by the amount that the strings are to be above the foundation (say, 2 ft), the rod may be held at each of the posts, and raised or lowered until the calculated rod reading is reached, at which time a pencil line is made on the post at the foot of the rod. These pencil lines will serve as guides in nailing the batter board horizontals. The top edges of all the horizontal batter boards are now at the same elevation.

To set the nails in the tops of the batter boards, set your instrument at each of the four corners and sight the two adjacent corners (for example, B and C from A). The nails are set on the lines so determined.

As a final step, check your work by stretching strings from nail to nail. These strings should lie entirely in one plane and should cross one another over points ABCD.

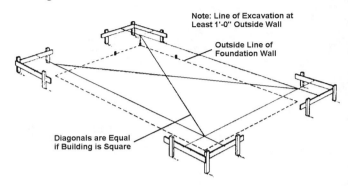

Figure 12.33

HAND LEVELS

Measuring Slopes

The hand level can be used to determine the slope of land in percent, or in feet of fall per hundred feet along the slope.

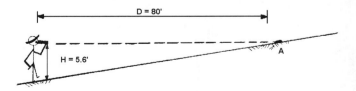

Figure 12.34

One man may measure slopes alone by the following procedure: First he must determine his eye height (H) above the ground by measuring from the ground up to his eye level. He then notes a definite object, such as a stone or stake on the ground, and moves down the slope until he sights directly on object A with the hand level. He then paces off the distance (D) from this point to the object and calculates the percentage of slope as shown.

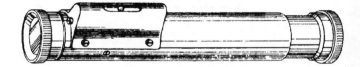

Figure 12.35. Slope = H/D = 5.6/80 = 0.07, or 7%.

If two men are working together, the level man stands midway down the slope to be measured. The rodman holds a level rod or marked stick at point A and takes a reading on the rod. Then he moves exactly 100 ft down the slope to B and takes a second reading. The difference between these two readings is the percentage of slope.

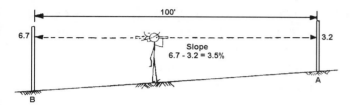

Figure 12.36

Slopes may be measured directly in percent with an Abney hand level. An Abney hand level differs from an ordinary hand level in that an arc is attached to the bubble tube. The Abney level is used as an ordinary hand level when the arc reading is set at 0°. To measure slopes it is first necessary to set the arc at 0° and determine eye height. Then the rodman moves up or down the slope and the levelman sights at the rod with the cross-hair at the reading for eye height. He swings the arc until the bubble is centered. The slope in percent is read directly on the arc of the hand level.

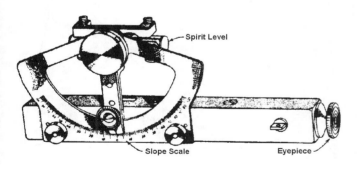

Figure 12.37

Locating Contour Lines

The hand level is useful in laying out contour guidelines. A contour line is a line of equal elevation inscribed on the Earth's surface. To lay out a contour line, the levelman determines his eye height by sighting on a level rod, a board, or a feature or mark on the rodman.

A point is selected to be located on a contour line. The levelman stands at the point, and the rodman moves 50 to 100 ft along the approximate contour line, and up or down the slope as directed by the levelman. When the point is located on the line, a stake is set by the rodman.

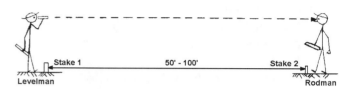

Figure 12.38

The rodman remains beside his stake, and the levelman moves past him another 50 to 100 ft, using the same distance as before. The levelman then sights back on the rodman and moves himself up or down the slope until he is on the contour line. A stake is set and the rodman moves past the level man as before. This procedure is continued until the entire contour line has been located.

PLANE TABLE AND PEEP-SIGHT ALIDADE

Plane Table Survey (See Figure 12.39)

Procedure

1. Make a visual inspection of the field to be surveyed.
    a. Observe boundary lines, special features, obstacles, etc.
2. Decide on a plan of attack for doing the job.
    a. Determine plane table locations
    b. Determine location and length of baseline, etc.
3. Stake out the field.
    a. Use sufficient stakes to outline irregularly shaped boundary lines.
4. Pace the field in two directions and determine the scale to be used.
    a. The scale should be established in relation to the paper size.

5. Establish a starting point A on the paper.
   a. Make allowances for the distance behind and sides of the table in relation to its location in the field.
6. Place a pin A in the ground directly below point A.
   a. Drop a plumb bob from point A and place a pin in the ground.
7. Measure (tape) a baseline a predetermined number of feet and place pin B in the ground.
   a. Length of the baseline should be as long as possible—limited only by the length of the field.
8. Using the peep sight alidade from point A, sight to pin B, and draw the baseline on paper to scale.
   a. Sight and measure the baseline accurately.
9. Sight to each stake from point A and draw a light line that extends off the paper.
10. Move the plane table and plumb point B to pin B.
    a. *Caution:* Be precise.

11. Sight back to pin A and align the baseline.
    a. Loosen the table top from the tripod and swivel the table until the baseline is realigned back to pin A, then tighten top.
12. Resight all the stakes from point B and draw lines off the paper.
    a. The intersecting point of lines from points A and B establish the location of each stake.
13. To check for accuracy, tape the distance between two stakes in the field and check against the scaled distance.
14. Calculate the acres of the field.
    a. Use the trapezoidal rule

$$\frac{\dfrac{h_o}{2} + (\text{sum of all h's}) + \dfrac{h_x}{2} \times d}{43,560}$$

plus triangles, rectangles, etc., where applicable.

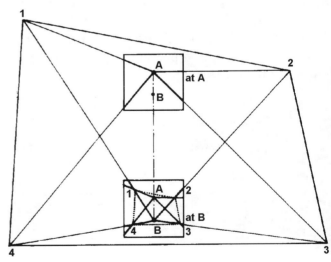

Figure 12.39

PRACTICE PROBLEM

Assume the field data shown was derived from a plane table survey of an oddly shaped field.

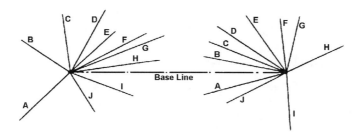

Figure 12.40

1. Using a straight edge, find the intersecting points. Then complete the borderline of the field.
2. Using the *trapezoidal rule* method, divide the field into segments and determine the length of each line.
   Note: Drawing is to scale.
3. Calculate the area of the field in *acres.*

SOLVING FOR SQUARE FEET OF AN IRREGULARLY SHAPED GREEN

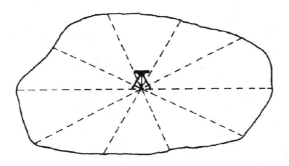

Figure 12.41

Procedure

1. Set up the level near the center of the green.
2. Below the attached plumb bob, place a marker in the ground.
3. Set the level graduated circle to zero degrees and set the pin at the edge of the green.
4. Tape the distance from the marker to the edge of the green.
5. Repeat measurements at 10–20° and/or at intervals as needed.
6. Draw the green to scale on paper using a 360° circle or protractor.
7. Solve for ft^2.

STUDENT PROBLEM

The data shown was obtained from a staked out field which included a parking area. The arrows give the degrees and distance (feet) to the stakes.

Problem

Solve for acres by each of the following methods:

    1. trapezoidal rule—see page 21.
    2. obtuse and/or acute triangles—see page 20.

    3. trigonometry: $A = \dfrac{bc \sin A}{2}$

Procedure

1. Tape a 360° circle onto a worksheet.
2. Determine the scale to fit the paper.
3. Measure and mark the distance of the stakes at the degrees shown.
4. Outline the shape of the field.
5. Solve for acres.

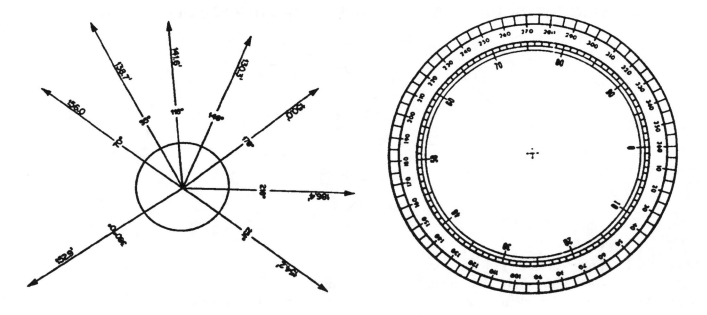

Figure 12.42

Important: Review the following example of a student Field Exercise and Lab Report.

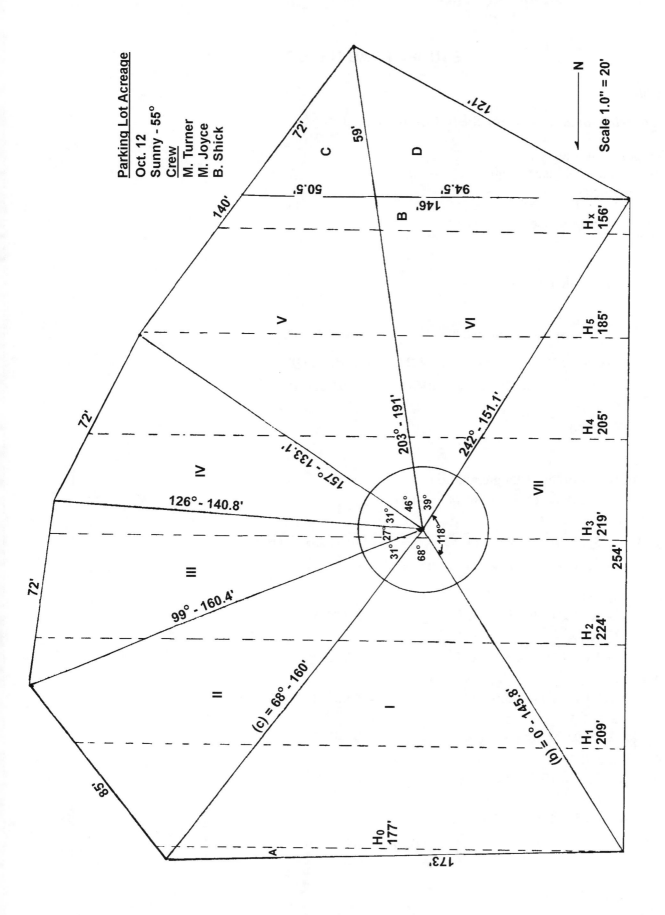

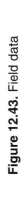

Figure 12.43. Field data

STUDENT LAB REPORT

Field Problem

Solve for *parking lot acreage* using the following methods:

1. Trapezodial rule
2. Acute and/or obtuse triangles
3. Trigonometry:

$$\text{Area} = \frac{\sin A \times \text{side (b)} \times \text{side (c)}}{2}$$

1. Trapezoidal rule method:

$$\text{area} = \frac{h_o}{2} + \text{(sum of numbered h's)} + \frac{h_x}{2} \times D + \text{(odd shapes)}$$

$$= 88 + (1039) + 78.2 \times 40 = 48,228 + (A, B, C, D)$$

$$= 48,228 + 435 + 2268.7 + 1441 + 2648 = 55,021 \text{ ft}^2$$

$$= \frac{55,021}{43,560} = 1.26 \text{ ac}$$

2. Acute and/or obtuse triangle method:

$$S = \frac{(A) + (B) + (C)}{2} = \frac{173 + 145.8 + 160}{2} = 239.4$$

a. $A = \sqrt{S(239.4)} \times \sqrt{S - A(173)} \times \sqrt{S - B(145.8)} \times \sqrt{S - C(160)}$

$$= 15.4 \times 8.1 \times 9.6 \times 8.9 = 10,658 \text{ ft}^2$$

b. $S = 202.7$
$A = 14.2 \times 10.8 \times 6.5 \times 6.5$ $\qquad = 6,479 \text{ ft}^2$

c. $S = 186.6$
$A = 13.6 \times 10.7 \times 5.1 \times 6.7$ $\qquad = 4,972 \text{ ft}^2$

d. $S = 172.9$
$A = 13.1 \times 10.0 \times 5.6 \times 6.3$ $\qquad = 4,622 \text{ ft}^2$

e. $S = 232.5$
$A = 15.2 \times 9.5 \times 9.9 \times 6.4$ $\qquad = 9,149 \text{ ft}^2$

f. $S = 230.5$
$A = 15.1 \times 10.5 \times 6.2 \times 8.9$ $\qquad = 8,749 \text{ ft}^2$

g. $S = 275.4$
$A = 16.5 \times 4.6 \times 11.1 \times 11.3$ $\qquad = 9,520 \text{ ft}^2$

$$\text{Total} = 54,149 \text{ ft}^2$$

$$= \frac{54,149}{43,560} = 1.24 \text{ ac}$$

3. Trigonometry method

    a. $\sin 68° = 0.9271$

    $$\frac{0.9271 \times 145.8 \times 160}{2} = 10,813.6 \text{ ft}^2$$

    b. $\sin 31° = 0.515$

    $$\frac{0.515 \times 160 \times 160.4}{2} = 6,608.4 \text{ ft}^2$$

    c. $\sin 27° = 0.4539$

    $$\frac{0.4539 \times 160.4 \times 140.8}{2} = 5,125.5 \text{ ft}^2$$

    d. $\sin 31° = 0.515$

    $$\frac{0.515 \times 140.8 \times 133.1}{2} = 4,825.6 \text{ ft}^2$$

    e. $\sin 46° = 0.7193$

    $$\frac{0.7193 \times 133.1 \times 191}{2} = 9,143.0 \text{ ft}^2$$

    f. $\sin 39° = 0.6293$

    $$\frac{0.6293 \times 191 \times 151.1}{2} = 9,080.8 \text{ ft}^2$$

    g. $\sin 118° = 0.8829$

    $$\frac{0.8829 \times 1.51.1 \times 145.8}{2} = 9,725.3 \text{ ft}^2$$

    Total $= 55,322.2 \text{ ft}^2$

    $$\frac{55,322.2}{43,560} = 1.27 \text{ ac}$$

PONDS

General

The demand for water has increased tremendously in recent years, and ponds are one of the most reliable and economical sources of water. On many golf courses ponds are serving a variety purposes, including irrigation, aesthetic value, fire protection, and recreation.

Ponds are an important source of irrigation water. However, pond capacity must be adequate to meet the turfgrass requirements and overcome unavoidable losses. For example, a 3-in. application of water on 1 ac requires 81,462 gal. The required storage capacity of a pond used for irrigation depends on a number of interrelated factors: turfgrass requirements, effective rainfall, evaporation losses, application efficiency, expected inflow, and seepage.

Figure 12.44

Important note: Most states and government entities have laws, rules, and regulations governing the installation of ponds.

Those responsible for planning and designing ponds must comply with all such laws and regulations. For details and assistance contact the U.S. Soil Conservation Service.

Water Storage Pondw

The discussion regarding storage ponds will be limited to:

1. Calculating the pond size based on given or known design conditions.
2. Method for *estimating* the storage capacity (gallons) of a pond—prismoidal formula.

Reasons for a Pond

1. To store up a large volume of water from low-yielding wells, springs, or other sources over a period of time that can be pumped out at a higher rate during irrigation.
2. Water may be pumped up from sources at lower elevations to a strategically placed pond at a higher elevation. This in turn will reduce the lift during irrigation, lowering the pump pressure, pump size, and operating cost.

Design Conditions Required

1. Water usage rate for your local area, in./wk.
2. Total number of acres—include fairways, greens, tees, and potential future expansion.
3. Shape of pond—rectangular, square, roundish, or other.
4. Side slopes.
5. Depth.
6. Number of weeks supply.
7. Safety factor—avoid pumping out all the water.

SAMPLE PROBLEM 1

Design Conditions

1. Water usage rate = 1 in./wk.
2. Number of acres = 24.
3. Pond shape = rectangular.
4. Side slopes = see following figures.
5. Depth = 12 ft.
6. Weeks supply = 1.
7. Safety factor = 20% (estimated).

Procedure

1. Solve for volume (V), ft³.
   Note: 1 ft³ = 7.48 or 7.5 gallons
   1 ac-in. = 27,154 gal

$$V = \frac{27,154 \times 24 \times 1.2}{7.5} = \frac{104,271 \text{ ft}^3}{782,033 \text{ gal}}$$

2. Solve for area (ft²) or length × width.

$$(L \times W) = \frac{V}{D} = \frac{104,271}{12} = 8689 \text{ ft}^2$$

   Next: Select a random length. In this case assume L = 136 ft.

3. Solve for width.

$$W = \frac{8689}{136} = 64.9 \text{ or } 64 \text{ ft}$$

4. Solve for pond design.

   Start at mid-depth, then add and subtract for side slopes as shown in Figure 13.45.

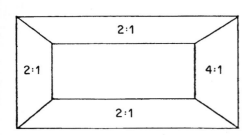

Figure 12.45a. Plan (Not to Scale).

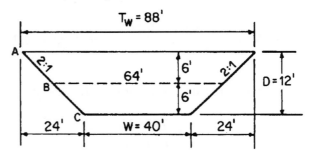

Figure 12.45b. Cross-Section (Not to Scale).

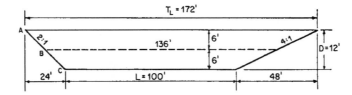

Figure 12.45c. Longitudinal Section (Not to Scale).

Estimating the Volume and Gallons of an Excavated or Existing Pond

After the dimensions and side slopes of the pond have been selected, it is usually necessary to prepare an estimate of the volume of excavation. Such an estimate determines the cost of the pond and serves as a basis for payment if the work is done by contract.

The volume of excavation required can be estimated with sufficient accuracy by use of the *prismoidal formula:*

$$V = \left[\frac{(A + 4B + C)}{6} \times D \right] \times 7.5 = \text{gal}$$

$$\text{or} \div 27 = \text{yd}^3$$

where: V = volume of excavation, in gallons
A = area of the excavation at the ground surface, ft^2
B = area of the excavation at the mid-depth point (1/2 D), ft^2
C = area of the excavation at the bottom of the pond, ft^2
D = average depth of the pond, ft

As an example, assume a pond with a depth, D, of 12 ft; a bottom width, W, of 40 ft, and a bottom length, L, of 100 ft as shown in the Figures 12.45a through c.

The side slope at the ramp end is 4:1 and the remaining slopes are 2:1. The volume excavation, V, is computed as follows:

Top length = 12(2) + 12(4) + 100 = 172 ft
Top width = 12(2)(2) + 40 = 88 ft
A = 88 × 172 = 15,136 ft^2
Mid-length = 6(2) + 6(4) + 100 = 136 ft
Mid-width = 6(2)(2) + 40 = 64 ft
4B = 4(64 × 136) = 34,816 ft^2
C = 40 × 100 = 4,000 ft^2
(A + 4B + C) = 53,952 ft^2

Then: $V = \dfrac{53.952}{6} \times 12 \times 7.5 = 809,280 \text{ gal}$

5. Solve for percent of accuracy of the design gallonage vs. estimated gallonage.

$$\text{accuracy} = \frac{782,033}{809,280} = 97\%$$

SAMPLE PROBLEM 2. POND DESIGN

Design Conditions

a. *roundish* shaped pond
b. average side slopes = 3:1
c. all other conditions are the same as Sample Problem 1

Procedure

1. Solve for pond volume (V):

$$V = \frac{27{,}154 \times 24 \times 1.2}{7.5} = 104{,}271 \text{ ft}^3$$

2. Solve for pond diameter (D):

where: V = Volume
π = 3.14
H = depth
r = radius
D = diameter

$$V = \pi r^2 \times H = r = \sqrt{\frac{V}{\pi H}}$$

$$r = \sqrt{\frac{104{,}271}{3.14 \times 12}} = 53 \text{ ft}$$

i.e., D = 106 ft

3. Solve for pond design:

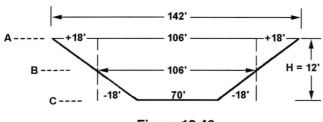

Figure 12.46

Estimating the volume of an existing pond.

Procedure: see above sketch.

$$V = \frac{(A + 4B + C)}{6} \times H$$

A = 3.14 × 71² = 15,829
B = 4 × 3.14 × 53 = 35,281
C = 3.14 × 35² = 3,847

$$V = \frac{54,957}{6} \times 12 = 109,913 \text{ ft}^3$$

Check percent of accuracy:

$$\frac{104,271}{109,913} = 95\%$$

STUDENT PROBLEMS

Storage Pond Design—*Rectangular Shape*

Design Conditions

a. total acres	= 110
b. water usage rate	= 1.25/wk (moderate climate)
c. number of weeks supply	= 1
d. depth (D)	= 10 ft
e. side slopes	= 4:1 (all sides)
f. safety factor	= 20% (estimated)

Procedure

1. Solve for volume (V), ft^3

$$V = \frac{gal/ac\text{-}in. \times usage\ rate \times total\ acres \times ft^2}{7.5}$$

2. Solve for surface area (ft^2) or length ´ width

$$(L \times W) = \frac{V}{D} = area\ ft^2$$

3. Select a random length. Assume L = 200 ft.
4. Solve for width:

$$W = \frac{ft^2}{assumed\ length}$$

5. Solve for: *longitudinal section* and *cross section*
   Start at mid-depth, then add and subtract for side slopes. Review sample problem.

Estimate the Volume of an Existing Pond

Design Conditions

a. Calculate the volume in ft^3 and gallons of the round-shaped pond shown below.

Procedure

1. Solve for the circular surface area (A), ft^2, at mid-depth (B) and bottom (C).
2. Use the prismoidal formula:

$$V = \frac{A + 4B + C}{6} \times D$$

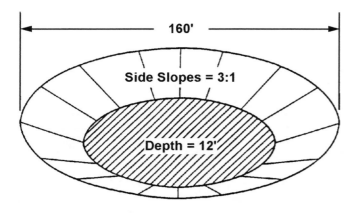

Figure 12.47

ACKNOWLEDGMENTS

1. "Elements of Surveying." Technical manual. TM 5-232, Department of the Army, USGPO.
2. "Engineering Field Manual Handbook for Conservation Practices." USDA, Soil Conservation Service.
3. "Drainage of Agricultural Land." USDA, Soil Conservation Service.
4. "Corrugated Plastic Tubing." Manual. Advanced Drainage Systems.
5. "Agricultural Engineering Yearbook." ASAE, R260.2, pp. 408–416.
6. "Corrugated Plastic Drainage Tubing." Circular 1078. University of Illinois, Urbana-Champaign, College of Agriculture.

Chapter 13

Tile Drainage Systems

EXPLANATION OF SOIL-WATER-PLANT TERMS AND THEIR RELATIONSHIPS

Soil Moisture

Water is held in the soil because of its natural attraction to the soil particles as well as to itself. Water is held in the form of a film around each particle of soil.

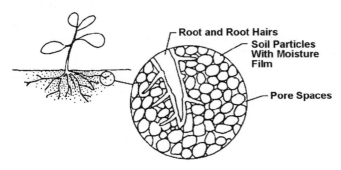

Figure 13.1

Forces Acting on Soil Moisture

Tension

Soil tension is a combination of *cohesion* (attraction of water to water) and *adhesion* (attraction of water to particles). Wet soils (thick moisture films) have less tension than dry soils (thin moisture films).

Capillarity

When soil particles are packed together, the spaces between the particles form tiny passages or channels through the soil. *Capillary action* is present due

to these channels. This is the action that lifts water up a sponge when one corner of it is placed in water.

How well capillary action works depends on the size of the channels. The smaller the channel, the better the capillary action.

Gravity

The force of gravity is constantly pulling down on the water in the soil. Gravity limits the height that water can rise by capillary action, and it limits the thickness of the film held by tension around each soil particle.

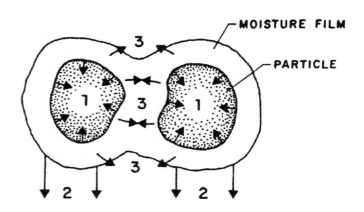

Figure 13.2. 1. Tension; 2. Capillarity; 3. Gravity.

Non-Saturated Flow

Just as magnetic forces attract and hold iron filings (up to the capacity of the magnet), water is attracted to dry soil particles (having high tension forces), and forms a moisture film around them. As the films get thicker, the particle tension forces are reduced and approach zero. At this point, water flows through the capillary passages to the adjacent particles having thin moisture films and higher tension forces. This movement of water through the soil is called *non-saturated flow*.

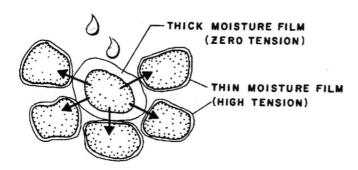

Figure 13.3

When water is added to one point of a dry soil, the wetting pattern is in all directions, but mostly downward. Because capillary action is stronger in fine soils having finer passages, more lateral movement results than in a coarser soil or sand. Clay soils are generally poor because drainage is slow and it is desirable to resume play as soon as possible after a rain. But, where there is an area of 100% sand, the water will drain right down and dry spots will show. Clearly, thorough mixing in correct proportions is necessary. A laboratory analysis of the soil is required to determine the correct soil mixture.

Perched Hydration Zone

If soil is being watered and the subsoil is wet, field capacity is exceeded very quickly, with the accompanying drainage (saturated flow).

Deep soil, to which water is added more slowly than the soil can move it in non-saturated flow, will not reach saturation. Saturation at the *surface* (puddling) occurs if water is added too fast. However, drainage will not occur unless the entire soil is saturated. Saturation starts to happen when moving water comes up against a barrier or soil stratum of coarser texture. Water cannot move through a stratum of coarser soil until a saturated region builds up above the barrier and hydraulic pressure is great enough to force the water through. The hydraulic pressure is proportional to the height of water in the saturated region.

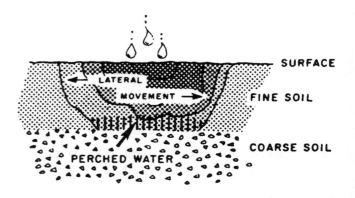

Figure 13.4

Perhaps this can be illustrated by a standard (1-in. × 3-in. × 6-in.) sponge. Lay the sponge flat in your hand and fill it to saturation. Now, without adding water, tip it on its side and water will run out. Again, tip the sponge on end and more water will run out. Each time the sponge was tipped, the hydraulic pressure increased, forcing water to flow from the sponge.

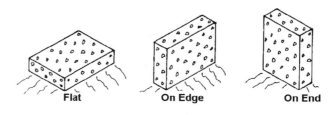

Figure 13.5

The base of the sponge is called the *effective perched water table* and the water above the base is called *perched water*.

Field Capacity

At the point where the films around the particles are thick, the tension of the soil is zero. The gravity force is now greater than the tension forces, pulling water downward from the particles. As the films get thinner, the tension will increase, and when the two forces become equalized (at a soil tension of about 1/3 atmosphere, or 5 psi), the downward flow ceases. This condition exists at the *water-holding capacity* or *field capacity* of the soil.

Field capacity is defined as the water retained in a soil against the pull of gravity.

Saturation

If more water is added, the field capacity is exceeded, and the water will fill the capillary pore spaces. Complete saturation is reached when all the pore spaces are filled. The addition of more water results in surface runoff and puddling.

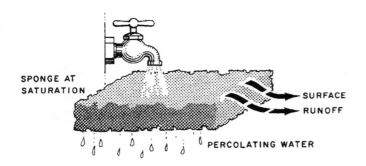

Figure 13.6

Saturated Flow or Drainage

The downward movement of soil water, caused by gravity, is *drainage,* or *saturated flow.* Drainage will not take place until the entire soil profile reaches zero tension.

Hydraulic Conductivity, Percolation Rate, or Permeability

This is the rate of moisture movement through the soil under saturated conditions.

SUBSURFACE DRAINAGE

General Information

The main purpose of subsurface drainage system is the removal of unwanted water.

Why Drain?

1. To maintain the water table at the proper level for healthiest plant growth.
2. Soil voids may be kept free of excess water, permitting the presence of air, which makes possible the key soil bacterial processes that liberate life-giving nitrogen in the plant root zone.
3. To keep equipment from bogging down.
4. It permits play to resume soon after heavy rains on golf courses and recreational areas.

Benefits of Proper Drainage

1. It deepens the soil root zone, which holds moisture in dry weather and creates an opportunity for the presence of plant food.

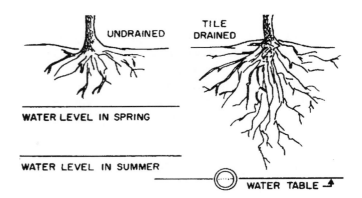

Figure 13.7

In undrained land crop roots spread out near the top of the soil in early wet months, and cannot reach the lower water table in dry periods. Observe the difference in tiled land, where original root growth extends into the newly created root zone.

2. It conserves the topsoil. When land is perpetually wet there is no escape for rainwater, except for it to run off over the surface, taking with it some of the topsoil.
3. It lengthens the growing season, as soils warm earlier.
4. It raises the soil temperature. Dry soil is much warmer than wet soil. Actually it takes five times as much heat to raise an equal volume of water 1° as it does to raise an equal volume of soil 1°. Germination of seeds occurs earlier.

5. It protects plants from frost heaving.
6. Increases the value of fertilizing. Rainwater will percolate down, carrying the fertilizer to the root zone.
7. Drain lines occupy no land surface and they do not harbor weeds.
8. Drain lines don't interfere with land usage and maintenance.
9. It reduces mosquito breeding areas.
10. It reduces pollution of waterways.

Subsurface Drainage

The underground drains take out only *excess* water, not water that plants can use. Water available to plants is held in the soil by capillarity, while the excess water flows by gravity into the drains.

1. Subsurface drainage is accomplished by means of perforated pipe, laid in a continuous line at a specified depth and grade so that free water entering the line through the perforations will flow out by gravity. It differs in principle from surface drainage in that water percolates into soil and is taken out by drains below the ground's surface. The principal purpose is to lower the groundwater level below the root zone of plants.
2. A subsurface drainage system consists of a drainage outlet, perforated pipe main, submain, and laterals. The function of the laterals is to remove the free water from the soil. The function of the submains and mains is to carry tile water to the drainage outlet. Unless otherwise stated herein, reference to perforated pipe will apply equally to tile.

Factors such as type of soil, design of pipe, filter material, and final outlet can have considerable effect on the efficiency of the system and must be taken into consideration.

SYSTEM LAYOUT AND LOCATION OF INTERCEPTORS

Although very few drainage problems are alike, below are some typical drainage layouts proven effective under the conditions indicated.

Subsurface Systems

In order to plan a subsurface drainage system, a pattern must be selected which fits the topography, sources of water, and other field conditions. Basic systems that may be selected are the random, multiple, parallel, herringbone, and interceptor.

Random Drains

The random system generally follows the largest natural depressions in the field, with laterals extending into individual wet areas. Often, such drains become outlets of a more complete system established on the higher areas of the field, as single lines, or even as complete systems. Their depth, location, and capacity should be considered with this possibility in mind.

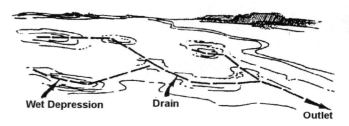

Figure 13.8. Random drains.

Multiple Drains in Depressional Areas

Closely spaced multiple drains are occasionally installed in depressional areas to remove both surface and subsurface water. This type of installation is easier to install than the *blind inlet,* and is usually just as effective. (See page 290.) A gravel envelope will improve infiltration into these multiple lines.

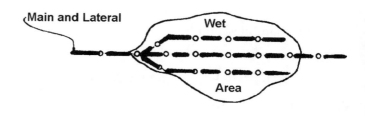

Figure 13.9

Herringbone Drains

Despite double drainage and the added cost of more numerous junctions, herringbone systems are sometimes advantageous in providing the extra drainage for heavier soil so often found in narrow depressions.

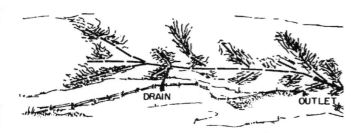

Figure 13.10. Herringbone drains.

Parallel or Gridiron

Parallel and gridiron systems should be so arranged that one main or sub-main serves as many laterals as possible. Drainage of level areas is accomplished by a system of laterals connecting to mains.

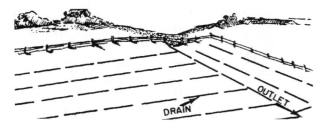

Figure 13.11. Gridiron or parallel drains.

Herringbone, gridiron, and parallel design systems are all quite popular and effective in reducing the groundwater to a level where it is not harmful.

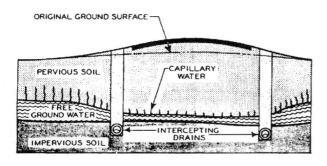

Figure 13.12

Intercepting Drains

Sloping water tables are found in slightly rolling, hilly, or mountainous places. The free groundwater, being subject to gravity, will flow in the direction of the slope along the impervious layer.

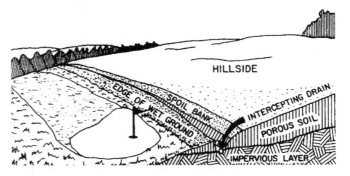

Figure 13.13. Intercepting Drains.

When this condition exists, it is possible for the free water to create a wetted area, or even reach the surface.

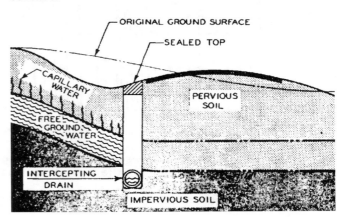

Figure 13.14

USE OF DRAIN LINE DESIGN CHARTS

To determine the size of drain line needed for a drainage system, it is necessary to carefully gather the design factors for a specific installation. A combination of either two or three of these known factors will be required to find the size of drain line from the drainage charts for ordinary tile, page 299, corrugated plastic drains, page 300. (See Figure 13.15.)

Design Factors

1. *Percentage of slope of the tile lines*—review percent slope, page 258.
2. *Drainage coefficient*—for complete systems as the parallel or gridiron.
3. *Water inflow rates*—in cfs for random or interceptor drains.
4. *Area of the watershed to be drained*—in acres.

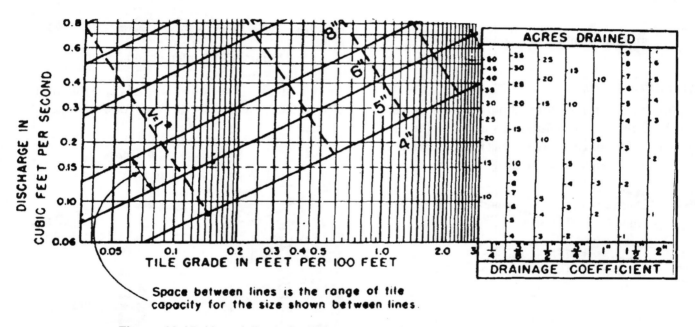

Figure 13.15. Use of tile drain design charts. (See complete chart on page 299.)

SAMPLE PROBLEMS

Example 1

Problem

To select the size of pipe required for the following conditions:

    a. discharge 100 gpm from a water-cooled condenser of a refrigeration system to a sewer line (or draining swimming pool).
    b. slope of the line = 1% for slope = 0.05%.

Solution

a. Convert gallons per minute to cubic feet per second.

$$\text{cfs} = \frac{\text{gpm}}{60\ (\text{min to sec}) \times 7.5 \left(\text{gal to ft}^3\right)}$$

$$\frac{100}{60 \times 0.75}\ 0.22\ \text{cfs}$$

b. From Tile Drainage Design Chart, find point of intersection of (0.22 cfs, 1% slope)
    Read—4-in. discharge pine required, and (0.22 cfs, 0.05% slope) = 5-in. pipe.

Example 2

Design drain lines for the following conditions:

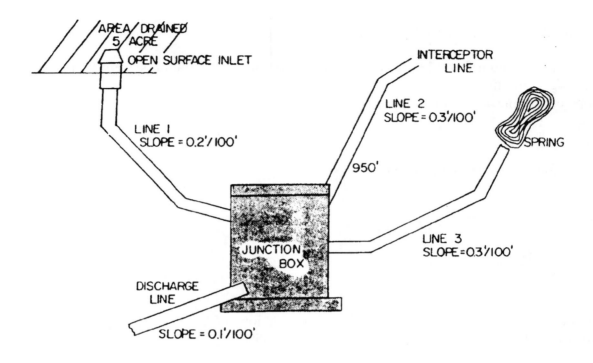

Figure 13.16

A. Line 1
 a. open surface inlet line to junction box
 b. watershed (acres drained) = 5 ac
 c. slope of drain line—2 ft/100 ft or 0.2%

Solution

1. drainage coefficient for water inletted directly into the surface inlet carried to junction box = 1 in. per day
2. area drained = 5 ac
3. slope of line = 0.2%

From Tile Drainage Design Chart find point of intersection. (D = 1–5 ac 0.2% slope) Read—6-in. tile = 0.21 cfs.

B. Line 2
950-ft-long interception line leading to junction box

type of soil = sandy loam
slope of drain line = 3 ft/100 ft, or 0.3%
slope of land = 3 to 4%

Solution

inflow rate selected = 0.1 cfs per 1000 ft of line

$$\text{inflow for 950 ft} = \frac{950}{1000} \times 0.1 = 0.095 \text{ cfs}$$

Note: For 2–5% slope, increase inflow rate by 10%. Total inflow rate = 0.095 + 0.0095 = 0.1045 cfs

From Tile Drainage Design Chart find point of intersection (0.1045 cfs – 0.3% slope) Read—4-in. drain line.

C. Line 3
a. tile line to carry 50 gpm from spring to junction box
b. slope = 0.3 ft/100 ft, or 0.3%

Solution

Convert gpm to cfs.

$$\text{cfs} = \frac{50}{60 \times 7.5} = 0.11 \text{ cfs}$$

From Tile Drainage Design Chart find point of intersection (0.11 cfs, 0.3% slope) Read—4-in. drain line.

D. Discharge Line
a. discharge line from junction box to carry off combined inflow of lines a,b,c
b. slope = 0.1 ft/100 ft, or 0.1% slope

Solution

total inflow = line 1 = 0.21 cfs
                       2 = 0.1045 cfs
                       3 = <u>0.11</u> cfs
              Total   0.4245 cfs

From Tile Drainage Design Chart find point of intersection (0.4245 cfs, 0.1% slope). Read—8-in. drain line.

Example 3

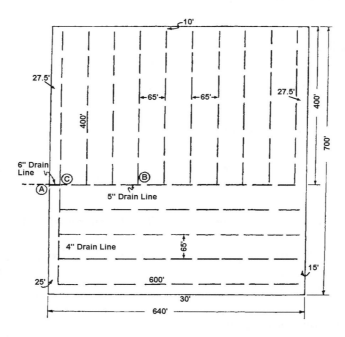

Figure 13.17

Given: Tract approximately 640 ft × 700 ft. Drainage area (from survey) = 10 ac.

soils (from soil borings, or soil map) silt loam
spacing—65 ft (Table 13.1)
drainage coefficient—3/8 in./day (Table 13.2)
use parallel system (from survey)

10 lines × 400 ft	= 4000	ft (from sketch)
4 lines × 600 ft	= 2400	ft
1 line 300 ft × 1/2	= 150	
1 line 600 ft × 1/2	= 300	
	6850	

65 ft × 6850 = 445,250 = 10 ac (approx.)

Grade = 0.0008 ft/ft (from survey)
(Some fields may require plotted profiles.)

Required: size of drain line at outlet.
  6-in. drain line (from chart)

Required: size of submain
  5-in. drain line will drain 8 ac = 348,480 ft^2 (from chart). 348,480/65 = 5400 ft (approx.)
  4-in. drain line will drain 4 ac = 174,240 (from chart). 174,240/65 = 2700
  Point B = 6 × 400 + 0.5 (6 × 65) = 2400 + 195 = 2595

  Therefore, use 5-in. drain line to point C. At point C, 10 × 400 + 0.5
  (9 × 65) = 4000 + 193 = 4193, which is less than 5400, and 5-in. drain line is okay.

Example 4

Given: Same as Example 3, except a 50-gpm spring also discharges into the drain line system.

Required: Size of the drain line main at outlet. From the point of intersection on Tile Drainage Chart, located in Example 3, continue horizontally across to left-hand vertical scale, which indicates the required discharge of the system for a 3/8-in. coefficient as 0.16 cfs. The equivalent discharge rate of a 50-gpm spring is equal to

$$\frac{50}{60 \ (\text{min to sec}) \ 7.5 \ \left(\text{gal to ft}^3\right)} = 0.11 \ \text{cfs}$$

Adding 0.11 cfs to 0.16 cfs = 0.27 cfs total discharge.

Opposite 0.27 cfs on the left-hand vertical scale, trace horizontally back across the chart to a point vertically above 0.08 grade on the horizontal scale. This point lies within diagonal spacing for an 8-in. drain line required at the outlet.

Example 5

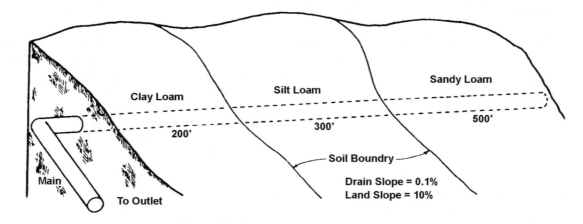

Figure 13.18. Interceptor drain line.

Given: 1000 ft interceptor drain line across 10% slope (from survey). Soils: top 500 ft sandy loam, next 300 ft silt loam, and the balance is clay loam (from soil borings or soil map).

   Grade: 0.001 ft/ft (from survey)
   (some fields may require plotted profiles)

Required: Size of drain line.
   Inflow for top 500 ft of line in sandy loam is 500 ft/1000 ft × 0.07 = 0.035 cfs
   Correcting for 10% land slope shows increased inflow of 0.035 × 0.20 = 0.007.
   Total inflow 0.035 + 0.007 = 0.42, say 0.04 cfs
   From Tile Drainage Chart point of intersection 0.04 cfs and 0.001 slope lies in 4-in. drain line size space.
   Inflow for next 300 ft of line in silt loam is 300 ft/1000 ft × 0.06 = 0.018 cfs
   Correcting for slope, 0.018 × 0.20 = 0.0036
     0.018 + 0.0036 = 0.0216, say 0.02 cfs
   Inflow for balance of line in clay loam is 200 ft/1000 ft × 0.04 = 0.008 cfs
   Correcting for slope 0.008 × 0.20 = 0.0016
     0.008 + 0.0016 = 0.0096, say 0.01
   Total inflow is 0.04 + 0.02 + 0.01 = 0.07.
   From Tile Drainage Chart, point of intersection 0.07 cfs and 0.001 slope lies in 4-in. drain line size space.
     Therefore, use 4-in. drain throughout the length of the line.

DETERMINING SIZE OF THE DRAINS

Factors to Consider for Drain Line Diameter Selection

Drainage requirements determined by crops. Different turfgrass species have widely differing tolerances for excess water, both as to amount and time. The designer of a drainage system recognizes these differences in species requirements by selecting the appropriate degree or intensity of drainage for the site. The drainage requirement is based on (1) the maximum duration and frequency of surface ponding, (2) the maximum height of the water table, or (3) the minimum rate at which the water table must be lowered. The local drainage guide indicates the drainage criteria required for various plant-soil combinations.*

Note: When surface water is admitted directly to the drain line through surface inlets, the entire watershed draining to the surface inlet should be used as the contributing area. An exception to this condition is where only a small amount of the runoff will be impounded at the location of the inlet and the remainder will flow away in a confined channel. The drain line should then be designed to carry away the impounded water in 8 to 12 hr depending upon the permeability of the soil, and lateral lines within the impounded area should be placed at about one-half the normal spacing. Further important guidelines are:

    a. 4- or 5-in. drain line may be used in mineral soils containing small amounts of fine sand or silt and where good alignment and uniform grade can be obtained.

    b. 6-in. drain line should be the minimum size for unstable soils, such as peat, muck, and quicksand.

    c. *No drain line less than 4 in. in diameter should be used.*

    d. Size of drain line required shall be determined from Tile Drainage Design Charts, and be based on slope, selected drainage coefficient, and area to be drained.

Drainage Coefficient *(not applicable to irrigated land)*

Underdrains should have sufficient capacity to remove excess water from minor surface depressions and the major portion of the root zone of the soil profile within a period of 24 hr after rainfall, and also to remove any seepage which enters the soil. The required amount of water to be removed is generally referred to as the *drainage coefficient,* expressed as inches of water depth to be removed uniformly over a prescribed area, or as an *inflow rate per length of drain.* Because of the difference in soil permeability, hazard, and value of the site, as well as the manner in which water may enter the underdrain, this coefficient should be modified to fit site conditions in accordance with local drainage guides established by the SCS.

Note: Drainage coefficient = the number of ac-in. of water to be removed per acre in 24 hr. In other words, DC = 1 is equivalent to removing 1 ac-in. (27,154 gal) per acre per day (24 hr).

DRAINAGE COEFFICIENT FOR GOLF COURSES

An exception to the basic rule of allowing 24 hr for water removal would be the case for recreational areas such as football fields and golf courses. Here, it is desirable to use the site as soon as possible following and even during a rainstorm, and flooding for any length of time cannot be tolerated. In such a case, a runoff rate similar to the rate of rainfall is justified. Often, a high rate—such as 1/3-in./hr or 8 in. in 24 hr (DC = 8)—is used in the design criteria to facilitate rapid water removal.

For *random, interceptor* and *single* drain lines, use the inflow rates found in the following table.

    • *Where no surface water is admitted directly to the drain line.* This applies to single drain lines on flat lands or across sloping lands, where seepage is intercepted. Size of drain line is based on the flow of gravitational water

---

* The local drainage guide may be obtained from the nearest Soil Conservation Service office.

Table 13.1. Inflow Rates for Random Lines[a]

Soil Texture	Unified Soil Classification	Inflow Rate of 1,000 ft of line in cfs
Coarse Sand and Gravel	GP, GW, SP, SW	0.15 to 1.00
Sandy Loam	SM, SC, GM, GC	0.07 to 0.25
Silt Loam	CL, ML	0.04 to 0.10
Clay and Clay Loam	CL, CH, MH	0.02 to 0.20

[a] For interceptor drain lines on sloping land, increase inflow rate as follows:

Land Slope (%)	Inflow Rate Increase (%)
2–5	10
5–12	20
12 and over	30

through soil, and shall be determined from the Tile Drainage Design Chart based on the inflow rates shown in Table 13.1.

• *Where surface water is to be admitted directly to the tile line through surface inlets,* the contributing subsurface flow shall be determined as above, and contributing surface flow added.

WORKSHEET AND PRACTICE PROBLEMS

A. 1. One acre = <u>43,560</u> ft^2
   2. One acre-foot = <u>43,560</u> ft^3
   3. One cubic foot = <u>7.48 or 7.50</u> gal.
   4. To convert cubic feet to gallons, multiply cubic feet by <u>7.5</u>.
   5. To convert gallons to cubic feet, divide by <u>7.5</u>.
   6. Show the equation for converting gallons per minute (gpm) to cubic feet per second (cfs).
   7. Show the equation for converting cfs to gpm.

B. Problems: *To be done on a separate sheet, showing procedure and all calculations.*

Problem 1

      1 cfs = (a) <u>450</u> gpm
      1 cfs = (b) <u>27,000</u> gph
      1 cfs = (c) <u>648,000</u> gpd

Problem 2

    Assume a drainage system was designed to remove 1/2 ac-in. of water (per acre) from a 6-ac fairway in one day (24 hr). Determine the following:

    a. gallons drained per 24 hours = <u>81,462.00</u>
    b. gallons drained per hour = <u>3,394.25</u>
    c. gallons drained per minute = <u>56.60</u>
    d. cubic feet per second = <u>0.126</u>

Problem 3

Assume that a swimming pool that measured 15 ft wide, 30 ft long, and 6 ft average depth was to be filled by a pump with a capacity output of 0.1 cfs. How many hours would it take to fill the pool?
  _7.5_ hr

a. total cubic ft = 15 × 30 × 6 = 7200 ft³
b. pump capacity = 0.1 cfs × 60 × 60 = 360 cfh
c. hours = 7200/360 = 7.5 hr

Problem 4

Assume a drainage system was installed in a fairway as shown.

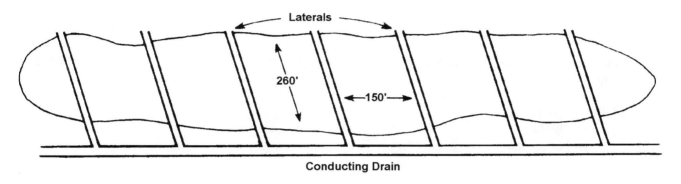

Figure 13.19

The drainage system design was based on the removal of one acre-inch of water per acre. Determine the following:

1. Area drained per lateral = _42,000 ft²_ = 0.964 ac
2. Total area drained = _294,000 ft²_ = _6.75_ ac
3. Gallons of water per acre inch = _27,154_
4. Gallons of water collected by the conducting drain = _183,290._
5. Cubic feet of water collected by the conducting drain = _24,439_ .

Problem 5

Assume that an interceptor drain was installed as shown.

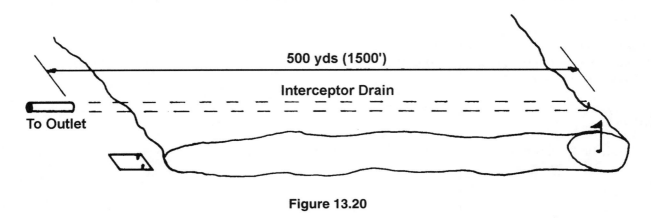

Figure 13.20

The water inflow rate as determined by the design engineer was based on 0.25 cfs per 1000 ft length. Determine the following:

1. total inflow = <u>0.375</u> cfs
2. total gpm = <u>169.00</u>
3. total gallons per day (24 hr) = <u>243,000</u>

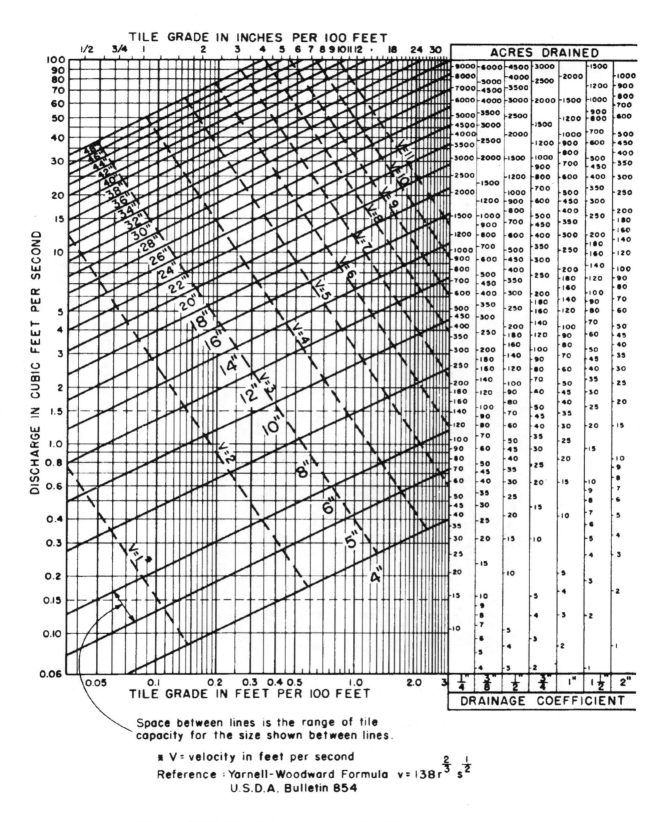

Figure 13.21. Tile drain chart. Agricultural Engineers Yearbook.

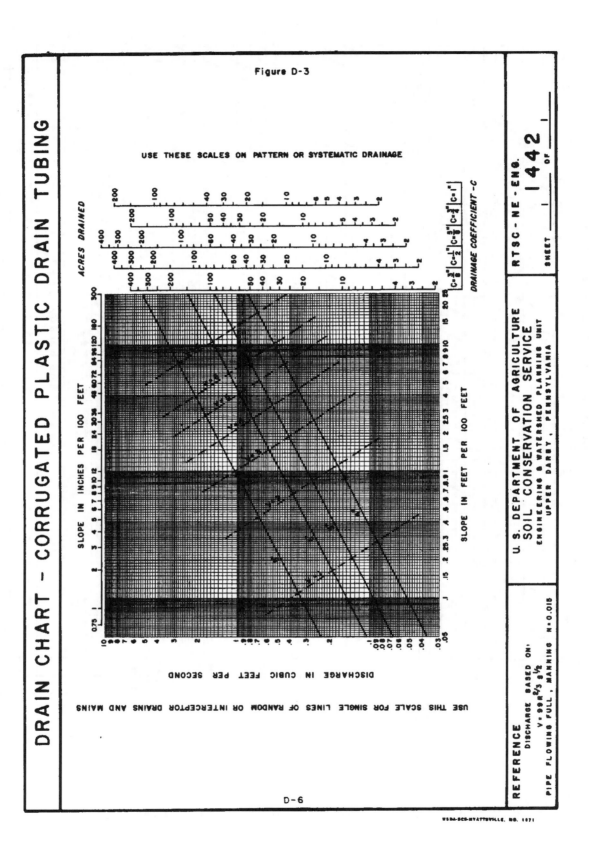

Figure 13.22. Drain chart—corrugated plastic drain tubing.

SAMPLE PROBLEMS

Tile Size Selection

Example 1

Problem

To select the size of pipe required for the following conditions:

a) Discharge 100 gpm from a water-cooled condenser of a refrigeration system to a sewer line (or draining swimming pool).

b) Slope of the line = 1%, for slope = 0.05%.

Solution

a) Convert gpm to cfs

$$\text{cfs} = \frac{\text{gpm}}{60 \ (\text{min to sec}) \times 7.5 \left(\text{gal to ft}^3\right)}$$

$$\frac{100}{60 \times 0.75} = 0.22 \ \text{cfs}$$

b) From Tile Drainage Design Chart, find point of intersection of (0.22 cfs, 1% slope)
Read—4-in. discharge pipe required, and (0.22 cfs, 0.05% slope) = 5-in. pipe.

Example 2

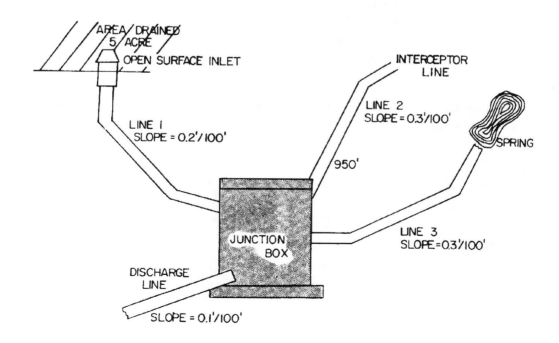

Figure 13.23

PRACTICE PROBLEMS

Problem 1

Given: type of crop = turf grass
       type of soil = mineral
       drainage coefficient—Use DC = 3/8
       percentage slope of conducting drain = 0.2%
       acres to be drained = 12 ac

Note: Use Tile Drainage Chart on page 299.

Find: 1. size of conducting drain = __6__ in.
     2. cfs = __0.19__ cfs
     3. velocity = __1.4__ fps

Problem 2

Given: Same conditions as above except % slope of conducting drain = 0.5%.

Find: 1. size of drain = __5__ in.
     2. cfs = __0.19__ cfs
     3. velocity = __1.90__ fps

Problem 3

As % slope increases:

    1. the tile size increases, or decreases, or stays the same?
    2. the cfs increases, or decreases, or stays the same?
    3. the velocity increases, or decreases, or stays the same?

Problem 4

Determine the maximum number of acres that can be drained for the following pipe sizes and drainage coefficients. Use a 0.25% slope:

    4-in. tile - DC = 1/4 = __11__ ac
                   DC = 1/2 = __6__ ac
    5-in. tile - DC = 1/4 = __20__ ac
                   DC = 1/2 = __10__ ac
    6-in. tile - DC = 1/4 = __32__ ac
                   DC = 3/8 = __22__ ac

Problem 5

Draining flow from a spring

Given: See sketch below

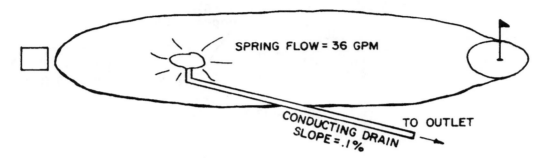

Figure 13.24

Find: 1. cfs = __0.08__ cfs
2. size of conducting drain = __5 in.__

Problem 6

Determine the drain line size for lines 1, 2, and 3.

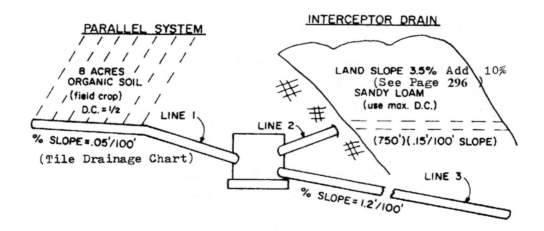

Figure 13.25

	cfs	Pipe Size (in.)
Line 1	0.170	8
Line 2	0.206	6
Line 3	0.380	5

Problem 7

Determine the drain line size for the sandy loam and the silt loam sections of the field.

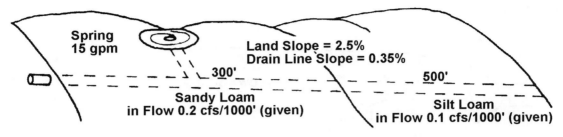

Figure 13.26

Note: $\text{cfs} = \dfrac{300 \text{ ft}}{1000 \text{ ft}} \times 0.2 + 10\%$

Find: 1. spring inflow = 0.033
   2. sandy loam (300 ft) inflow = _0.066_ cfs
   3. silt loam (500 ft) inflow = _0.055_ cfs
        Total inflow = _0.15_ cfs
   4. drain line size = _5_ in. dia.

Problem 8

Green Drainage
   Assume the green shown below was to be drained.

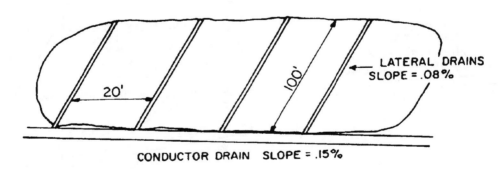

Figure 13.27

To assure rapid drainage after a heavy downpour, a drainage coefficient of 8 in. for 24 hr (DC = 8 in.) was selected. Solve the following: (*Note:* 27,154 × 8 = 217,232 gal/24 hr)

$$\frac{217{,}232}{24 \times 60} = 150.86$$

Find: 1. green acreage = _0.18_
     2. gallons per minute for 1 ac = _____
     3. gallons per minute for green = _27.2_
     4. gallons per minute for 1 lateral drain = _6.80_
     5. cubic feet per second for 1 lateral = _0.015_
     6. cubic feet per second for conductor drain = _0.06_
     7. lateral pipe size = _4 in. min. size_
     8. conductor pipe size = _4 in._

Problem 9

Green and approach measured about 22,000 ft². (Use 1/2 ac.)

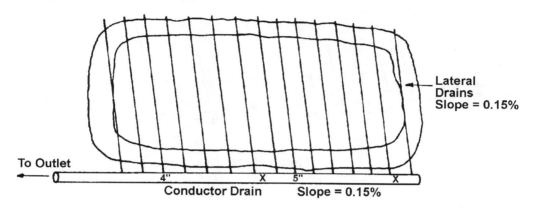

Figure 13.28

For the golf course green, use DC = 8.

Find: 1. total gallons for 1 ac = _217,232_ 24 hr
     2. total gallons for green (1/2 ac) = _108,616_ 24 hr
     3. gallons per hour (green) = _4526_ hr
     4. cubic feet per second inflow to conducting drain = _0.17_ cfs
     5. cubic feet per second inflow per lateral = _0.011_ cfs
     6. What is the maximum cfs:
        for a 4-in. tile—0.15% slope _0.08_ cfs = 7.25 lateral?
        for a 5-in. tile—0.15% slope _0.16_ cfs = 0.11 = 14.5 lateral?
     7. On the sketch, indicate by an ⊗ where the conduction drain size should increase from 4 in. to 5 in. and from 5 in. to 6 in.

Problem 10

Fairway Drainage. (Use Drain Chart for *Corrugated Plastic Tubing* on page 300.)

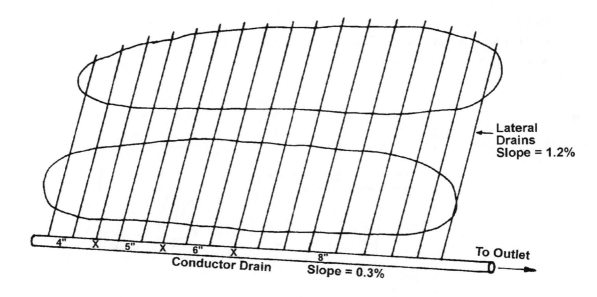

Figure 13.29

Given: drainage coefficient DC = 1
lateral length = 300 ft
lateral spacing = 100 ft

Find: 1. total acres being drained = __11.70__ ac
2. acres drained per lateral = __0.67__ ac
3. total gallons inflow into conductor drain per 24 hr = __317,702__
4. total gallons per lateral per 24 hr = __18,686__
5. cubic feet per second for conductor drain = __0.49__
6. cubic feet per second for lateral = __0.03__
7. lateral drain size = __4 in.__
8. By ⊗ indicate where conductor drain size increases from 4 in. to 5 in. and from 5 in. to 6 in.
   Note: max. inflow:

$$4 \text{ in.} = \frac{0.09}{0.03} = 3 \text{ laterals}$$

$$5 \text{ in.} = \frac{0.17}{0.03} = 6 \text{ laterals}$$

$$6 \text{ in.} = \frac{6.10}{0.03} = 9 \text{ laterals}$$

$$8 \text{ in.} = \frac{0.58}{0.03} = 19 \text{ laterals}$$

GROUNDWATER

Water present in the ground may be confined, free, or perched. Confined water usually presents no problem. Both the free and perched groundwater, however, are subject to capillary action in which water rises toward the surface, thereby changing the characteristics of the subsoil. This in turn reduces the soil's load carrying capacity and may cause frost heaving. Also, accumulation of water may cause subgrade softening.

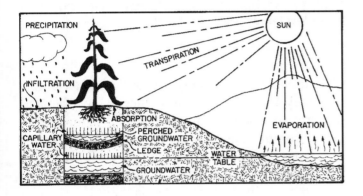

Figure 13.30

Water Table

The water table may be defined as the upper surface of the saturated zone of free, unconfined groundwater. The soil moisture content for a significant height above a water table is substantially greater than field capacity. For this reason, plant root growth is affected by a water table much more than the height of the water table alone indicates.

Another important feature of water tables is their fluctuation, both seasonal and short-term. Water tables are seldom static. They respond to additions and depletions of groundwater from natural or artificial causes. Sources such as distant influent seepage from precipitation and stream flow are seasonal, and their effects on the wet area may be delayed for months or even years. Direct precipitation and irrigation-percolation wastes, of course, may change the water table height almost immediately.

The Water Table and the Capillary Fringe

Close to the water table, the quantity of capillary water held in a granular material is greater than field capacity. The amount of water held at a given point depends on the distance above the water table, as well as on the soil pore sizes and shapes. This form of capillary water is sometimes called fringe water. Just above the water table, fringe water completely fills the capillary pores.

Level water tables occur in floodplains, level sandy shoreline or swampy areas. The groundwater has little or no flow, but its level can rise or fall with the seasons, and with either rainfall or drought.

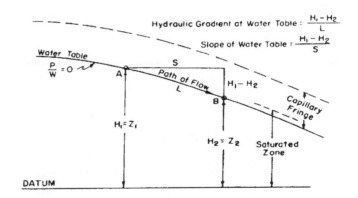

Figure 13.31

Hydraulic Gradient

The hydraulic gradient is defined as the difference in hydraulic head at two points divided by the distance (L) *measured along the path of flow,* as illustrated in Figure 13.31.

Difference Between Hydraulic Gradient and Slope of the Water Table

Subdrainage is accomplished by placing, below the water table, an artificial channel (drain line) in which the hydraulic head is less than it is in the soil to be drained. (See Figure 13.32.)

Note: Holes face downward when installing a drain line system.

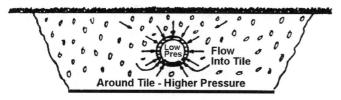

Figure 13.32

Drain Line Drawdown

Two factors determine the rate at which water moves toward the drain; these are directly related to *depth* and *spacing* of drain lines.

1. The potential for flow increases as the hydraulic head increases, i.e., *depth of drain line.*

2. The potential for flow decreases by distance along the path of flow, i.e., *spacing of drain line.*

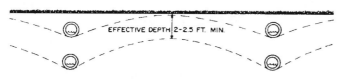

Figure 13.33

In uniformly structured soils, the drawdown is directly related to tubing depth. The principal limitation on depth is governed by the economics of deeper installation vs. less depth with narrower spacing. Only in exceptional cases would a depth in excess of 6 ft be recommended.

Spacing

The drawdown curves, and hence the optimum spacing and depth, are directly related to the *permeability* of the soil in question. In addition, the various strata and barriers present must be considered. (See Figure 13.34.)

Permeability, Hydraulic Conductivity, and Percolation Rate

This is the rate of moisture movement through the soil under saturated conditions.

Permeability, hydraulic conductivity, and drainability are all practically synonymous, and are

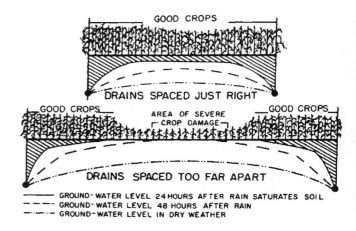

Figure 13.34. How spacing of tile drains affects ground-water level and crop damage.

often used interchangeably. Regardless of terminology, they are of utmost importance in any drainage investigation and are directly related to an intelligent recommendation of spacing, depth, and sizing of various drain lines in a system. (See Table 13.2.)

When a system of parallel, lateral drain lines is used, the spacing between laterals should be according to the local drainage guide for the applicable soil. Table 13.3 can also be used as a general guide.

The design engineer should specify the spacing for individual sites within the above limits based on soil conditions, turfgrass species, and local experiences.

DRAINAGE FOR GOLF COURSES— RECREATIONAL AREAS

An exception to the basic water removal rule is the depth used for drains in recreational area drainage. Because of the shallow-rooted characteristics of turfgrasses, drainage is needed only in the top foot of soil. This, plus the need for rapid water removal, *dictates a drain depth of 1 to 2 ft, with close spacing of lines.*

Minimum Depth

a) *Mineral Soils.* The designer should recognize that turfgrass growth improves as the drain line depths

Table 13.2

Soil Separate	Diameter (range in mm)	Permeability	Percolation Rate (in./hr)
Clay	Below 0.002	Very slow	Less than 0.05
Silt	0.05–0.002	Slow	0.05–0.2
Very fine sand	0.10–0.05	Moderately slow	0.2–0.8
Fine sand	0.25–0.10	Moderate	0.8–2.5
Medium sand	0.5–0.25	Moderately rapid	2.5–5.0
Coarse sand	1.0–0.5	Rapid	5.0–10.0
Very coarse sand	2.0–1.0	Very rapid	10.0 and over

Table 13.3

Soil	Permeability	Spacing Range
Clay and clay loam	Very slow	30 ft to 70 ft apart
Silt loam	Slow to moderately slow	60 ft to 100 ft apart
Sandy loam	Moderate to rapid	100 ft to 300 ft apart
Muck	—	80 ft to 100 ft apart
	(Space at Maximum Initially)	

Golf Course Greens and Recreational Areas, See Drainage for Golf Courses, page 308.

is increased from 24 to 42 in. and more. It is therefore desirable for the depth of laterals to average 3 to 4 ft. This may be reduced to 2.5 ft with laterals spaced approximately 3 ft apart, under the following conditions:

1. for slowly permeable soil
2. where there is a layer of extremely tight soil, sand, or large stones that prohibits greater depth
3. in depressional or impounded areas
4. where outletting is a problem
5. where usage conditions require rapid drainage of upper horizons of the soil profile.

The very minimum of soil cover to protect the drain line from breakage by heavy machinery should be 2 ft of soil over the drain line.

b) *Organic Soils.* Organic soils should not be drained until after initial subsidence has occurred. To produce initial subsidence, open ditches should be constructed in deep muck and peat load to carry off free water and the area should be allowed to stand or be partially cultivated for a period of from 3 to 5 years before installing drain lines. Subsidence will continue after the initial stage and may be as much as 0.1 ft/yr due to oxidation under cultivation, compaction, shrinkage and wind erosion. Therefore, the minimum working depth of drain line should not be less than 4 ft. Control drainage should always be given consideration. The subsidence rate increases as the average depth to the water table during the growing season increases. The very minimum of cover to protect the drain line in organic soils from breakage by heavy machinery should be 2.5 ft of soil over the drain lines. If control drainage is not provided to hold subsidence to a minimum, this depth of cover should be increased to 3 ft.

c) When it is impossible or uneconomical to secure the minimum cover specified in (a) and (b) above under waterways and road ditches, at the outlet end of the main, or near structures, the drain lines should be replaced with metal pipe or other high-strength continuous rigid pipe.

d) Mains should be laid deep enough so that laterals can be joined at the approximate center of the main drain line.

SUBDRAIN MATERIALS

Materials for Drain Lines

For many years after subsurface drainage was first introduced in the United States, most of the drains were made of ceramic tile. *Tile drain* became the most common term used to refer to a subsurface drain, and it is still used in many parts of the country to refer to subsurface drains manufactured from all kinds of materials.

Materials now commonly used for drains include ceramic tile, concrete, bituminized fiber, plastics, asbestos cement, aluminum alloy, and steel. Some of the products manufactured from these materials are made in such a way that drainage water will enter the conduit through the joint between adjacent sections of the drain, and some are made with perforations through which water enters the drain.

Factors involved in the selection of the drain to be used include: climatic conditions, chemical characteristics of the soil, depth requirements, and installation cost. In northern areas freezing and thawing conditions must be considered in selecting the drain. Where acids or sulfates exist in the soil or drainage water, drains that will resist these conditions must be selected. The method of installation and the depth requirements influence the selection of the type and strength of the drain. The drain must sustain the loads to which it will be subjected. The cost of the drain—including the cost of transportation from the manufacturer to the place of installation—is a big factor in selection of drains.

Standards and Specifications

Specifications covering the physical requirements and testing methods for use in quality control of drains have been developed by several organizations and groups interested in standardization and maintenance of high quality for manufactured products. These include specifications and standards of the American Society for Testing and Materials (ASTM); product standards published by the National Bureau of Standards, U.S. Department of Commerce; and Federal Specifications of the Federal Supply Service, General Services Administration.

These specifications and standards are revised as necessary to reflect improved methods of manufacture, new materials, and changed requirements for use of the product. The latest revision, which is indicated by its date, should be used. Specifications for materials which are approved for use in installations for which the Soil Conservation Service is technically responsible are listed in current Engineering Practice Standards of the Soil Conservation Service.

Newly developed products, for which specifications are written by the manufacturer and approved by the Soil Conservation Service, may also be used pending development of standard specifications by one of the organizations listed above.

Clay Drain Tile

Clay tile may fail due to the freezing and thawing reversals in northern areas, where frost penetrates the ground to the depth of the tile or where the tile is stockpiled on the ground during the winter before being installed. Damages due to such exposure may be serious. The absorption of water by tile is a good index of its resistance to freezing and thawing. Tile manufactured from shale usually has lower absorption rates and is more resistant to freezing and thawing than tile made from surface clays. The quality of clay tile is also dependent upon the manufacturing process. Many plants use de-airing equipment, which increases the density of the extruded material. Generally, color and salt glazing are not reliable indicators of the quality of tile. Clay tiles are not affected by acids or sulfates. Low temperatures normally will not affect the use of clay tile, provided the tiles are properly selected for absorption and care is taken in handling and storage of the tile during freezing weather.

Concrete Drain Tile

Concrete tile of high density is not affected by freezing and thawing reversals and freezing temperatures. It will be adversely affected, however, if exposed to the action of acids or sulfates unless the tile has been manufactured to meet these situations. Concrete tile is not recommended for extremely acid or sulfate conditions.

Bituminized Fiber Pipe

Bituminized fiber pipe may have a homogeneous or laminated wall structure. The pipe comes in various lengths from 4 to 20 ft. The pipes are connected by tapered couplings to form a watertight joint, or with a sleeve coupling for butt joints. This pipe, without perforations, may be used to protect the outlet of drains from other material. However, the out-of-bank projection of bituminized fiber pipe must be kept short—not more than 3 to 4 diameters in length—to maintain its shape against softening and collapse by exposure to the heat of the sun. Precautions must also be taken against any point loading against the pipe walls by a stone or buried log, which could cause failure by cold flow of the bituminous material.

CORRUGATED PLASTIC POLYETHYLENE DRAIN LINE

The use of corrugated plastic polyethylene tubing for subsurface drainage has increased rapidly since 1967. (See Figure 13.35.)

Advantages

Ease of handling. The material is light, ranging in weight from 72 to 85 lb for a 250-ft roll of 4-in. tubing. It is flexible and requires less labor for handling during installation than concrete or clay drain tile.

lengths
6-in. diameter

drain guard
2,4,6,8-in. diameter

coil
2,3,4,5,6-in. diameter

sections
10,12,15,18-in. diameter

Figure 13.35

Better alignment in unstable soils. The long lengths of plastic tubing result in better alignment in unstable soils such as mucks, peats, and wet sands. Joint openings are eliminated.

Ease of hauling on soft fields. The light weight of plastic tubing makes it easier to distribute material on soft, wet fields.

Lower cost installation. The use of plastic tubing may result in lower total cost of drainage systems.

Disadvantages

Reduced strength at high temperatures. The temperature of plastic tubing can reach 120° to 140°F when strung out in a field on a hot, bright day. The strength of 4-in. tubing is reduced 50% when the temperature is raised from 70° to 120°F. For this reason, precautions must be taken to prevent the impact of sharp, heavy objects or excessive pull on the tubing during installation. The tubing regains strength when its temperature returns to that of the trench bottom. The relative strength of 4-in. tubing at various temperatures is shown in the following table.

Temperature (°F)	Relative Strength (%)	Percent Change from 70°F
35	170	+70
50	139	+39
70	100	0
85	79	−21
100	62	−38
120	49	−51
140	43	−57

Reduced flexibility at low temperatures. Although the strength of the tubing increases as the temperature is decreased, flexibility decreases. If the tubing is rapidly uncoiled at low temperatures, it may be stressed excessively and crack.

Check with the manufacturer concerning their recommendation for handling the material under either hot or cold conditions.

Reduced strength by stretching. Stretch that may occur during installation will cause some decrease in strength. This stretch is influenced by the temperature of the tubing at the time it is installed, the amount and duration of drag encountered when the tubing feeds through the installation equipment, and the stretch-resistance characteristics of the tubing. In general, increased stretch is associated with low-strength tubings. Stretch should not exceed 5%. The relative strength of 4-in. tubing at 73°F for various stretch lengths is shown here.

Percent Stretch	Relative Strength (%)	Percent Decrease from % Stretch
0	100	0
2	95	5
4	91	9
5	89	11
6	87	13
8	83	17
10	76	24

Flotation in water. Plastic tubing will float in shallow depths of water. Once the tubing floats during installation, it is difficult to get backfill material around and over the tubing without getting the material underneath the tubing, causing misalignment. The tubing should be held in place and covered immediately when water is present.

Sizes Available

Tubing commonly used in turf drainage is 4 in. in diameter or larger. The smaller sizes are available in coils of various lengths, depending on the diameter of the tubing. The larger sizes are produced in short, straight sections. Tubing smaller than 4 in. is available for special applications.

Strength Requirements

In contrast to rigid materials such as clay and concrete drain tile, corrugated plastic drain tubing is a flexible material. Failures occur in rigid material by cracking, and in flexible material by deflection or collapse. Flexible plastic drain tubing gains most of its soil load-carrying capacity by support from the soil at the sides of the tubing. A load on the top of the tubing causes the sidewalls to bulge outward against the soil, as shown in the figure below. This

soil resists the bulging, and its effect is to give flexible tubing more load-carrying ability. Plastic drain tubing must have sufficient strength to withstand the soil load without excessive deflection, collapse on the top, or failure of the sidewalls.

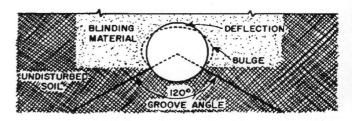

Figure 13.36

How Do Corrugations Affect Flow?

A 4-in. corrugated plastic drain line can carry about 75% as much water as a well-aligned 4-in. clay or concrete drain tile on the same grade. This reduction in capacity is not significant for most lateral lines, since they seldom flow full. Four inch laterals spaced 50 ft apart at a 0.1% grade seldom flow full unless they are more than 2,000 ft long. Main lines are designed for full flow, however, and the reduction in flow due to corrugations must be considered.

SPECIAL TECHNIQUES NEEDED TO INSTALL CORRUGATED PLASTIC TUBING

Groove. A specially shaped groove is required in the trench bottom. This groove provides improved alignment and side support against bulging. It can be cut or formed in a number of ways. A typical groove giving 120° support is shown Figure 13.36. A gravel envelope is recommended when it is impossible to form a groove.

Blinding. In *cohesive,* well-structured soils, water enters freely and there is no sediment inflow. Soil aggregates and large soil particles bridge across drain openings. The practice of blinding with topsoil is designed to put stable soil next to the drain for this purpose. For naturally stable soils, no envelope is needed, but for *non-cohesive,* structureless, fine-grain soils, envelopes are required.

Blinding and backfilling. Proper blinding and backfilling of plastic drain tubing are essential. Granular topsoil for blinding around and 4 in. above the tubing is recommended to permit free water movement into the drain. Some soil compaction on both sides of the tubing is desirable to provide good side support, which will reduce bulging. For spatial conditions, coarse sand to fine gravel are good blinding materials.

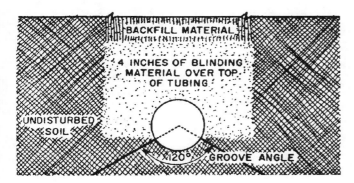

Figure 13.37

No stones or other hard objects should be allowed in the blinding material, since they apply point loads and may cause the tubing to collapse. Blinding holds the tubing in place during backfilling to ensure proper alignment, and provides protection for the tubing during the backfilling operation, when the impact of rocks and hard clods could damage it. Even with good blinding, large stones or clods and heavy loads should not be dropped over the tubing.

A number of different blinding methods have proved acceptable. Some contractors place such importance on blinding that they apply selected material around and over the tubing by hand. Tubing should be held in place in the trench until secured by blinding. This operation is especially important when water is in the trench and when the air temperature is below 45°F. When the tubing feels warm (98°F), delay backfilling until the tubing temperature reaches the soil temperature.

Envelopes in Unstable or Non-Cohesive Soils

In unstable or non-cohesive soils where a groove cannot be cut into the bottom of the trench, an en-velope must be used. Conditions where envelopes or stabilizing materials are needed include: (1) soils that easily fill a drain with sediment, such as fine and medium sands in the size ranges of 0.05 to 1.0 mm; (2) soils that do not provide a stable foundation, such as saturated sands in a quick condition; and (3) soils that tend to seal or clog drain openings and limit water entry into the drain.

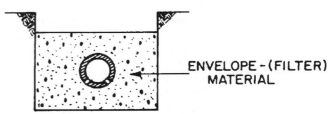

Figure 13.38

A filter, as it is commonly defined, will be doomed to failure over a period of time, as the voids in the filter material around the drain filled and a layer of fine sediment (filter cake) builds up. This seals off the flow of groundwater.

Filter

The term *filter,* as applied to soil mechanics (drainage), is the placement of permeable materials around drains for the following purposes:

1. To prevent fine-grained particles in the surrounding soil from being carried into the drain with the groundwater.
2. To allow an unrestricted flow of groundwater to enter the drain.

A drain is not successful unless both of these conditions are satisfied.

Envelope

An *envelope* is the placement of a permeable material around drains for the following purposes:

1. To improve the flow of groundwater into the drain.
2. To improve the bedding conditions by stabilizing the soil around the drain.
3. To maintain alignment of the drain.

Although a difference in the purposes of a filter and an envelope does in fact exist, in practice they are frequently used interchangeably among soil engineers.

Filter Materials

Studies have been made to determine the filter medium requirements to prevent infiltration of fines and filter media into underdrain pipes. One of the well-known and accepted studies is that conducted by the U.S. Waterways Experiment Station at Vicksburg, Mississippi. Conclusions reached were: Material approximately the same size and grade as concrete sand (AASHO Spec.) can be used as a filter medium, and is in fact recommended by many authorities.

Design of Sand and Gravel Envelopes

One of the basic functions of an envelope is to improve the permeability in the zone around the drain line. For this reason the first requirement is that the envelope material have a permeability higher than that of the base material. All of the envelope material should pass the 1.5-in. sieve, 90% of the material should pass the 3/4-in. sieve, and not more than 10% of the material should pass the No. 60 sieve. The graduation of the envelope material is not important, as it does not serve the purpose of a filter. Where sand and gravel filters are used, the thickness of the filter material around the drain should be 3 in. or more. (Figure 13.39A.)

In those cases where the top of the drain is covered with a plastic strip, the requirement for a filter above the drain may be waived. This design will reduce the amount of filter material required, and may be the most economical in areas where sand and gravel are expensive. Experience to date indicates that this is the best type of filter installation, as shown in Figure 13.39B.

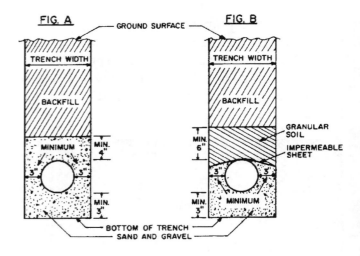

Figure 14.39

Determination of Need for Filters and Envelopes

In designing a drainage system, one of the first considerations is to determine if a filter is needed. Referring to the table below, it will be noted that soils are separated into three groups (Unified Classification System) according to the need for a filter. The first group always needs a filter, the second group may or may not need filters, and the third group seldom needs filters, except under unusual conditions.

Determination of the need for envelopes must be secondary to that for filters, as filters often will serve the same purpose. The need for an envelope should be considered in those cases where a filter is not specified, and where flexible pipe is used for the drain. Referring to Table 13.4, it will be noted that envelopes may be used in all cases where filters are not recommended.

Table 13.4. A Classification to Determine the Need for Drain Filters or Envelopes, and Minimum Velocities in Drains

Unified Soil Classification	Soil Description	Filter Recommendation	Envelope Recommendation	Recommendation for Minimum Drain Velocity
SP (fine)	Poorly graded sands, gravelly sands.	Filter needed	Not needed where sand and gravel filter is used, but may be needed with flexible drain tubing and other type filters.	None
SM (fine)	Silty sands, poorly graded sand-silt mixture.			
ML	Inorganic silts and very fine sands, rock flour, silty or clayey fine sands with slight plasticity.			
MH	Inorganic silts, micaceous or diatomaceous fine sandy or silty soils, elastic silts.			
GP	Poorly graded gravels, gravel-sand mixtures, little or no fines.	Subject to local on-site determination.	Not needed where sand and gravel filter is used, but may be needed with flexible drain tubing and other type filters.	With filter—none Without filter—1.40 ft/sec
SC	Clayey sands, poorly graded sand-clay mixtures.			
GM	Silty gravels, poorly graded gravel-sand silt mixtures.			
SM (coarse)	Silty sands, poorly graded sand-silt mixtures.			
GC	Clayey gravels, poorly graded gravel-sand-clay mixtures.	None	Optional. May be needed with flexible drain tubing.	None—for soils with little or no fines. 1.40 fps for soils with appreciable fines.
CL	Inorganic clays of low to medium plasticity, gravelly clays, sandy clays, silty clays, lean clays.			
SP,GP (coarse)	Same as SP and GP above.			
GW	Well-graded gravels, gravel-sand mixtures, little or no fines.			
SW	Well-graded sands, gravelly sands, little or no fines.			
CH	Inorganic, fat clays.			
OL	Organic silts and organic silt-clays of low plasticity.			
OH	Organic clays of medium to high plasticity.			
Pt	Peat			

Source: U.S. Department of Agriculture, Soil Conservation Service, Engineering Division—Drainage Section

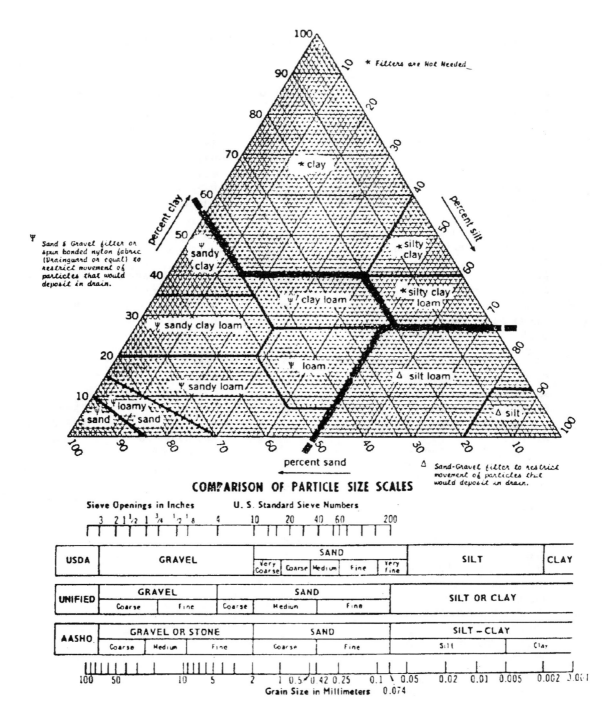

Figure 13.40. Guide for use of spun, bonded nylon fabric to protect corrugated drainage tubing.

GOLF COURSE GREENS

General

For an ideal round of par 72, about 36 shots are made on the green and about 18 shots to the green. The putting green figures into about three-quarters of the total strokes. It is little wonder why golfers demand excellent turf and playing surfaces on these tremendously important areas.

The superintendent should have a good understanding of green construction; this is basic to good drainage practices.

Good drainage on greens is essential: (1) to allow play to continue uninterrupted from puddles of water or from soggy greens, (2) to provide good aeration, which will encourage deep roots and aid in preventing root diseases.

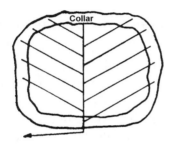

Figure 13.41a. Herringbone pattern.

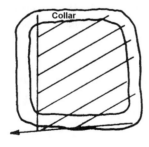

Figure 13.41b. Gridiron pattern.

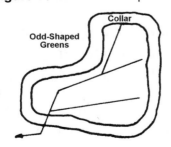

Figure 13.41c. Random pattern.

Drainage

Drain lines of at least 4-in. diameter should be so spaced that water will not have to travel more than 10 ft to reach a drain line. Any suitable pattern or drain line arrangement may be used, but the herringbone, the gridiron, or the random arrangements will fit most situations.

Cut ditches or trenches into the subgrade so the drain lines slope uniformly. Do not place drain line deeper than is necessary to obtain the desired amount of slope. Drain lines should have a minimum fall of 0.5%. Steeper grades can be used, but there will seldom be a need for drain line grades steeper than 3% to 4% on a putting green.

Drain line may be agricultural clay tile, concrete, plastic, or perforated asphalt paper composition. Agricultural tile joints should be butted together with no more than 0.25 in. of space between joints. The tops of tile joints should be covered with asphalt paper, fiberglass composition, or with plastic spacers or covers designed for this purpose. The covering prevents gravel from falling into the tile.

Drain line should be laid on a firm bed of 0.5 in. to 1 in. deep gravel or crushed stone to reduce possible wash of subgrade soil up into the drain lines by fast water flow. If the subgrade consists of undisturbed soil, so that washing is unlikely, it is permissible to lay the drain lines directly on the bottom of the trench.

After the drain line is laid, the trenches should be backfilled with gravel, and care should be taken not to displace the covering over the joints.

Specifications* for Green Construction and Drainage

The specifications for a method of putting green construction and drainage, as reported by the USGA Green Section Staff in the March–April, 1993 issue of the USGA Green Section Record. (See Figure 13.42.)

* Complete information and details for putting green construction may be obtained by writing to: Golf House, Far Hills, NJ 07931.

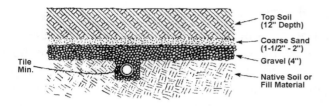

Figure 13.42. Profile of a properly constructed green.

USGA Standard Specifications for Drainage of Golf Course Greens:

Subgrade:
- The subgrade need not conform to the proposed finished grade; the purpose of the subgrade is to facilitate water movement to the drainage system. However, the surface of the gravel drainbed layer must conform to the proposed finished grade.

Drainage:
- Drainage trenches shall be a minimum of 8 in. (20 cm) deep.
- Drainage lines shall be not more than 15 ft (5 m) apart.
- A perimeter drain line (smile drain) shall be installed along the low end of the gradient, usually along the front of the green.

Gravel drainbed:
- Both crushed stone and pea gravel are acceptable for use as the drainage bed.
- Gravel materials suspected of lacking mechanical stability to withstand common construction traffic should be checked with the LA Abrasion test—ASTM procedure C-131 (value should not exceed 40).
- Gravel materials of questionable weathering stability should be tested using the sulfate soundness test—ASTM procedure C-88 (not more than 12% loss by weight).
- Slight change in particle size distribution for gravel where the intermediate layer is used. Previously, 100% required between 3/8 in. and 1/4 in. (9 mm and 6 mm); now minimum 65% required between 3/8 in. and 1/4 in., plus limits on percentages and sizes above and below this range.

Intermediate layer:
- The acceptable particle size range has been expanded. Now, 90% must fall between 1 mm and 4 mm (previously 1–2 mm).
- Need not be included if the properly sized gravel is used.

Particle Size Description of Gravel and Intermediate Materials

Material	Description
Gravel: Intermediate layer is used	Not more than 10% of the particles greater than 1/2 in. At least 65% of the particles between 1/4 in. and 3/8 in. Not more than 10% of the particles less than 2 mm
Intermediate layer material	At least 90% of the particles between 1 mm and 4 mm

Gravel Drainbed and Coarse Sand Base

The entire subgrade should be covered with a layer of clean washed gravel or crushed stone to a minimum thickness of 4 in. The preferred material for this purpose is washed gravel of 1/4 in. to 3/8 in. diameter particle size. If particles of any other size are present, they should be screened out. *This is important for the proper functioning of the perched hydration zone.*

A 1-1/2-in. layer of coarse sand is then spread over the entire gravel base. This sand should be within a range of five to seven diameters of the gravel. In other words, if 1/4-in. gravel (about 6 mm) is used, then the particles of the overlying layer of sand should not be less than 1 mm in diameter. In order to prevent movement of the sand into the gravel, the maximum allowable discrepancy appears to be five to seven diameters.

Fairway Drainage—Isolated Wet Spots or Small Simple Projects

Particularly in the isolated wet areas that affect play or traffic—as in front of greens, in bunkers and on tees, the excess water can be removed by installing a good drainage system. The following system and list of guidelines is offered.

Checklist

1. Determine outlet. Water must flow out of the line. The drain line should never be dead-ended.
2. Intercept lines must be installed across the slope of the land, or perpendicular to flow.
3. Drain lines must run downhill or downgrade.
4. Dig the drainage ditch—12 in. wide and 18 in. to 24 in. deep. Remove the spoils from the site as digging progresses.
5. Backfill as shown.

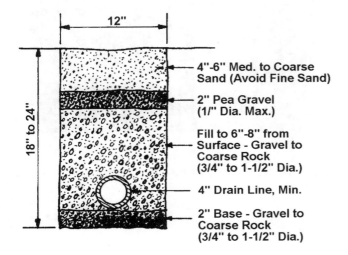

Figure 13.43

6. Do not sod over drainage lines. Allow the scar to gradually cover over, with grass growing in from both sides. The drain lines are more effective if allowed to heal.
7. It is critical to resod immediately. Use only sod with a sand base. Avoid clay-silt base sod that may seal off drainage from the surface.
8. Rope off the area. Keep mowers and cart traffic off until the backfill has stabilized.

Manual Grade Control

Manual grade control with visual sight bars and targets can be used for normal trenching operations. Stakes to govern alignment, direction, and grade should be set 30 m (50 ft) or less on curves for all drain lines to be constructed. They should also be set at all intersections of mains and points of grade change. Grade stake elevations should be determined *with a level.*

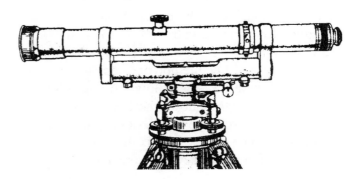

Figure 13.44

Surveyor's Level—For Placement of Drain Lines

Particularly when the grade approaches the lower limits of 1.0% or less, the surveyor's level should be used. For short runs, as on golf course greens, it could be used either directly or for establishing a measuring line (a taut string between two stakes) set to grade. The depth of the trench, preparation of the trench bottom, and the placement of the drain lines can be accurately gauged with a measuring stick or surveyor's rod.

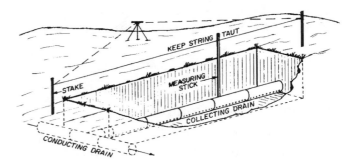

Figure 13.45

Grade Stake Elevations

For long runs, grade stakes should be set every 50 ft if drain line is installed from a string line. If string is used, a weight should be tied to each end and hung over the support, so that the string is tight at all times. If the target method is used for a tiling machine, grade stakes can be set at 100-ft intervals. Always have three targets set ahead of the machine. Grade stake lines usually are offset from 2 to 4 ft from the centerline of the drain. A method of setting grade for a hand-dug trench is shown in Figure 13.46.

The gage-and-line method of establishing grade for a drain line drain is as follows: A cut of 3 ft 11 in. is indicated at the hub in the first station. Subtracting this cut from the length of the gage stick (6 ft), the difference of 2 ft 1 in. is the height for setting the top of the crossbar above the hubs. The gage stick at the first station shows that the trench has been dug to the correct grade elevation at that point. The gage stick at the second station shows that more digging is required for the cut of 3 ft 10 in.

Figure 13.47. A popular laser system shown prior to installation setup.

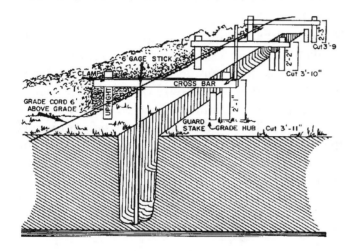

Figure 13.46

Automatic Grade Control

For high-speed trenching, a trencher utilizing optical and/or electronic devices (as lasers) capable of following the designated grade line is recommended.

Width of the Trench

To prevent overloading in deep and wide trenches, it is sometimes feasible to construct, with a trenching machine or by hand, a subditch in the bottom of a wide ditch which has been excavated by a bulldozer, dragline, steam shovel, or backhoe. The width of the subditch measured at the top of the drain line

Figure 13.48. A modern high-speed trencher. Two workers can operate this trencher, laying plastic tubing to grade.

determines the load, and the width of the excavation above this point is unimportant. Therefore, any feasible method may be employed for removing the material above the top of the drain line. (See Figure 13.49.)

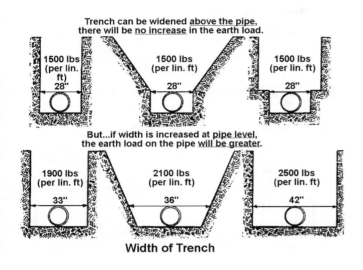

Trench can be widened <u>above the pipe</u>, there will be <u>no increase</u> in the earth load.

1500 lbs (per lin. ft) 28"

1500 lbs (per lin. ft) 28"

1500 lbs (per lin. ft) 28"

But...if width is increased at <u>pipe level</u>, the earth load on the pipe <u>will be greater</u>.

1900 lbs (per lin. ft) 33"

2100 lbs (per lin. ft) 36"

2500 lbs (per lin. ft) 42"

Width of Trench

Figure 13.49

PLANNING AN INSTALLATION

The Outlet

A starting point in planning a subsurface drainage system is the location of the outlet. Drain lines may outlet by gravity into natural or artificial channels or tile outlet mains. Any of these are suitable, provided they are deep enough and of sufficient capacity to take away all the drainage water from the drain line. *Before proceeding with the design of the system, the planner should be satisfied with the adequacy of the outlet.*

At the outflow end of the drain line, be sure the water it carries has someplace to go. A drain line should never dead-end. It must daylight, connect to another flowing line, dump into a large dry well, or in some way ensure the movement of water out of the line. (See Figure 13.50.)

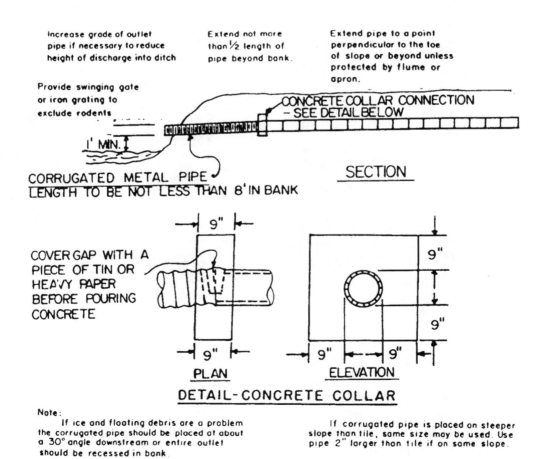

Increase grade of outlet pipe if necessary to reduce height of discharge into ditch

Extend not more than ½ length of pipe beyond bank.

Extend pipe to a point perpendicular to the toe of slope or beyond unless protected by flume or apron.

Provide swinging gate or iron grating to exclude rodents

CONCRETE COLLAR CONNECTION – SEE DETAIL BELOW

1' MIN.

CORRUGATED METAL PIPE LENGTH TO BE NOT LESS THAN 8' IN BANK

SECTION

COVER GAP WITH A PIECE OF TIN OR HEAVY PAPER BEFORE POURING CONCRETE

9"

9"

9"

9"

9"

9"

PLAN

ELEVATION

DETAIL-CONCRETE COLLAR

Note:
   If ice and floating debris are a problem the corrugated pipe should be placed at about a 30° angle downstream or entire outlet should be recessed in bank.

   If corrugated pipe is placed on steeper slope than tile, same size may be used. Use pipe 2" larger than tile if on same slope.

Figure 13.50. Tile outlet erosion protection.

DRAIN OUTLETS—EROSION PROTECTION

Capacity and depth of subsurface mains. When existing mains are used as the outlet, they should be in good condition, free of failures, and working properly. The main should have sufficient available capacity, based on the area served, to handle the proposed drain line system, and the depth should be sufficient to permit the new drain system to be laid at the depth specified and provide at least 2 ft of cover for protection. (See Figure 13.51.)

Outlet by pumping. Where a gravity outlet is not available, the possibilities of an outlet by pumping may be considered. This is a special problem that should be investigated by an engineer.

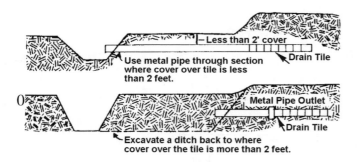

Figure 13.51. Grating end of tile outlet pipe.

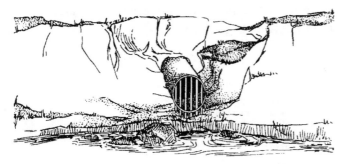

Figure 13.52. Methods for handling shallow depths at tile outlet.

2. *Surface inlets.* Surface inlets should be used in low areas where surface drainage cannot be provided. These must be properly constructed to prevent washouts and silting of the line. Surface inlets should be avoided whenever possible. If silt is a particular hazard, use a blind inlet, or place a silt trap immediately downstream from the inlet at a

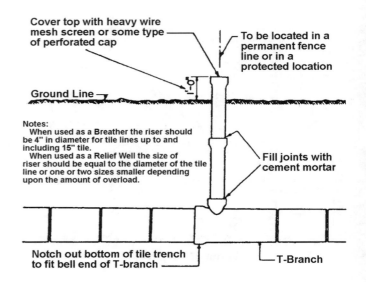

Figure 13.55

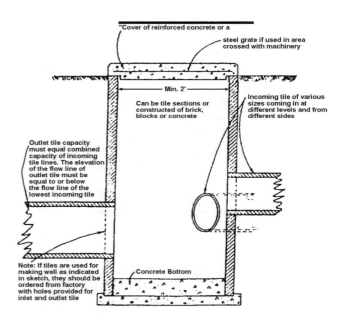

Figure 13.56

convenient location. It is good practice to locate the inlet on laterals or at least 20 ft to one side of the main, since inlets are a frequent source of trouble.

All open surface inlets shall be provided with some type of grating to exclude coarse debris.

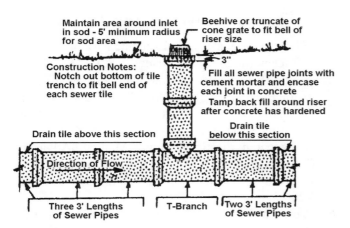

Figure 13.53. Surface inlet to tile line.

3. *Blind inlet.* If silt is a particular hazard, use a *blind inlet,* or place a silt trap immediately downstream from the surface inlet at a convenient location.

Figure 13.54. Blind inlet to tile line.

4. *Vents and relief valves.* Vents or breathers should be used to reduce vacuum in the line. Breathers should be used where the drain line changes abruptly from a flat grade to a steep grade. (See Figure 13.55.)

5. *Junction boxes.* Junction boxes should be used where more than two tile mains join, or where two or more laterals or mains join at different elevations. These boxes also serve as visual inspection ports to monitor for proper function. (See Figure 13.56.)

MISCELLANEOUS REQUIREMENTS FOR SUBSURFACE DRAINAGE SYSTEM

Alignment

All drain lines shall be laid true to line and grade. (See Figure 13.57.)

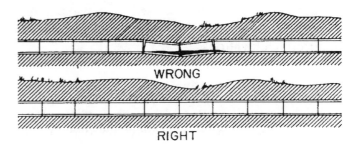

Figure 13.57

Changes in Horizontal Direction

There should be a gradual curve of the tile trench on a radius that the trenching machine can dig and still maintain grade. (See Figure 13.58.)

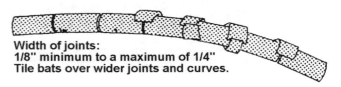

Figure 13.58

Manufactured bends or Y and T fittings should be used so that the change in direction is a smooth curve.

Note: Manufactured connections or branches for joining two drain lines should be used where available. If such connections are not available, the junction shall be chipped and fitted, and the connection sealed with mortar. Laterals should be connected into main drain lines at the approximate midpoint on handmade connections.

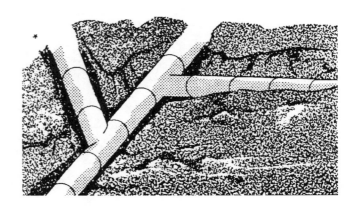

Figure 13.59

Junction boxes should be used when two or more mains or submains join, or when several laterals join at different elevations.

Crossing Waterways and Roads

Special precautions should be taken where drain lines are placed under waterways or roads. If the cover is 30 in. or less, they should either be encased in concrete or metal pipe, or extra-strength sewer tile with cemented joints should be used.

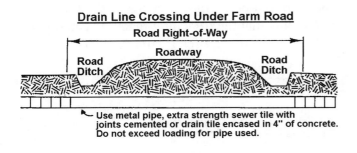

Drain Line Crossing Under Farm Road

Use metal pipe, extra strength sewer tile with joints cemented or drain tile encased in 4" of concrete. Do not exceed loading for pipe used.

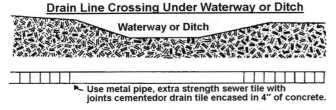

Drain Line Crossing Under Waterway or Ditch

Use metal pipe, extra strength sewer tile with joints cementedor drain tile encased in 4" of concrete.

Figure 13.60. Drain line crossing and shallow tile outlets.

ACKNOWLEDGMENTS

1. "Elements of Surveying." Technical Manual. TM 5-232, Department of the Army, U.S. GPO.
2. "Engineering Field Manual Handbook for Conservation Practices." USDA Soil Conservation Service.
3. "Drainage of Agricultural Land." USDA Soil Conservation Service.
4. "Corrugated Plastic Tubing." Manual. Advanced Drainage Systems.
5. "Agricultural Engineering Yearbook." ASAE, R260.2. pp. 408–416.
6. "Corrugated Plastic Drainage Tubing." Circular 1078. University of Illinois, Urbana–Champaign, College of Agriculture.

APPENDICES

Appendix 1

Standard Irrigation System Symbols

The following list of symbols shows those most often used on irrigation plans. They are intended to be used to streamline and clarify the interpretation of the plan. (*Source:* Buckner Irrigation Systems Design Manual)

Standard Irrigation System Symbols

Symbol	Description
WM	Domestic Water Meter
RP	Reduced Pressure Backflow Preventer
PT	Pressure Backflow Preventer
DC	Double Check Backflow Preventer
VB	Atmospheric Vacuum Breaker
VB	Manual Valve/Vacuum Breaker
MS	Moisture Sensor
PR	Pressure Regulator
FC	Flow Control Valve
	Check Valve
S 2	Strainer, Size Indicated
	Air or Pressure Relief Valve
DF	Drinking Fountain
POC	Point of Connection
P	Pump
HB	Hose Bibb
	Quick Coupling
	ADV Drain Valve, Automatic
DV	Drain Valve, Manual
	MAV-1-1/2 Valve, Manual Angle Size Indicated
1" A3	Valve, Automatic, Size and Designation Indicated
M C	Controller Master, Designation Indicated

Symbol	Description
⅝ 5	Controller, Satellite Designation Indicated
C 4	Controller, Standalone Designation Indicated
–DW–	Water Line, Domestic
–RW–	Water Line, Reclaimed
▬ ▬ ▬	Mainline, Existing, Size Indicated
▪▪▪▪▪	Lateral Line, Existing, Size Indicated
▬▬▬	Mainline, New, Size Indicated
———	Lateral Line, New, Size Indicated
▰▰▰	Wire Path, High and Low Voltage, Sizes Indicated
⌐===⌐	Pipe Sleeve, Existing
	Pipe Size, New
1 S-2	Detail Number with Sheet Number
	Emitter
○	Large Rotor, Pop-Up, Full-Circle
	Large Rotor, Pop-Up, Part-Circle
○●	Small Rotor, Pop-Up, Full-Circle
	Small Rotor, Pop-Up, Part-Circle
●○	Spray Head, Pop-Up, Full-Circle
▽∨	Bubbler, Pop-Up, Part-Circle
▽ ∜	Bubbler, Riser Mount

A.S.P.E. Standard Plumbing and Piping Symbols

Gate Valve			Side Branch Connection	
Globe Valve			Pipe End Cap	
Angle Valve			Plugged Tee	
Hose Valve			Concentric Reducer	
Swing Check Valve			Eccentric Reducer	
Non-Slam Check Valve			Union	
Backflow Preventer			Yard Cleanout	
Gas Cock, Gas Stop			Valve in Riser	
Solenoid Valve			Valve in Yard Box	
Pressure Reducing Valve			Expansion Joint	
Pressure Relief Valve			Flexible Connector	
Flow Switch			Cold Water	
Pressure Switch			Storm Drain	
Fire Hydrant			Sanitary Sewer	
Wye Strainer			Fire Water Supply	
Basket Strainer			Gas, Low Pressure	
Water Hammer Arrestor			Gas, Medium Pressure	
Auto Air Vent			Gas, High Pressure	
Hose Bibb			Industrial Waste	
Recessed Hose Bibb			Lawn Sprinkler Main	
Top Branch Connection			Lawn Sprinkler Lateral	
Bottom Branch Connection			Flow, In Direction of Arrow	

Note: These symbols are those suggested by the American Society of Plumbing Engineers. They are intended for your use in interpreting engineering drawings as you prepare the irrigation plans.

Appendix 2

Goulds Conversion Charts
(Technical Data)

English measures—unless otherwise designated, are those used in the United States.

Gallon—designates the U.S. gallon. To convert into the Imperial gallon, multiply the U.S. gallon by 0.83267. Likewise, the word ton designates a short ton, 2,000 pounds.

Properties of water—it freezes at 32°F, and is at its maximum density at 39.2°F. In the multipliers using the properties of water, calculations are based on water at 39.2°F in a vacuum, weighing 62.427 lb/ft^3, or 8.345 lb/U.S. gallon.

MULTIPLY	BY	TO OBTAIN	MULTIPLY	BY	TO OBTAIN
Acres	43,560	Square feet	Cubic inches	16.39	Cubic centimeters
Acres	4047	Square meters	Cubic inches	5.787×10^{-4}	Cubic feet
Acres	1.562×10^3	Square miles	Cubic inches	1.639×10^{-5}	Cubic meters
Acres	4840	Square yards	Cubic inches	2.143×10^{-5}	Cubic yards
Atmospheres	76.0	Cms. of mercury	Cubic inches	4.329×10^{-3}	Gallons
Atmospheres	29.92	Inches of mercury	Cubic inches	1.639×10^{-2}	Liters
Atmospheres	33.90	Feet of water	Cubic inches	0.03463	Pints (liq.)
Atmospheres	10,332	Kgs./sq. meter	Cubic inches	0.01732	Quarts (liq.)
Atmospheres	14.70	Lbs./sq. inch	Cubic yards	764,554.86	Cubic centimeters
Atmospheres	1,058	Tons/sq. ft.	Cubic yards	27	Cubic feet
Barrels-Oil	42	Gallons-Oil	Cubic yards	46,656	Cubic inches
Barrels-Beer	31	Gallons-Beer	Cubic yards	0.7646	Cubic meters
Barrels-Whiskey	45	Gallons-Whiskey	Cubic yards	202.0	Gallons
Barrels/Day-Oil	0.02917	Gallons/Min-Oil	Cubic yards	764.5	Liters
Bags or sacks-cement	94	Pounds-cement	Cubic yards	1616	Pints (liq.)
			Cubic yards	807.9	Quarts (liq.)
Board feet	144 in.$^2 \times 1$ in.	Cubic inches	Cubic yards/min.	0.45	Cubic feet/sec.
BTU/min.	12.96	Foot-lbs./sec.	Cubic yards/min.	3.336	Gallons/sec.
BTU/min.	0.02356	Horsepower	Cubic yards/min.	12.74	Liters/sec.
BTU/min.	0.01757	Kilowatts	Fathoms	6	Feet
BTU/min.	17.57	Watts	Feet	30.48	Centimeters
Centimeters	0.3937	Inches	Feet	12	Inches
Centimeters	0.01	Meters	Feet	0.3048	Meters
Centimeters	10	Millimeters	Feet	1/3	Yards
Cubic feet	2.832×10^{-4}	Cubic cms.	Feet of water	0.0295	Atmospheres
Cubic feet	1728	Cubic inches	Feet of water	0.8826	Inches of mercury
Cubic feet	0.02832	Cubic meters	Feet of water	304.8	Kgs./sq. meter
Cubic feet	0.03704	Cubic yards	Feet of water	62.43	Lbs./sq. ft.
Cubic feet	7.48052	Gallons	Feet of water	0.4335	Lbs./sq. inch
Cubic feet	28.32	Liters	Feet/min.	0.5080	Centimeters/sec.
Cubic feet	59.84	Pints (liq.)	Feet/min.	0.01667	Feet/sec.
Cubic feet	29.92	Quarts (liq.)	Feet/min.	0.01829	Kilometers/hr.
Cubic feet/min.	472.0	Cubic cms./sec.	Feet/min.	0.3048	Meters/min.
Cubic feet/min.	0.1247	Gallons/sec.	Feet/min.	0.01136	Miles/hr.
Cubic feet/min.	0.4719	Liters/sec.	Feet/sec.	30.48	Centimeters/sec.
Cubic feet/min.	62.43	Lbs of water/min.	Feet/sec.	1,097	Kilometers/hr.
Cubic feet/sec.	0.646317	Millions gals./day	Feet/sec.	0.5924	Knots
Cubic feet/sec.	448,831	Gallons/min.	Feet/sec.	18.29	Meters/min.

MULTIPLY	BY	TO OBTAIN
Feet/sec.	0.6818	Miles/hr.
Feet/sec.	0.01136	Miles/min.
Feet/sec./sec.	30.48	Cms./sec./sec.
Feet/sec./sec.	0.3048	Meters/sec./sec.
Foot-pounds	1.286×10^{-3}	British Thermal Units
Foot-pounds	5.050×10^{-7}	Horsepower-hrs.
Foot-pounds	3.240×10^{-4}	Kilogram-calories
Foot-pounds	0.1383	Kilogram-meters
Foot-pounds	3.766×10^{-7}	Kilowatt-hours
Gallons	3785	Cubic centimeters
Gallons	0.1337	Cubic feet
Gallons	231	Cubic inches
Gallons	3.785×10^{-3}	Cubic meters
Gallons	4.951×10^{-3}	Cubic yards
Gallons	3,785	Liters
Gallons	8	Pints (liq.)
Gallons	4	Quarts (liq.)
Gallons-Imperial	1.20095	U.S. gallons
Gallons-U.S.	0.83267	Imperial gallons
Gallons water	8,345	Pounds of water
Gallons/min.	2.228×10^{-3}	Cubic feet/sec.
Gallons/min.	0.06308	Liters/sec.
Gallons/min.	8.0208	Cu. ft./hr.
Gallons/min.	2271	Meters^3/hr.
Grains/U.S. gal.	17.118	Parts/million
Grains/U.S. gal.	142.86	Lbs./million gal.
Grains/imp. gal.	14.254	Parts/million
Grams	15.43	Grains
Grams	001	Kilograms
Grams	1000	Milligrams
Grams	0.03527	Ounces
Grams	2.205×10^{-3}	Pounds
Horsepower	42.44	BTU/min.
Horsepower	33,000	Foot-lbs./min.
Horsepower	550	Foot-lbs./sec.
Horsepower	1,014	Horsepower (metric)
Horsepower	0.7457	Kilowatts
Horsepower	745.7	Watts
Horsepower (boiler)	33,493	BTU/hr.
Horsepower (boiler)	9,809	Kilowatts
Horsepower-hours	2546	BTU
Horsepower-hours	1.98×10^6	Foot-lbs.
Horsepower-hours	2.737×10^5	Kilogram-meters
Horsepower-hours	0.7457	Kilowatt-hours
Inches	2.540	Centimeters
Inches of mercury	0.03342	Atmospheres
Inches of mercury	1.133	Feet of water
Inches of mercury	345.3	Kgs./sq. meter
Inches of mercury	70.73	Lbs./sq. ft.
Inches of mercury (32°F)	0.491	Lbs./sq. inch
Inches of water	0.002458	Atmospheres
Inches of water	0.07355	Inches of mercury
Inches of water	25.40	Kgs./sq. meter

MULTIPLY	BY	TO OBTAIN
Inches of water	0.578	Ounces/sq. inch
Inches of water	5.202	Lbs./sq. foot
Inches of water	0.03613	Lbs./sq. inch
Kilograms	2.205	Lbs.
Kilograms	1.102×10^{-3}	Tons (short)
Kilograms	10^3	Grams
Kiloliters	10^3	Liters
Kilometers	10^5	Centimeters
Kilometers	3281	Feet
Kilometers	10^3	Meters
Kilometers	0.6214	Miles
Kilometers	1094	Yards
Kilometers/hr.	27.78	Centimeters/sec.
Kilometers/hr.	54.68	Feet/min.
Kilometers/hr.	0.9113	Feet/sec.
Kilometers/hr.	5399	Knots
Kilometers/hr.	16.67	Meters/min.
Kilowatts	56.907	BTU/min.
Kilowatts	4.425×10^4	Foot-lbs./min.
Kilowatts	737.6	Foot-lbs./sec.
Kilowatts	1.341	Horsepower
Kilowatts	10^3	Watts
Kilowatt-hours	3414.4	BTU
Kilowatt-hours	2.655×10^6	Foot-lbs.
Kilowatt-hours	1.341	Horsepower-hrs.
Kilowatt-hours	3.671×10^5	Kilogram-meters
Liters	10^3	Cubic centimeters
Liters	0.03531	Cubic feet
Liters	61.02	Cubic inches
Liters	10^{-3}	Cubic meters
Liters	1.308×10^{-3}	Cubic yards
Liters	0.2642	Gallons
Liters	2.113	Pints (liq.)
Liters	1.057	Quarts (liq.)
Liters/min.	5.886×10^{-4}	Cubic ft./sec.
Liters/min.	4.403×10^{-3}	Gals./sec.
Lumber Width (in.) × Thickness (in.) ⁄ 12	Length (ft.)	Board feet
Meters	100	Centimeters
Meters	3.281	Feet
Meters	39.37	Inches
Meters	10^{-3}	Kilometers
Meters	10^3	Millimeters
Meters	1.094	Yards
Miles	1.609×10^5	Centimeters
Miles	5280	Feet
Miles	1.609	Kilometers
Miles	1760	Yards
Miles/hr.	44.70	Centimeters/sec.
Miles/hr.	88	Feet/min.
Miles/hr.	1.467	Feet/sec.
Miles/hr.	1.609	Kilometers/hr.
Miles/hr.	0.8689	Knots
Miles/hr.	26.82	Meters/min.
Miles/min.	2682	Centimeters/sec.
Miles/min.	88	Feet/sec.

MULTIPLY	BY	TO OBTAIN
Miles/min.	1.609	Kilometers/min.
Miles/min.	60	Miles/hr.
Ounces	16	Drams
Ounces	437.5	Grains
Ounces	0.0625	Pounds
Ounces	28.3495	Grams
Ounces	2.835×10^{-5}	Tons (metric)
Parts/million	0.0584	Grains/U.S. gal.
Parts/million	0.07015	Grains/Imp. gal.
Parts/million	8.345	Lbs./million gal.
Pounds	16	Ounces
Pounds	256	Drams
Pounds	7000	Grains
Pounds	0.0005	Tons (short)
Pounds	453.5924	Grams
Pounds of water	0.01602	Cubic feet
Pounds of water	27.68	Cubic inches
Pounds of water	0.1198	Gallons
Pounds of water/min.	2.670×10^{-4}	Cubic ft./sec.
Pounds/cubic foot	0.01602	Grams/cubic cm.
Pounds/cubic foot	16.02	Kgs./cubic meters
Pounds/cubic foot	5.787×10^{-4}	Lbs./cubic inch
Pounds/cubic inch	27.68	Grams/cubic cm.
Pounds/cubic inch	2.768×10^{4}	Kgs./cubic meter
Pounds/cubic inch	1728	Lbs./cubic foot
Pounds/foot	1.488	Kgs./meter
Pounds/inch	1152	Grams/cm.
Pounds/sq. foot	0.01602	Feet of water
Pounds/sq. foot	4.882	Kgs./sq. meter
Pounds/sq. foot	6.944×10^{-3}	Pounds/sq. inch
Pounds/sq. inch	0.06804	Atmospheres
Pounds/sq. inch	2.307	Feet of water
Pounds/sq. inch	2.036	Inches of mercury
Pounds/sq. inch	703.1	Kgs./sq. meter
Quarts (dry)	67.20	Cubic inches
Quarts (liq.)	57.75	Cubic inches
Square feet	2.296×10^{-5}	Acres
Square feet	929.0	Square centimeters
Square feet	144	Square inches
Square feet	0.09290	Square meters
Square feet	3.587×10^{-4}	Square miles
Square feet	1/9	Square yards
$\dfrac{1}{\text{Sq. ft./gal./min.}}$	8.0208	Overflow rate (ft./hr.)
Square inches	6.452	Square centimeters
Square inches	6.944×10^{-3}	Square feet
Square inches	645.2	Square millimeters
Square kilometers	247.1	Acres
Square kilometers	10.76×10^{6}	Square feet
Square kilometers	10^{6}	Square meters
Square kilometers	0.3861	Square miles
Square kilometers	1.196×10^{6}	Square yards
Square meters	2.471×10^{4}	Acres
Square meters	10.76	Square feet
Square meters	3.861×10^{-7}	Square miles
Square meters	1.196	Square yards
Square miles	640	Acres
Square miles	27.88×10^{6}	Square feet
Square miles	2.590	Square kilometers
Square miles	3.098×10^{6}	Square yards
Square yards	2.066×10^{-4}	Acres
Square yards	9	Square feet
Square yards	0.8361	Square meters
Square yards	3.228×10^{-7}	Square miles
Temp. (°C.)+273	1	Abs. temp. (°C.)
Temp. (°C.)+17.78	1.8	Temp. (°F.)
Temp. (°F.)+460	1	Abs. temp. (°F.)
Temp. (°F.)−32	5/9	Temp. (°C.)
Tons (metric)	10^{3}	Kilograms
Tons (metric)	2205	Pounds
Tons (short)	2000	Pounds
Tons (short)	32,000	Ounces
Tons (short)	907.1843	Kilograms
Tons (short)	0.89287	Tons (long)
Tons (short)	0.90718	Tons (metric)
Tons of water/24 hrs	83.333	Pounds water/hr.
Tons of water/24 hrs.	0.16643	Gallons/min.
Tons of water/24 hrs.	1.3349	Cu. ft./hr.
Watts	0.05686	B.T.U./min.
Watts	44.25	Foot-lbs./min.
Watts	0.7376	Foot-lbs./sec.
Watts	1.341×10^{-3}	Horsepower
Watts	0.01434	Kg.-calories/min.
Watts	10^{-3}	Kilowatts
Watt-hours	3.414	BTU
Watt-hours	2655	Foot/lbs.
Watt-hours	1.341×10^{-3}	Horsepower-hrs.
Watt-hours	0.8604	Kilogram-calories
Watt-hours	367.1	Kilogram-meters
Watt-hours	10^{-3}	Kilowatt-hours
Yards	91.44	Centimeters
Yards	3	Feet
Yards	36	Inches
Yards	0.9144	Meters

Appendix 3

Glossary

The following terms and definitions are generally accepted throughout the irrigation industry.

They are provided to enable the user to better understand and use the contents of this book.

A

Acre-Foot: The volume of water that will cover one acre to a depth of one foot—325,850 U.S. gallons (approximate).

Acre-Inch: Volume of water that will cover one acre one inch deep—27,154 U.S. gallons (approximate).

Adhesion: In irrigation, the property of attraction that soil particles have for water.

Aeration Irrigation: A sprinkler irrigation concept based on low application rates, usually not more than one-half the soil's intake rate; it allows water and air to enter the soil in more balanced proportions; said to improve soil structures and reduce compaction problems.

Algae: Any of the group of chiefly aquatic nonvascular plants (as seaweeds, pond scum, stoneworts) with chlorophyll, often masked by a brown or red pigment.

Alkali: A soluble salt obtained from the ashes of plants and consisting largely of potassium or sodium carbonates.

Angle Valve: A valve from which water flows out at a 90° angle from the direction which it enters.

Anti-Siphon Device: A device to protect domestic water from possible contaminated water in sprinkler system lines. Also called a vacuum breaker.

Application Rate: The rate (inches or gallons per week) at which water is applied to the turf or ornamentals. *Also:* In sprinkler irrigation, refers to the amount of water applied to a given area in one hour—usually measured in inches/hour.

Arc: The degrees of coverage of a sprinkler from one side of throw to the other. A 90° arc would be quarter-circle coverage. Likewise a 180° arc would be identified as a half-circle coverage.

As-Built Plan: A complete plan of an installed irrigation system designating valve, sprinkler, and controller locations, and routing of pipe and control wire. The plan includes all changes to the original design necessitated during the systems installation.

Atmosphere: A pressure of 14.71 pounds per square inch, a column height of 76.39 centimeters of mercury (29.92 inches of mercury), or a column height of 34.01 feet (1,036 centimeters) of water at sea level when the temperature is 69.8°F.

Atmospheric Vacuum Breaker: An anti-siphon device which uses a disc float assembly to seal off the atmospheric vent when an irrigation lateral line is pressurized. The disc float falls, opening the atmospheric vent, and allows air to enter the system when the lateral line is shut off or when the pressure drops to atmospheric levels or below (vacuum).

Automatic Control Valve: A valve in a sprinkler system which is activated by an automatic controller through the use of electric control wire.

Automatic System: An irrigation system which will irrigate in accordance to a preset program.

Available Water: The amount of water held in the plant root zone between field capacity and the wilting point.

B

Backflow: The unwanted reverse flow of liquids in a piping system.

Backflow Preventer: A mechanical device which allows water to flow in only one direction, but does not allow the water to flow in a reverse direction. There are several types of backflow preventers, namely:

1. vacuum breaker
   a. atmospheric vacuum breaker
   b. combination atmospheric vacuum breaker and sprinkler control valve
   c. pressure vacuum breaker
   d. hose connection vacuum breaker
2. double check valve assembly
3. reduced-pressure principle backflow preventer

The choke or backflow preventer to be used will depend on the degree of hazard and the particular piping arrangement involved.

Battery (of Sprinklers): A group of sprinklers controlled by one valve.

Booster (Pump): A pump which has a pressurized suction and is designed to raise the existing pressure of the water in the irrigation mainline to a desired level.

Brackish Water: Water polluted or contaminated by organic matter, salts, or acids, or a combination thereof.

C

Caliche: A soil deposit of fine silts or clays impregnated with crystalline salts, such as sodium nitrate or sodium chloride. May be a zone of calcium or mixed carbonates in soils found in semi-arid regions.

Cam Drive: A method of creating rotational movement in a sprinkler by means of two solid pieces of material (cams) interacting repeatedly.

Cavitation: The formation and collapse of vapor bubbles that form in a pumping situation when the pressure in the suction line or pump is reduced to the vapor pressure of the water; may cause pumping inefficiency and damage to pump.

Check Valve: A valve which permits water to flow in one direction only. Check valves rely on seat weight or spring force to remain closed against reverse flows.

Chloride: A compound of chlorine with another element, especially a salt or ester of hydrochloric acid.

Clay: Soil particles that are visible only with magnification of 100 power or more; plate-like in shape (less than 0.002 millimeter in diameter).

Coefficient of Uniformity: A numerical expression which is an index of the uniformity of water application to a given area within a specific geometric arrangement of sprinklers.

Cohesion: The attraction that water molecules have for each other.

Combination Atmospheric Vacuum Breaker: A combination atmospheric vacuum breaker and sprinkler control valve is a device which combines, in one body, an atmospheric vacuum breaker and a tightly closing shutoff valve (located upstream of the vacuum breaker portion). This device is an atmospheric vacuum breaker.

Consumptive Irrigation Water Requirement: The amount of water that must be applied by irrigation to supplement stored moisture, or that furnished by precipitation, in order to supply the total moisture required.

Consumptive Use: Combined total of moisture withdrawn from the soil by the plant (transpiration) plus that evaporated from the soil, plus the amount of water intercepted by the plant foliage. The term evapotranspiration is preferred. May be expressed as acre-inches or as acre-feet.

Controller: The automatic timing device and its enclosure. The controller signals the automatic valves to open and close on a preset program.

Coverage: General terms referring to the manner in which water is applied to the ground, with respect to the spacing between sprinklers.

Cycle, Irrigation: Refers to one complete operation of a controller station.

D

Deep-Well Turbine Pump: A mixed-flow pump used in deep wells. The pump is set below the water level and power to the turbine is transmitted by a line shaft inside the water column pipe.

Design: To recommend the use of sprinkler system components which make up the total system. A design can be an informal field plot location or a formal engineering drawing.

Designer: The individual responsible for specifying which sprinkler system components to use and how they should be used. Also responsible for analyzing the performance and operation of the irrigation system.

Diaphragm Valve: The type of globe or angle pattern valve which utilizes a diaphragm to aid the control of liquid flow through the valve.

Direct Burial Wire: Thermoplastic-coated copper wire capable of being installed in soil without being placed in conduit. Direct burial wire is designated "Type UF" (underground feeder).

Direct-Drive (Pump): The power unit and drive mechanism are connected so that operating revolutions per minute (rpm) are equal.

Distribution, Water: The manner in which a sprinkler applies water to the turf or ground.

Distribution Curve (Sprinkler): A curve showing the rate of water application by a sprinkler at various points along the radius.

Distribution Pattern: The pattern of water application by a sprinkler over the area the sprinkler covers.

Domestic Water: Water which is meant for human consumption. Potable or drinkable water. It can be used for irrigation systems as long as protection is provided against contamination.

Double Check Valve Assembly: Generally refers to a type of backflow preventer composed of two in-line positive seating check valves. The DC assembly also includes two shutoff valves and approved ball valve test cocks.

Downthrust (Pump): The sum of the weights of the line shaft, bowls, and impellers, plus the hydraulic thrust of the impellers of a deep-well turbine pump.

Drain Valve: A valve used to drain water from a line. It can be a manual drain valve in mainlines or an automatic drain valve in non-pressurized lateral lines.

Drawdown (Pump): Difference between the static water level of a well and water level maintained after a period of consistent pumping. Expressed in feet or meters.

E

Electric Valve: Automatic valve usually controlled by 24- to 30-volt (AC) current. Valves using other voltages are also available, generally on special order.

Elevation Gain (Pressure): Increase in static pressure due to water being stored below its original source. Expressed as head, feet, or psi pressure. Pressure increase is calculated as 0.43 psi gain per foot of elevation difference.

Elevation Loss (Pressure): Decrease in static pressure due to water being stored above its original source. Expressed as head, feet, or psi pressure. Pressure decrease (loss) is calculated as 0.43 psi loss per foot of elevation difference

Emitter (Trickle): A device used in trickle (drip) irrigation to reduce the water pressure within the lateral line to nil before discharging the water to the soil.

Evaporation: The loss of water from a surface, soil or otherwise, into the atmosphere in the form of vapor.

Evapotranspiration (ET): Refers to the total water utilized by the plant through transpiration, evaporation, and by intercepting water on the plant's foliage.

F

Field Capacity: Amount of moisture held by a soil after drainage following a water application. This figure may be stated as a percentage on a dry weight basis or in inches of moisture per foot of depth.

Float Valve: A valve which utilizes an external floating ball to open or close the valve. Float valves are often used as reservoir level control valves.

Flooding: Excess water on turf resulting from long watering cycles and/or runoff from higher elevations.

Flow: The movement of fluids through pipe fittings and valves or other vessels.

Flow Control Valve: A valve which modulates in order to maintain a predetermined liquid flow rate without drastically altering the pressure.

Flow Switch: A device in a pumping station which controls the pumps' output based on demand for water (gpm).

Friction Loss: The loss of pressure caused by water flowing in a pipe system. Pressure loss due to turbulence produced by water flow against the inside wall of the pipe.

Friction loss is a function of the pipe inside diameter, wall surface roughness, and the velocity of the water flow.

G

Gear Head: The gear drives used to change the direction of power input or to change the speed of the power shaft and drive shaft in a pump plant.

Globe Valve: A valve whose body is constructed in a globe pattern. The globe valve is available in manual, hydraulic, and electric solenoid–actuated configurations.

gpm: Abbreviation for gallons per minute.

Gravity Flow: Flow of water in a pipe on a descending path.

Groundwater: Water found below the soil surface. Usually does not include the water flowing in underground streams.

Gypsum: A widely distributed mineral consisting of hydrous calcium sulfate that is used as a soil amendment to counteract alkaline soil conditions.

H

Head Feet: A measure of pressure in feet of water. Equivalent to 0.43 psi per foot of water.

Head-to-Head Spacing: The placement of sprinklers such that the radius of throw of one strikes the adjacent sprinkler locations. Also known as 100% and head-to-head coverage.

Hose Connection Vacuum Breaker: A device consisting of a positive seating check valve and an atmospheric vent, biased to a normally open position. The device is designed specifically for use on a hose threaded outlet. Although designed to protect primarily against back-siphonage, it will also protect against low head pressure backflow.

Hydraulic Conductivity: Refers to the readiness of a soil to permit fluid flow through it; both soil and fluid characteristics determine the extent of this property.

I

Impact Drive: A method of providing rotational movement to a sprinkler through the use of a weighted or spring-loaded arm being pushed away from the sprinkler by the water stream and returning to collide with the sprinkler to force a movement.

Infiltration Rate: Generally used in relation to sprinkler irrigation. It is the rate at which the soil will take in measured water, measured in inches per hour.

Irrigation Efficiency: The percentage of irrigation water that is actually stored in the soil and available for consumptive use by the turfgrass species when compared to the total amount provided.

Irrigation Frequency: The amount of time that can be allowed between irrigation, during periods of peak water use. In turf applications, the frequency may be expressed in days or hours.

Irrigation Requirement: The quantity of water required to be added to the turf, shrubs, etc., to satisfy the evaporation, transpiration, and other uses of water in the soil and general environment. The irrigation requirement is usually expressed in inches, and equals the net water required divided by the irrigation efficiency.

Irrigation System: A complete set of system components which include the water source, the water distribution network, and the general irrigation equipment.

J

Jockey Pump: A small pump in a pumping station used to supply small demands or replenish a hydropneumatic pressure tank.

L

Lateral: The low-pressure pipeline downstream of the control valve on which the sprinkler devices are located.

Leaching: The removal of harmful soluble salts from the plant root zone by an extra-heavy application of water. The undesirable salts are carried by gravitational water to a point below or out of the root zone.

Loam: Soil which has a relatively even mixture of different grades of sand, silt, and clay.

Loop: A piping network which allows more than one path for water to flow from the supply to the point of demand.

M

Main: A large pipe sized to carry the water for the irrigation system. Usually sprinklers are not connected directly to the main. Main pipelines are usually under pressure at all times unless drained during the winter.

Manual Drain Valve: A manual valve located at a low point in the irrigation piping which allows the laterals, valves, and mainlines to be drained during winterization of the system.

Manual System: A system in which control valves are opened manually rather than by automatic controls.

Master Valve: A normally closed valve installed at the supply point of the main which opens only when the automatic system is activated

Miner's Inch: A unit of water flow measurement commonly used in several western states. It is the rate of flow through a 1-in.-square vertical opening when the water is used with a pressure head of 4 in. to 7 in. An ordinary miner's inch is 9 gpm; its legal definition is generally set at 11.22 gpm. In the states of Arizona, California, Montana, Nevada, and Oregon, 40 miner's inches are equivalent to 1 cfs (cubic foot per second); in Idaho, Kansas, Nebraska, New Mexico, North Dakota, South Dakota, and Utah 50 miner's inches are required to equal 1 cfs; and in Colorado, a cfs equals 38.4 miner's inches.

Mist Irrigation: A sprinkler concept involving the application of water in the form of extremely small drops, or mist, to a plant.

N

Net Irrigation Water Requirement: The amount of irrigation water that must be stored in the root zone to meet the consumptive use requirement of the plant, exclusive of water supplied by rain.

Normally Closed Valve: An automatic valve through which no water will flow unless externally activated by hydraulic or electrical forces. A normally closed valve will fail to the closed position if external activation power is lost.

Normally Open Valve: An automatic valve through which water will flow unless external forces are applied to close the valve. Normally open valves will fail to the open position if external activation forces are removed.

O

Operating Cycle: Refers to one complete run of a controller through all programmed controller stations.

Operating Pressure: The pressure at which a system of sprinklers operates. Static pressure less pressure losses. Usually indicated at the base or nozzle of a sprinkler.

Orifices: Openings in pipe, tubing, and nozzles.

Overlap: The coincidence of coverage by more than one sprinkler into a common area. The amount of overlap is expressed as a percentage of the radius or spacing of the sprinklers.

P

Peak Consumptive Use: The average daily rate at which moisture is used during the growing season at periods when evapotranspiration is at the highest level. Peak consumptive use may be expressed in inches of water per day.

Percolation: The movement of water through the soil.

Permanent Wilting Point: Refers to the moisture content of the soil at which plants can no longer obtain enough water to meet transpiration requirements.

Permeability: The quality of soil that permits water and air to move through it.

Polyethylene Pipe: A black flexible pipe commonly used in trickle and sprinkler irrigation systems. PE pipe for irrigation usage is manufactured with controlled inside and outside diameters.

Potable Water: Water which is meant for human consumption. Domestic or drinkable water. It can be used for irrigation systems as long as protection is provided to prevent contamination to the domestic supply.

Precipitation Rate: The rate at which sprinklers apply water to the turf. Usually figured for a pattern at a given spacing. Expressed in inches/hour.

Pressure: The force per unit area measured in pounds per square inch, psi, or head feet.

Pressure Loss: The loss of pressure energy under flow conditions caused by pipe friction, elevation, and directional changes within the irrigation system.

Pressure Regulator: A device which regulates the available pressure to a preset maximum under static or flow conditions.

Pressure Relief Valve: A valve which will open when the pressure exceeds a preset limit to relieve and reduce the pressure.

Pressure Vacuum Breaker: A pressure vacuum breaker is a device consisting of either one or two positive seating check valves and an internally force-loaded disc float assembly, downstream of the valves, installed as a unit between two tightly closing shutoff valves, and fitted with properly located test cocks. The disc float assembly is force-loaded (generally by means of a spring) to a normally open position, and allows air to enter the piping system when the line pressure drops to 1 psi or below. Since the disc float is force-loaded, this device can be installed on the pressure side of a shutoff valve (i.e., shutoff valves may be located downstream). However, it is designed to prevent backsiphonage only, and is not effective against backflow due to backpressure.

Program: The watering plan or schedule.

Programming: The act of devising and applying to the controllers a plan or procedure for irrigating the plant material.

psi: Abbreviation for pounds per square inch, the standard pressure measurement of water.

Pump Start Circuit: The feature on some automatic controllers which allows a connection to be made through a relay with the pump starter so that the starter will be energized when the watering cycle begins.

PVC Pipe: Unplasticized polyvinyl chloride pipe. A semi-rigid plastic pipe in general use in irrigation systems.

Q

Quick-Coupling System: A sprinkler system which uses quick-coupling valves, keys, and impact heads. The valves are permanently installed, while the keys and sprinklers are manually moved from valve to valve.

Quick-Coupling Valve: The valve used in a quick-coupling system which is activated by inserting a coupling key.

R

Rain Shutdown Switch: A device which will stop the watering program when a preset amount of rain falls.

Rate of Application: The rate at which water is applied to the ground by the sprinklers within a pattern, sometimes referred to as precipitation rate.

Reduced-Pressure Backflow Preventer: A device consisting of two positive seating check valves, and an automatically operating pressure differential relief valve integrally located between the two check valves, installed as a unit between two tightly closing shutoff valves, and fitted with properly located test cocks. During normal operation, the pressure in the zone between the two check valves is maintained at a lower pressure than the supply pressure. If the zone pressure starts to approach the supply pressure, the differential pressure relief valve will automatically maintain a differential of not less than 2 psi between the supply pressure and the zone between the two check valves by discharging to the atmosphere. This device is effective against backflow caused by backpressure and backsiphonage and is used to protect the water from substances which are hazardous to health (high hazard).

Remote Control Unit: Can be a controller or manually controlled device used to activate automatic remote control valves.

Remote Control Valve: An automatic valve which is activated by an automatic controller or manual remote control unit through use of hydraulic or electric control lines.

Repeat Cycle: The programming of an automatic controller to repeat an irrigation cycle automatically for those controller stations so set.

Riser: Usually refers to a length of pipe affixed to a lateral line or submain for the purpose of supporting a valve or sprinkler head.

Runoff: Water which is not absorbed by the soil and turf to which it is applied. Runoff occurs when water is applied at too great a rate or when there is a severe slope.

S

Saddle: A type of fitting which clamps on over the pipe. A hole is drilled through the pipe to deliver water to the outlet of the saddle.

Saline Water: Water containing a high concentration of dissolved salts, usually stated in parts per million.

Sand: Soil particles that can be seen with the naked eye and felt as an individual grain. They vary in size from fine to coarse.

Sand Filter: A device installed in a pipeline to remove sand or silt from water by allowing it to settle out of the flowing stream.

Sandy Loam: A soil made up predominantly of sand particles, but having enough silt and clay particles to make it cohesive.

Silt: Soil particles that appear and feel flour-like; a grain of silt requires approximately 4-power magnification before it can be seen; may be either rough or angular in shape, just as sand.

Sleeve (Tubing): Pipe through which hydraulic tubing or electrical wiring is run for added protection or for ease in replacing tubing or wire when running under paved areas.

Slip Fitting: A fitting that is solvent-welded on PVC or ABS pipe.

Sodium Chloride: A chemical compound, crystalline in nature when dissolved, commonly referred to as table salt.

Sodium Sulfate: A compound of sodium and sulfur—a salt.

Soil Compaction: Compression of soil particles which may cause the water intake rate of a particular soil to be reduced.

Soil-Moisture Tension: The measure of force with which water is held in the soil by adhesion and cohesion (expressed in terms of atmospheres), against the forces exerted by a plant's root system and evaporative processes.

Soil Profile: The makeup of the root zone section of the soil, which may be composed of layers of various textured particles.

Solvent: A material which causes a partial dissolving of PVC pipe and fittings so that a chemical fusion can occur between the pipe and fitting.

Solvent Welding: The act of chemically fusing pipe and fittings together using solvent.

Spacing: The distance between sprinkler heads.

Specific Capacity: The water production of a well measured in gpm per foot of drawdown when the pump is being operated at its peak efficiency.

Sprinkler: A hydraulic mechanical device which discharges water through a nozzle or an orifice.

Static Pressure: The pressure (psi) in a closed system, without any water movement.

Station: A position on the controller which indicates control of automatic valves.

Stop-a-Matic Valve (Sprinkler): A spring-loaded check valve used beneath a sprinkler to prevent water from draining out of that sprinkler. The check valve feature may also be built into the sprinkler device.

Surge: An energy wave in pipelines caused by sudden opening or closing of valves.

Swing Check Valve: A valve which allows water flow in only one direction. Closure against a backflow is provided by the weight of the pendulum-action seat.

Swing Joint: A threaded connection of pipe and fittings between the pipe and sprinkler which allows movement to be taken up in the threads rather than as a sheer force on the pipe. Also used to raise or lower sprinklers to a final grade without plumbing changes.

T

Tensiometers: Devices for measuring the moisture content of the soil that work on the principal that a partial vacuum is created in a closed tube when water moves out through a porous ceramic tip to the surrounding soil. The tension causing the movement of water is measured on a vacuum gauge.

Tension: Energy used in moving moisture from a soil or exerted by soil particles to hold moisture. The higher the moisture content of a soil, the lower the tension, and vice versa.

Transpiration: The process by which a plant removes water from the soil throughout itself to its leaves and transpires moisture to the atmosphere.

U

Uniformity of Application: A general term designating how uniform the application of the sprinkler is over the area it is covering while in pattern.

V

Vacuum Breaker: A type of backflow prevention device which prevents the reverse flow of water from a potentially contaminated source to the potable water supply, by allowing air to enter the supply line to break the vacuum or siphon condition.

Vacuum Pump: A type of pump used to move fluids at low pressure, or to prime larger pumps; operates on the principle of reducing pressure in the direction of desired movement.

Valve-in-Head: Indicates that the automatic control valve, electric or hydraulic, is an integral part of the sprinkler assembly.

Valve-under-Head: A sprinkler in which there is a separate automatic valve under each sprinkler. The valve is a separate and independent component of the system.

Velocity (of Water): The speed at which water travels, expressed in feet per second (fps, or ft/sec).

W

Wall-to-Wall Coverage: Indicates complete coverage of the area to be irrigated from one border to the other.

Water Hammer: A shock wave created by a fast-closing valve. Also referred to as a surge of pressure.

Water Pressure: Pressure which water exerts, as measured in pounds per square inch or in head feet.

Water Ram: A shock wave set up by introducing water under high pressure into an air-filled pipe.

Wire Gauge—Size Standards for Wire: The larger the gauge number, the smaller the wire.

Appendix 4

Friction Loss Characteristics

FRICTION LOSS CHARACTERISTICS
PVC CLASS 125 IPS PLASTIC PIPE
(1120, 1220) SDR 32.5 C = 150
PSI LOSS PER 100 FEET OF PIPE (PSI/100 FT)

Sizes 1" thru 5"
Flow GPM 1 thru 600

SIZE	1.00	1.25	1.50	2.00	2.50	3.00	3.50	4.00	5.00	SIZE
OD	1.315	1.660	1.900	2.375	2.875	3.500	4.000	4.500	5.563	OD
ID	1.211	1.548	1.784	2.229	2.699	3.284	3.754	4.224	5.221	ID
WALL THK	0.052	0.056	0.058	0.073	0.088	0.108	0.123	0.138	0.171	WALL THK

Flow G.P.M.	1.00 Vel. F.P.S.	1.00 P.S.I. Loss	1.25 Vel. F.P.S.	1.25 P.S.I. Loss	1.50 Vel. F.P.S.	1.50 P.S.I. Loss	2.00 Vel. F.P.S.	2.00 P.S.I. Loss	2.50 Vel. F.P.S.	2.50 P.S.I. Loss	3.00 Vel. F.P.S.	3.00 P.S.I. Loss	3.50 Vel. F.P.S.	3.50 P.S.I. Loss	4.00 Vel. F.P.S.	4.00 P.S.I. Loss	5.00 Vel. F.P.S.	5.00 P.S.I. Loss	Flow G.P.M.
1	0.27	0.02	0.17	0.01	0.12	0.00													1
2	0.55	0.06	0.34	0.02	0.25	0.01	0.16	0.00											2
3	0.83	0.13	0.51	0.04	0.38	0.02	0.24	0.01											3
4	1.11	0.22	0.68	0.07	0.51	0.03	0.32	0.01	0.22	0.00									4
5	1.39	0.33	0.85	0.10	0.64	0.05	0.41	0.02	0.28	0.01									5
6	1.66	0.46	1.02	0.14	0.76	0.07	0.49	0.02	0.33	0.01									6
7	1.94	0.62	1.19	0.19	0.89	0.09	0.57	0.03	0.39	0.01	0.26	0.00							7
8	2.22	0.79	1.36	0.24	1.02	0.12	0.65	0.04	0.44	0.02	0.30	0.01							8
9	2.50	0.98	1.53	0.30	1.15	0.15	0.73	0.05	0.50	0.02	0.34	0.01							9
10	2.78	1.19	1.70	0.36	1.28	0.18	0.82	0.06	0.56	0.02	0.37	0.01	0.28	0.00					10
11	3.06	1.42	1.87	0.43	1.41	0.22	0.90	0.07	0.61	0.03	0.41	0.01	0.31	0.01					11
12	3.33	1.67	2.04	0.51	1.53	0.25	0.98	0.09	0.67	0.03	0.45	0.01	0.36	0.01	0.27	0.00			12
14	3.89	2.22	2.38	0.67	1.79	0.34	1.14	0.11	0.78	0.05	0.52	0.02	0.40	0.01	0.32	0.01			14
16	4.45	2.85	2.72	0.86	2.05	0.43	1.31	0.15	0.89	0.06	0.60	0.02	0.46	0.01	0.36	0.01			16
18	5.00	3.54	3.06	1.07	2.30	0.54	1.47	0.18	1.00	0.07	0.68	0.03	0.52	0.01	0.41	0.01			18
20	5.56	4.31	3.40	1.30	2.56	0.65	1.64	0.22	1.12	0.09	0.75	0.03	0.57	0.02	0.45	0.01			20
22	6.12	5.14	3.74	1.56	2.82	0.78	1.80	0.26	1.23	0.10	0.83	0.04	0.63	0.02	0.50	0.01			22
24	6.67	6.04	4.08	1.83	3.07	0.92	1.97	0.31	1.34	0.12	0.90	0.05	0.69	0.02	0.54	0.01	0.35	0.00	24
26	7.23	7.00	4.42	2.12	3.33	1.06	2.13	0.36	1.45	0.14	0.98	0.05	0.75	0.03	0.59	0.02	0.38	0.01	26
28	7.78	8.03	4.76	2.43	3.58	1.22	2.29	0.41	1.56	0.16	1.05	0.06	0.81	0.03	0.64	0.02	0.41	0.01	28
30	8.34	9.13	5.10	2.76	3.84	1.39	2.46	0.47	1.68	0.18	1.13	0.07	0.86	0.04	0.68	0.02	0.44	0.01	30
35	9.73	12.14	5.95	3.68	4.48	1.84	2.87	0.62	1.96	0.25	1.32	0.09	1.01	0.05	0.80	0.03	0.52	0.01	35
40	11.12	15.55	6.81	4.71	5.12	2.36	3.28	0.80	2.24	0.31	1.51	0.12	1.15	0.06	0.91	0.04	0.59	0.01	40
45	12.51	19.34	7.66	5.86	5.76	2.94	3.69	0.99	2.52	0.39	1.70	0.15	1.30	0.08	1.02	0.04	0.67	0.02	45
50	13.91	23.50	8.51	7.12	6.40	3.57	4.10	1.21	2.80	0.48	1.89	0.18	1.44	0.10	1.14	0.05	0.74	0.02	50
55	15.30	28.04	9.36	8.49	7.05	4.26	4.51	1.44	3.08	0.57	2.08	0.22	1.59	0.11	1.25	0.06	0.82	0.02	55
60	16.69	32.94	10.21	9.98	7.69	5.00	4.92	1.69	3.36	0.67	2.26	0.26	1.73	0.13	1.37	0.08	0.89	0.03	60
65	18.08	38.21	11.06	11.57	8.33	5.80	5.33	1.96	3.64	0.77	2.45	0.30	1.88	0.16	1.48	0.09	0.97	0.03	65
70	19.47	43.83	11.91	13.27	8.97	6.65	5.74	2.25	3.92	0.89	2.64	0.34	2.02	0.18	1.60	0.10	1.04	0.04	70
75			12.76	15.08	9.61	7.56	6.15	2.56	4.20	1.01	2.83	0.39	2.17	0.20	1.71	0.11	1.12	0.04	75
80			13.62	17.00	10.25	8.52	6.56	2.88	4.48	1.14	3.02	0.44	2.31	0.23	1.82	0.13	1.19	0.05	80
85			14.47	19.02	10.89	9.53	6.98	3.23	4.76	1.27	3.21	0.49	2.46	0.26	1.94	0.14	1.27	0.05	85
90			15.32	21.14	11.53	10.60	7.39	3.59	5.04	1.41	3.40	0.54	2.60	0.28	2.05	0.16	1.34	0.06	90
95			16.17	23.37	12.17	11.71	7.80	3.96	5.32	1.56	3.59	0.60	2.75	0.31	2.17	0.18	1.42	0.06	95
100			17.02	25.69	12.81	12.88	8.21	4.36	5.60	1.72	3.78	0.66	2.89	0.35	2.28	0.19	1.49	0.07	100
110			18.72	30.65	14.10	15.37	9.03	5.20	6.16	2.05	4.16	0.79	3.18	0.41	2.51	0.23	1.64	0.08	110
120					15.38	18.06	9.85	6.11	6.72	2.41	4.53	0.93	3.47	0.48	2.74	0.27	1.79	0.10	120
130					16.66	20.94	10.67	7.09	7.28	2.79	4.91	1.08	3.76	0.56	2.97	0.32	1.94	0.11	130
140					17.94	24.02	11.49	8.13	7.84	3.20	5.29	1.23	4.05	0.64	3.20	0.36	2.09	0.13	140
150					19.22	27.30	12.31	9.24	8.40	3.64	5.67	1.40	4.34	0.73	3.43	0.41	2.24	0.15	150
160							13.13	10.41	8.96	4.10	6.05	1.58	4.63	0.82	3.65	0.46	2.39	0.17	160
170							13.96	11.65	9.52	4.59	6.43	1.77	4.92	0.92	3.88	0.52	2.54	0.19	170
180							14.78	12.95	10.08	5.10	6.80	1.96	5.21	1.02	4.11	0.58	2.69	0.21	180
190							15.60	14.31	10.64	5.64	7.18	2.17	5.50	1.13	4.34	0.64	2.84	0.23	190
200							16.42	15.74	11.20	6.20	7.56	2.39	5.79	1.25	4.57	0.70	2.99	0.25	200
225							18.47	19.57	12.60	7.72	8.51	2.97	6.51	1.55	5.14	0.87	3.36	0.31	225
250									14.00	9.38	9.45	3.61	7.23	1.88	5.71	1.06	3.74	0.38	250
275									15.40	11.19	10.40	4.31	7.96	2.25	6.28	1.27	4.11	0.45	275
300									16.80	13.15	11.34	5.06	8.68	2.64	6.86	1.49	4.49	0.53	300
325									18.20	15.25	12.29	5.87	9.40	3.06	7.43	1.72	4.86	0.62	325
350									19.60	17.49	13.24	6.73	10.13	3.57	8.00	1.98	5.23	0.71	350
375											14.18	7.65	10.85	3.99	8.57	2.25	5.61	0.80	375
400											15.13	8.62	11.58	4.50	9.14	2.53	5.98	0.90	400
425											16.07	9.65	12.30	5.03	9.71	2.83	6.36	1.01	425
450											17.02	10.72	13.02	5.59	10.29	3.15	6.73	1.12	450
475											17.96	11.85	13.75	6.18	10.86	3.48	7.10	1.24	475
500											18.91	13.03	14.47	6.80	11.43	3.83	7.48	1.37	500
550													19.92	8.11	12.57	4.57	8.23	1.63	550
600															13.72	5.37	8.98	1.91	600

(Continued)

Note: Shaded areas of chart indicate velocities over 5' per second. Use with Caution.

FRICTION LOSS CHARACTERISTICS
PVC CLASS 125 IPS PLASTIC PIPE
(1120, 1220) SDR 32.5 C = 150
PSI LOSS PER 100 FEET OF PIPE (PSI/100 FT)

Sizes 6" thru 12"
Flow GPM 1 thru 600

SIZE	6.00	8.00	10.00	12.00	SIZE
OD	6.625	8.625	10.750	12.750	OD
ID	6.217	8.095	10.088	11.966	ID
WALL THK	0.204	0.265	0.331	0.392	WALL THK

Flow G.P.M.	Velocity F.P.S.	P.S.I. Loss	Velocity F.P.S.	P.S.I. Loss	Velocity F.P.S.	P.S.I. Loss	Velocity F.P.S.	P.S.I. Loss	Flow G.P.M.
1									1
2									2
3									3
4									4
5									5
6									6
7									7
8									8
9									9
10									10
11									11
12									12
14									14
16									16
18									18
20									20
22									22
24									24
26									26
28									28
30									30
35	0.36	0.00							35
40	0.42	0.01							40
45	0.47	0.01							45
50	0.52	0.01							50
55	0.58	0.01							55
60	0.63	0.01							60
65	0.68	0.01							65
70	0.73	0.02							70
75	0.79	0.02	0.46	0.00					75
80	0.84	0.02	0.49	0.01					80
85	0.89	0.02	0.52	0.01					85
90	0.95	0.02	0.56	0.01					90
95	1.00	0.03	0.59	0.01					95
100	1.05	0.03	0.62	0.01					100
110	1.16	0.04	0.68	0.01					110
120	1.26	0.04	0.74	0.01					120
130	1.37	0.05	0.80	0.01	0.52	0.00			130
140	1.47	0.06	0.87	0.02	0.56	0.01			140
150	1.58	0.06	0.93	0.02	0.60	0.01			150
160	1.68	0.07	0.99	0.02	0.64	0.01			160
170	1.79	0.08	1.05	0.02	0.68	0.01			170
180	1.90	0.09	1.12	0.02	0.72	0.01			180
190	2.00	0.10	1.18	0.03	0.76	0.01			190
200	2.11	0.11	1.24	0.03	0.80	0.01	0.56	0.00	200
225	2.37	0.13	1.40	0.04	0.90	0.01	0.64	0.01	225
250	2.63	0.16	1.55	0.04	1.00	0.02	0.71	0.01	250
275	2.90	0.19	1.71	0.05	1.10	0.02	0.78	0.01	275
300	3.16	0.23	1.86	0.06	1.20	0.02	0.85	0.01	300
325	3.43	0.26	2.02	0.07	1.30	0.02	0.92	0.01	325
350	3.69	0.30	2.17	0.08	1.40	0.03	0.99	0.01	350
375	3.95	0.34	2.33	0.09	1.50	0.03	1.06	0.01	375
400	4.22	0.39	2.49	0.11	1.60	0.04	1.13	0.02	400
425	4.48	0.43	2.64	0.12	1.70	0.04	1.21	0.02	425
450	4.75	0.48	2.80	0.13	1.80	0.05	1.28	0.02	450
475	5.01	0.53	2.95	0.15	1.90	0.05	1.35	0.02	475
500	5.27	0.58	3.11	0.16	2.00	0.06	1.42	0.02	500
550	5.80	0.70	3.42	0.19	2.20	0.07	1.56	0.03	550
600	6.33	0.82	3.73	0.23	2.40	0.08	1.70	0.03	600

(Continued)

Note: Shaded areas of chart indicate velocities over 5' per second. **Use with Caution.**

FRICTION LOSS CHARACTERISTICS
PVC CLASS 160 IPS PLASTIC PIPE
(1120, 1220) SDR 26 C = 150
PSI LOSS PER 100 FEET OF PIPE (PSI/100 FT)

Sizes 1″ thru 5″. Flow GPM 1 thru 1250.

SIZE	1.00	1.25	1.50	2.00	2.50	3.00	3.50	4.00	5.00	SIZE
OD	1.315	1.660	1.900	2.375	2.875	3.500	4.000	4.500	5.563	OD
ID	1.195	1.532	1.754	2.193	2.655	3.230	3.692	4.154	5.133	ID
WALL THK	0.060	0.064	0.073	0.091	0.110	0.135	0.154	0.173	0.214	WALL THK

Flow G.P.M.	1.00 Vel F.P.S.	1.00 P.S.I. Loss	1.25 Vel F.P.S.	1.25 P.S.I. Loss	1.50 Vel F.P.S.	1.50 P.S.I. Loss	2.00 Vel F.P.S.	2.00 P.S.I. Loss	2.50 Vel F.P.S.	2.50 P.S.I. Loss	3.00 Vel F.P.S.	3.00 P.S.I. Loss	3.50 Vel F.P.S.	3.50 P.S.I. Loss	4.00 Vel F.P.S.	4.00 P.S.I. Loss	5.00 Vel F.P.S.	5.00 P.S.I. Loss	Flow G.P.M.
1	0.28	0.02	0.17	0.01	0.13	0.00													1
2	0.57	0.06	0.34	0.02	0.26	0.01	0.16	0.00											2
3	0.85	0.14	0.52	0.04	0.39	0.02	0.25	0.01											3
4	1.14	0.23	0.69	0.07	0.53	0.04	0.33	0.01	0.23	0.00									4
5	1.42	0.35	0.86	0.11	0.66	0.05	0.42	0.02	0.28	0.01									5
6	1.71	0.49	1.04	0.15	0.79	0.08	0.50	0.03	0.34	0.01	0.23	0.00							6
7	1.99	0.66	1.21	0.20	0.92	0.10	0.59	0.03	0.40	0.01	0.27	0.01							7
8	2.28	0.84	1.39	0.25	1.06	0.13	0.67	0.04	0.46	0.02	0.31	0.01							8
9	2.57	1.05	1.56	0.31	1.19	0.16	0.76	0.05	0.52	0.02	0.35	0.01	0.26	0.00					9
10	2.85	1.27	1.73	0.38	1.32	0.20	0.84	0.07	0.57	0.03	0.39	0.01	0.29	0.01					10
11	3.14	1.52	1.91	0.45	1.45	0.23	0.93	0.08	0.63	0.03	0.43	0.01	0.32	0.01					11
12	3.42	1.78	2.08	0.53	1.59	0.28	1.01	0.09	0.69	0.04	0.46	0.01	0.35	0.01	0.28	0.00			12
14	3.99	2.37	2.43	0.71	1.85	0.37	1.18	0.12	0.81	0.05	0.54	0.02	0.41	0.01	0.33	0.01			14
16	4.57	3.04	2.78	0.91	2.12	0.47	1.35	0.16	0.92	0.06	0.62	0.02	0.47	0.01	0.37	0.01			16
18	5.14	3.78	3.12	1.13	2.38	0.58	1.52	0.20	1.04	0.08	0.70	0.03	0.53	0.02	0.42	0.01			18
20	5.71	4.59	3.47	1.37	2.65	0.71	1.69	0.24	1.15	0.09	0.78	0.04	0.59	0.02	0.47	0.01			20
22	6.28	5.48	3.82	1.64	2.91	0.85	1.86	0.29	1.27	0.11	0.86	0.04	0.65	0.02	0.52	0.01	0.34	0.00	22
24	6.85	6.44	4.17	1.92	3.18	1.00	2.03	0.34	1.38	0.13	0.93	0.05	0.71	0.03	0.56	0.02	0.37	0.01	24
26	7.42	7.47	4.51	2.23	3.44	1.15	2.20	0.39	1.50	0.15	1.01	0.06	0.77	0.03	0.61	0.02	0.40	0.01	26
28	7.99	8.57	4.86	2.56	3.71	1.32	2.37	0.45	1.62	0.18	1.09	0.07	0.83	0.04	0.66	0.02	0.43	0.01	28
30	8.57	9.74	5.21	2.91	3.97	1.50	2.54	0.51	1.73	0.20	1.17	0.08	0.89	0.04	0.70	0.02	0.46	0.01	30
35	9.99	12.96	6.08	3.87	4.64	2.00	2.96	0.68	2.02	0.27	1.36	0.10	1.04	0.05	0.82	0.03	0.54	0.01	35
40	11.42	16.59	6.95	4.95	5.30	2.56	3.39	0.86	2.31	0.34	1.56	0.13	1.19	0.07	0.94	0.04	0.61	0.01	40
45	12.85	20.63	7.82	6.16	5.96	3.19	3.81	1.08	2.60	0.42	1.75	0.16	1.34	0.09	1.06	0.05	0.69	0.02	45
50	14.28	25.07	8.69	7.49	6.63	3.88	4.24	1.31	2.89	0.52	1.95	0.20	1.49	0.10	1.18	0.06	0.77	0.02	50
55	15.71	29.91	9.56	8.93	7.29	4.62	4.66	1.56	3.18	0.62	2.15	0.24	1.64	0.12	1.30	0.07	0.85	0.02	55
60	17.14	35.14	10.43	10.49	7.95	5.43	5.09	1.83	3.47	0.72	2.34	0.28	1.79	0.15	1.41	0.08	0.92	0.03	60
65	18.57	40.76	11.29	12.17	8.62	6.30	5.51	2.12	3.76	0.84	2.54	0.32	1.94	0.17	1.53	0.09	1.00	0.03	65
70	19.99	46.76	12.16	13.96	9.28	7.23	5.93	2.44	4.06	0.96	2.73	0.37	2.09	0.19	1.65	0.11	1.08	0.04	70
75			13.03	15.86	9.94	8.21	6.36	2.77	4.34	1.09	2.93	0.42	2.24	0.22	1.77	0.12	1.16	0.04	75
80			13.90	17.88	10.60	9.25	6.78	3.12	4.63	1.23	3.12	0.47	2.39	0.25	1.89	0.14	1.23	0.05	80
85			14.77	20.00	11.27	10.35	7.21	3.49	4.91	1.38	3.32	0.53	2.54	0.28	2.00	0.16	1.31	0.06	85
90			15.64	22.23	11.93	11.51	7.63	3.88	5.20	1.53	3.51	0.59	2.69	0.31	2.12	0.17	1.39	0.06	90
95			16.51	24.58	12.59	12.72	8.05	4.29	5.49	1.69	3.71	0.65	2.84	0.34	2.24	0.19	1.47	0.07	95
100			17.38	27.03	13.26	13.99	8.48	4.72	5.78	1.86	3.91	0.72	2.99	0.37	2.36	0.21	1.54	0.08	100
110			19.12	32.24	14.58	16.69	9.33	5.63	6.36	2.22	4.30	0.86	3.29	0.45	2.60	0.25	1.70	0.09	110
120					15.91	19.61	10.18	6.61	6.94	2.61	4.69	1.01	3.59	0.52	2.83	0.30	1.85	0.11	120
130					17.24	22.74	11.02	7.67	7.52	3.03	5.08	1.17	3.89	0.61	3.07	0.34	2.01	0.12	130
140					18.56	26.09	11.87	8.80	8.10	3.47	5.47	1.34	4.19	0.70	3.31	0.39	2.16	0.14	140
150					19.89	29.64	12.72	10.00	8.68	3.94	5.86	1.52	4.48	0.79	3.54	0.45	2.32	0.16	150
160							13.57	11.27	9.26	4.45	6.25	1.71	4.78	0.89	3.78	0.50	2.47	0.18	160
170							14.42	12.61	9.83	4.97	6.64	1.92	5.08	1.00	4.01	0.56	2.63	0.20	170
180							15.27	14.02	10.41	5.53	7.03	2.13	5.38	1.11	4.25	0.63	2.78	0.22	180
190							16.11	15.49	10.99	6.11	7.43	2.35	5.68	1.23	4.49	0.69	2.94	0.25	190
200							16.96	17.03	11.57	6.72	7.82	2.59	5.98	1.35	4.72	0.76	3.09	0.27	200
225							19.08	21.19	13.02	8.36	8.79	3.22	6.73	1.68	5.31	0.95	3.48	0.34	225
250									14.47	10.16	9.77	3.91	7.48	2.04	5.91	1.15	3.87	0.41	250
275									15.91	12.12	10.75	4.67	8.23	2.44	6.50	1.37	4.25	0.49	275
300									17.36	14.24	11.73	5.49	8.97	2.86	7.09	1.61	4.64	0.58	300
325									18.81	16.51	12.70	6.36	9.72	3.32	7.68	1.87	5.03	0.67	325
350											13.68	7.30	10.47	3.81	8.27	2.15	5.41	0.77	350
375											14.66	8.29	11.22	4.33	8.86	2.44	5.80	0.87	375
400											15.64	9.35	11.97	4.88	9.45	2.75	6.19	0.98	400
425											16.62	10.46	12.72	5.46	10.04	3.07	6.58	1.10	425
450											17.59	11.62	13.46	6.07	10.63	3.42	6.96	1.22	450
475											18.57	12.85	14.21	6.70	11.23	3.78	7.35	1.35	475
500											19.55	14.13	14.96	7.37	11.82	4.15	7.74	1.48	500
550													16.46	8.80	13.00	4.96	8.51	1.77	550
600													17.95	10.33	14.18	5.82	9.29	2.08	600
650													19.45	11.99	15.36	6.75	10.06	2.41	650
700															16.55	7.75	10.83	2.77	700
750															17.73	8.80	11.61	3.14	750
800															18.91	9.92	12.38	3.54	800
850																	13.16	3.96	850
900																	13.93	4.41	900
950																	14.71	4.87	950
1000																	15.48	5.36	1000
1050																	16.25	5.86	1050
1100																	17.03	6.39	1100
1150																	17.80	6.94	1150
1200																	18.58	7.51	1200
1250																	19.35	8.10	1250

Note: Shaded areas of chart indicate velocities over 5′ per second. Use with Caution.

(Continued)

FRICTION LOSS CHARACTERISTICS
PVC CLASS 160 IPS PLASTIC PIPE
(1120, 1220) SDR 26 C = 150
PSI LOSS PER 100 FEET OF PIPE (PSI/100 FT)
Sizes 6" thru 12"
Flow GPM 1 thru 5000

SIZE	6.00		8.00		10.00		12.00	
OD	6.625		8.625		10.750		12.750	
ID	6.115		7.961		9.924		11.770	
WALL THK	0.225		0.332		0.413		0.490	
Flow G.P.M.	Velocity F.P.S.	P.S.I. Loss	Velocity F.P.S.	P.S.I. Loss	Velocity F.P.S.	P.S.I. Loss	Velocity F.P.S.	P.S.I. Loss
1								
2								
3								
4								
5								
6								
7								
8								
9								
10								
11								
12								
14								
16								
18								
20								
22								
24								
26								
28								
30	0.38	0.00						
35	0.43	0.01						
40	0.49	0.01						
45	0.54	0.01						
50	0.54	0.01						
55	0.60	0.01						
60	0.65	0.01						
65	0.70	0.01						
70	0.76	0.02	0.45	0.00				
75	0.81	0.02	0.48	0.01				
80	0.87	0.02	0.51	0.01				
85	0.92	0.02	0.54	0.01				
90	0.98	0.03	0.57	0.01				
95	1.03	0.03	0.61	0.01				
100	1.09	0.03	0.64	0.01				
110	1.20	0.04	0.70	0.01				
120	1.30	0.05	0.77	0.01				
130	1.41	0.05	0.83	0.01	0.53	0.00		
140	1.52	0.06	0.90	0.02	0.57	0.01		
150	1.63	0.07	0.96	0.02	0.62	0.01		
160	1.74	0.08	1.02	0.02	0.66	0.01		
170	1.85	0.09	1.09	0.02	0.70	0.01		
180	1.96	0.10	1.15	0.03	0.74	0.01		
190	2.07	0.11	1.22	0.03	0.78	0.01		
200	2.18	0.12	1.28	0.03	0.82	0.01	0.58	0.00
225	2.45	0.14	1.44	0.04	0.93	0.01	0.66	0.01
250	2.72	0.18	1.60	0.05	1.03	0.02	0.73	0.01
275	3.00	0.21	1.77	0.06	1.13	0.02	0.80	0.01
300	3.27	0.25	1.93	0.07	1.24	0.02	0.88	0.01
325	3.54	0.29	2.09	0.08	1.34	0.03	0.95	0.01
350	3.81	0.33	2.25	0.09	1.44	0.03	1.03	0.01
375	4.09	0.37	2.41	0.10	1.55	0.04	1.10	0.02
400	4.36	0.42	2.57	0.12	1.65	0.04	1.17	0.02
425	4.63	0.47	2.73	0.13	1.76	0.04	1.25	0.02
450	4.90	0.52	2.89	0.14	1.86	0.05	1.32	0.02
475	5.18	0.58	3.05	0.16	1.96	0.05	1.39	0.02
500	5.45	0.63	3.21	0.18	2.07	0.06	1.47	0.03
550	6.00	0.76	3.54	0.21	2.27	0.07	1.61	0.03
600	6.54	0.89	3.86	0.25	2.48	0.08	1.76	0.04

Flow G.P.M.	Velocity F.P.S.	P.S.I. Loss	Velocity F.P.S.	P.S.I. Loss	Velocity F.P.S.	P.S.I. Loss	Velocity F.P.S.	P.S.I. Loss
650	7.09	1.03	4.18	0.29	2.69	0.10	1.91	0.04
700	7.63	1.18	4.50	0.33	2.89	0.11	2.06	0.05
750	8.18	1.34	4.82	0.37	3.10	0.13	2.20	0.06
800	8.72	1.51	5.15	0.42	3.31	0.14	2.35	0.06
850	9.27	1.69	5.47	0.47	3.52	0.16	2.50	0.07
900	9.81	1.88	5.79	0.52	3.72	0.18	2.65	0.08
950	10.36	2.08	6.11	0.58	3.93	0.20	2.79	0.09
1000	10.91	2.29	6.43	0.63	4.14	0.22	2.94	0.09
1050	11.45	2.50	6.75	0.69	4.34	0.24	3.09	0.10
1100	12.00	2.73	7.08	0.76	4.55	0.26	3.23	0.11
1150	12.54	2.96	7.40	0.82	4.76	0.28	3.38	0.12
1200	13.09	3.20	7.72	0.89	4.97	0.30	3.53	0.13
1250	13.63	3.45	8.04	0.96	5.17	0.33	3.68	0.14
1300	14.18	3.72	8.36	1.03	5.38	0.35	3.82	0.15
1350	14.72	3.98	8.69	1.10	5.59	0.38	3.97	0.16
1400	15.27	4.26	9.01	1.18	5.79	0.40	4.12	0.18
1450	15.82	4.55	9.33	1.26	6.00	0.43	4.27	0.19
1500	16.36	4.84	9.65	1.34	6.21	0.46	4.41	0.20
1550	16.91	5.15	9.97	1.43	6.42	0.49	4.56	0.21
1600	17.45	5.46	10.30	1.51	6.62	0.52	4.71	0.23
1650	18.00	5.78	10.62	1.60	6.83	0.55	4.85	0.24
1700	18.54	6.11	10.94	1.69	7.04	0.58	5.00	0.25
1750	19.09	6.44	11.26	1.78	7.24	0.61	5.15	0.27
1800	19.63	6.79	11.58	1.88	7.45	0.64	5.30	0.28
1850			11.90	1.98	7.66	0.68	5.44	0.30
1900			12.23	2.08	7.87	0.71	5.59	0.31
1950			12.55	2.18	8.07	0.75	5.74	0.33
2000			12.87	2.29	8.28	0.78	5.89	0.34
2100			13.51	2.50	8.69	0.86	6.18	0.37
2200			14.16	2.73	9.11	0.93	6.47	0.41
2300			14.80	2.96	9.52	1.01	6.77	0.44
2400			15.45	3.20	9.94	1.10	7.06	0.48
2500			16.09	3.46	10.35	1.18	7.36	0.52
2600			16.73	3.72	10.77	1.27	7.65	0.55
2700			17.38	3.98	11.18	1.36	7.95	0.59
2800			18.02	4.26	11.59	1.46	8.24	0.64
2900			18.66	4.55	12.01	1.56	8.54	0.68
3000			19.31	4.84	12.42	1.66	8.83	0.72
3100			19.95	5.15	12.84	1.76	9.12	0.77
3200					13.25	1.87	9.42	0.81
3300					13.67	1.98	9.71	0.86
3400					14.08	2.09	10.01	0.91
3500					14.49	2.20	10.30	0.96
3600					14.91	2.32	10.60	1.01
3700					15.32	2.44	10.89	1.07
3800					15.74	2.57	11.19	1.12
3900					16.15	2.69	11.48	1.17
4000					16.57	2.82	11.78	1.23
4100					16.98	2.96	12.07	1.29
4200					17.39	3.09	12.36	1.35
4300					17.81	3.23	12.66	1.41
4400					18.22	3.37	12.95	1.47
4500					18.64	3.51	13.25	1.53
4600					19.05	3.66	13.54	1.59
4700					19.47	3.81	13.84	1.66
4800					19.88	3.96	14.13	1.73
4900							14.43	1.79
5000							14.72	1.86

(Continued)

Note: Shaded areas of chart indicate velocities over 5' per second. Use with Caution.

FRICTION LOSS CHARACTERISTICS
PVC CLASS 200 IPS PLASTIC PIPE
(1120, 1220) SDR 21 C = 150
PSI LOSS PER 100 FEET OF PIPE (PSI/100)

Sizes 1" thru 5"
Flow GPM 1 thru 1200

SIZE	1.00	1.25	1.50	2.00	2.50	3.00	3.50	4.00	5.00	SIZE
OD	1.315	1.660	1.900	2.375	2.875	3.500	4.000	4.500	5.563	OD
ID	1.189	1.502	1.720	2.149	2.601	3.166	3.620	4.072	5.033	ID
WALL THK	0.063	0.079	0.090	0.113	0.137	0.167	0.190	0.214	0.265	WALL THK

Flow GPM	1.00 Vel F.P.S.	1.00 P.S.I. Loss	1.25 Vel F.P.S.	1.25 P.S.I. Loss	1.50 Vel F.P.S.	1.50 P.S.I. Loss	2.00 Vel F.P.S.	2.00 P.S.I. Loss	2.50 Vel F.P.S.	2.50 P.S.I. Loss	3.00 Vel F.P.S.	3.00 P.S.I. Loss	3.50 Vel F.P.S.	3.50 P.S.I. Loss	4.00 Vel F.P.S.	4.00 P.S.I. Loss	5.00 Vel F.P.S.	5.00 P.S.I. Loss	Flow GPM
1	0.28	0.02	0.18	0.01	0.13	0.00													1
2	0.57	0.07	0.36	0.02	0.27	0.01	0.17	0.00											2
3	0.86	0.14	0.54	0.04	0.41	0.02	0.26	0.01	0.18	0.00									3
4	1.15	0.24	0.72	0.08	0.55	0.04	0.35	0.01	0.24	0.01									4
5	1.44	0.36	0.90	0.12	0.68	0.06	0.44	0.02	0.30	0.01									5
6	1.73	0.51	1.08	0.16	0.82	0.08	0.53	0.03	0.36	0.01	0.24	0.00							6
7	2.02	0.67	1.26	0.22	0.96	0.11	0.61	0.04	0.42	0.01	0.28	0.01							7
8	2.30	0.86	1.44	0.28	1.10	0.14	0.70	0.05	0.48	0.02	0.32	0.01							8
9	2.59	1.07	1.62	0.34	1.24	0.18	0.79	0.06	0.54	0.02	0.36	0.01	0.28	0.00					9
10	2.88	1.30	1.80	0.42	1.37	0.22	0.88	0.07	0.60	0.03	0.40	0.01	0.31	0.01					10
11	3.17	1.56	1.98	0.50	1.51	0.26	0.97	0.09	0.66	0.03	0.44	0.01	0.34	0.01					11
12	3.46	1.83	2.17	0.59	1.65	0.30	1.06	0.10	0.72	0.04	0.48	0.02	0.37	0.01	0.29	0.00			12
14	4.04	2.43	2.53	0.78	1.93	0.40	1.23	0.14	0.84	0.05	0.56	0.02	0.43	0.01	0.34	0.01			14
16	4.61	3.11	2.89	1.00	2.20	0.52	1.41	0.17	0.96	0.07	0.65	0.03	0.49	0.01	0.39	0.01			16
18	5.19	3.87	3.25	1.24	2.48	0.64	1.59	0.22	1.08	0.09	0.73	0.03	0.56	0.02	0.44	0.01			18
20	5.77	4.71	3.61	1.51	2.75	0.78	1.76	0.26	1.20	0.10	0.81	0.04	0.62	0.02	0.49	0.01	0.32	0.00	20
22	6.34	5.62	3.97	1.80	3.03	0.93	1.94	0.32	1.32	0.12	0.89	0.05	0.68	0.02	0.54	0.01	0.35	0.01	22
24	6.92	6.60	4.34	2.12	3.30	1.09	2.12	0.37	1.44	0.15	0.97	0.06	0.74	0.03	0.59	0.02	0.38	0.01	24
26	7.50	7.65	4.70	2.46	3.58	1.27	2.29	0.43	1.56	0.17	1.05	0.07	0.80	0.03	0.63	0.02	0.41	0.01	26
28	8.08	8.78	5.06	2.82	3.86	1.46	2.47	0.49	1.68	0.19	1.13	0.07	0.87	0.04	0.68	0.02	0.45	0.01	28
30	8.65	9.98	5.42	3.20	4.13	1.66	2.65	0.56	1.80	0.22	1.22	0.09	0.93	0.04	0.73	0.02	0.48	0.01	30
35	10.10	13.27	6.32	4.26	4.82	2.20	3.09	0.75	2.11	0.29	1.42	0.11	1.08	0.06	0.86	0.03	0.56	0.01	35
40	11.54	17.00	7.23	5.45	5.51	2.82	3.53	0.95	2.41	0.38	1.62	0.14	1.24	0.08	0.98	0.04	0.64	0.02	40
45	12.98	21.14	8.13	6.78	6.20	3.51	3.97	1.19	2.71	0.47	1.83	0.18	1.40	0.09	1.10	0.05	0.72	0.02	45
50	14.42	25.70	9.04	8.24	6.89	4.26	4.41	1.44	3.01	0.57	2.03	0.22	1.55	0.11	1.23	0.06	0.80	0.02	50
55	15.87	30.66	9.94	9.83	7.58	5.09	4.85	1.72	3.31	0.68	2.23	0.26	1.71	0.14	1.35	0.08	0.88	0.03	55
60	17.31	36.02	10.85	11.55	8.27	5.97	5.30	2.02	3.61	0.80	2.44	0.31	1.86	0.16	1.47	0.09	0.96	0.03	60
65	18.75	41.77	11.75	13.40	8.96	6.93	5.74	2.35	3.92	0.93	2.64	0.36	2.02	0.19	1.59	0.10	1.04	0.04	65
70			12.65	15.37	9.65	7.95	6.18	2.69	4.22	1.06	2.84	0.41	2.17	0.21	1.72	0.12	1.12	0.04	70
75			13.56	17.47	10.34	9.03	6.62	3.06	4.52	1.21	3.05	0.46	2.33	0.24	1.84	0.14	1.20	0.05	75
80			14.46	19.68	11.03	10.18	7.06	3.44	4.82	1.36	3.25	0.52	2.49	0.27	1.96	0.15	1.28	0.05	80
85			15.37	22.02	11.72	11.39	7.50	3.85	5.12	1.52	3.45	0.59	2.64	0.30	2.09	0.17	1.36	0.06	85
90			16.27	24.48	12.41	12.66	7.95	4.28	5.42	1.69	3.66	0.65	2.80	0.34	2.21	0.19	1.44	0.07	90
95			17.18	27.06	13.10	13.99	8.39	4.74	5.72	1.87	3.86	0.72	2.95	0.37	2.33	0.21	1.53	0.08	95
100			18.08	29.76	13.79	15.39	8.83	5.21	6.03	2.06	4.07	0.79	3.11	0.41	2.46	0.23	1.61	0.08	100
110			19.89	35.50	15.17	18.36	9.71	6.21	6.63	2.45	4.47	0.94	3.42	0.49	2.70	0.28	1.77	0.10	110
120					16.54	21.57	10.60	7.30	7.23	2.88	4.88	1.11	3.73	0.58	2.95	0.33	1.93	0.12	120
130					17.92	25.02	11.48	8.47	7.84	3.34	5.29	1.29	4.04	0.67	3.19	0.38	2.09	0.13	130
140					19.30	28.70	12.36	9.71	8.44	3.84	5.69	1.47	4.35	0.77	3.44	0.43	2.25	0.15	140
150							13.25	11.04	9.04	4.36	6.10	1.68	4.67	0.87	3.69	0.49	2.41	0.18	150
160							14.13	12.44	9.64	4.91	6.51	1.89	4.98	0.98	3.93	0.55	2.57	0.20	160
170							15.01	13.91	10.25	5.50	6.91	2.11	5.29	1.10	4.18	0.62	2.73	0.22	170
180							15.90	15.47	10.85	6.11	7.32	2.35	5.60	1.22	4.42	0.69	2.89	0.25	180
190							16.78	17.10	11.45	6.75	7.73	2.60	5.91	1.35	4.67	0.76	3.06	0.27	190
200							17.66	18.80	12.06	7.43	8.14	2.85	6.22	1.49	4.92	0.84	3.22	0.30	200
225							19.87	23.38	13.56	9.24	9.15	3.55	7.00	1.85	5.53	1.04	3.62	0.37	225
250									15.07	11.23	10.17	4.31	7.78	2.25	6.15	1.27	4.02	0.45	250
275									16.58	13.39	11.19	5.15	8.56	2.68	6.76	1.51	4.42	0.54	275
300									18.09	15.74	12.21	6.05	9.34	3.15	7.38	1.78	4.83	0.63	300
325									19.60	18.25	13.22	7.01	10.11	3.65	7.99	2.06	5.23	0.74	325
350											14.24	8.05	10.89	4.19	8.61	2.36	5.63	0.84	350
375											15.26	9.14	11.67	4.76	9.22	2.69	6.03	0.96	375
400											16.28	10.30	12.45	5.37	9.84	3.03	6.44	1.08	400
425											17.29	11.53	13.23	6.01	10.45	3.39	6.84	1.21	425
450											18.31	12.81	14.01	6.68	11.07	3.77	7.24	1.34	450
475											19.33	14.16	14.78	7.38	11.68	4.16	7.65	1.48	475
500													15.56	8.11	12.30	4.58	8.05	1.63	500
550													17.12	9.68	13.53	5.46	8.86	1.95	550
600													18.68	11.37	14.76	6.42	9.66	2.29	600
650															15.99	7.44	10.46	2.65	650
700															17.22	8.54	11.27	3.04	700
750															18.45	9.70	12.07	3.46	750
800															19.68	10.93	12.88	3.90	800
850																	13.69	4.36	850
900																	14.49	4.85	900
950																	15.30	5.38	950
1000																	16.10	5.89	1000
1050																	16.91	6.45	1050
1100																	17.71	7.03	1100
1150																	18.52	7.64	1150
1200																	19.32	8.26	1200

Note: Shaded areas of chart indicate velocities over 5' per second. Use with Caution.

(Continued)

FRICTION LOSS CHARACTERISTICS
PVC CLASS 200 IPS PLASTIC PIPE
(1120, 1220) SDR 21 C = 150
PSI LOSS PER 100 FEET OF PIPE (PSI/100)
Sizes 6" thru 12"
Flow GPM 1 thru 5000

SIZE	6.00	8.00	10.00	12.00
OD	6.625	8.625	10.750	12.750
ID	5.993	7.805	9.728	11.538
WALL THK	0.316	0.410	0.511	0.606

Flow G.P.M.	Velocity F.P.S.	P.S.I. Loss	Velocity F.P.S.	P.S.I. Loss	Velocity F.P.S.	P.S.I. Loss	Velocity F.P.S.	P.S.I. Loss
1								
2								
3								
4								
5								
6								
7								
8								
9								
10								
11								
12								
14								
16								
18								
20								
22								
24								
26								
28								
30	0.34	0.00						
35	0.39	0.01						
40	0.45	0.01						
45	0.51	0.01						
50	0.56	0.01						
55	0.62	0.01						
60	0.68	0.01						
65	0.73	0.02	0.43	0.00				
70	0.79	0.02	0.46	0.01				
75	0.85	0.02	0.50	0.01				
80	0.90	0.02	0.53	0.01				
85	0.96	0.03	0.56	0.01				
90	1.02	0.03	0.60	0.01				
95	1.07	0.03	0.63	0.01				
100	1.13	0.04	0.66	0.01				
110	1.24	0.04	0.73	0.01				
120	1.36	0.05	0.80	0.01	0.51	0.00		
130	1.47	0.06	0.87	0.02	0.56	0.01		
140	1.59	0.07	0.93	0.02	0.60	0.01		
150	1.70	0.08	1.00	0.02	0.64	0.01		
160	1.81	0.08	1.07	0.02	0.68	0.01		
170	1.93	0.09	1.13	0.03	0.73	0.01		
180	2.04	0.11	1.20	0.03	0.77	0.01		
190	2.15	0.12	1.27	0.03	0.81	0.01	0.58	0.00
200	2.27	0.13	1.33	0.04	0.86	0.01	0.61	0.01
225	2.55	0.16	1.50	0.04	0.97	0.02	0.68	0.01
250	2.83	0.19	1.67	0.05	1.07	0.02	0.76	0.01
275	3.12	0.23	1.84	0.06	1.18	0.02	0.84	0.01
300	3.40	0.27	2.00	0.07	1.29	0.03	0.91	0.01
325	3.69	0.31	2.17	0.09	1.40	0.03	0.99	0.01
350	3.97	0.36	2.34	0.10	1.50	0.03	1.07	0.01
375	4.25	0.41	2.51	0.11	1.61	0.04	1.14	0.02
400	4.54	0.46	2.67	0.13	1.72	0.04	1.22	0.02
425	4.82	0.52	2.84	0.14	1.83	0.05	1.30	0.02
450	5.11	0.57	3.01	0.16	1.94	0.05	1.37	0.02
475	5.39	0.63	3.18	0.18	2.04	0.06	1.45	0.03
500	5.67	0.70	3.34	0.19	2.15	0.07	1.53	0.03
550	6.24	0.83	3.68	0.23	2.37	0.08	1.68	0.03
600	6.81	0.98	4.01	0.27	2.58	0.09	1.83	0.04

Flow G.P.M.	Velocity F.P.S.	P.S.I. Loss	Velocity F.P.S.	P.S.I. Loss	Velocity F.P.S.	P.S.I. Loss	Velocity F.P.S.	P.S.I. Loss
650	7.38	1.14	4.35	0.31	2.80	0.11	1.99	0.05
700	7.95	1.30	4.68	0.36	3.01	0.12	2.14	0.05
750	8.51	1.48	5.02	0.41	3.23	0.14	2.29	0.06
800	9.08	1.67	5.35	0.46	3.44	0.16	2.45	0.07
850	9.65	1.87	5.69	0.52	3.66	0.18	2.60	0.08
900	10.22	2.07	6.02	0.57	3.88	0.20	2.75	0.09
950	10.79	2.29	6.36	0.63	4.09	0.22	2.91	0.09
1000	11.35	2.52	6.69	0.70	4.31	0.24	3.06	0.10
1050	11.92	2.76	7.03	0.76	4.52	0.26	3.21	0.11
1100	12.49	3.01	7.36	0.83	4.74	0.28	3.37	0.12
1150	13.06	3.27	7.70	0.90	4.95	0.31	3.52	0.13
1200	13.63	3.53	8.03	0.98	5.17	0.33	3.67	0.15
1250	14.19	3.81	8.37	1.05	5.38	0.36	3.83	0.16
1300	14.76	4.10	8.70	1.13	5.60	0.39	3.98	0.17
1350	15.33	4.39	9.04	1.22	5.82	0.42	4.13	0.18
1400	15.90	4.70	9.37	1.30	6.03	0.45	4.29	0.19
1450	16.47	5.02	9.71	1.39	6.25	0.48	4.44	0.21
1500	17.03	5.34	10.04	1.48	6.46	0.51	4.59	0.22
1550	17.60	5.68	10.38	1.57	6.68	0.54	4.75	0.23
1600	18.17	6.02	10.71	1.66	6.89	0.57	4.90	0.25
1650	18.74	6.37	11.05	1.76	7.11	0.60	5.05	0.26
1700	19.31	6.73	11.38	1.86	7.32	0.64	5.21	0.28
1750	19.87	7.11	11.72	1.97	7.54	0.67	5.36	0.29
1800			12.05	2.07	7.76	0.71	5.51	0.31
1850			12.39	2.18	7.97	0.75	5.66	0.33
1900			12.72	2.29	8.19	0.78	5.82	0.34
1950			13.06	2.40	8.40	0.82	5.97	0.36
2000			13.39	2.52	8.62	0.86	6.12	0.38
2100			14.06	2.75	9.05	0.94	6.43	0.41
2200			14.73	3.00	9.48	1.03	6.74	0.45
2300			15.40	3.26	9.91	1.12	7.04	0.49
2400			16.07	3.53	10.34	1.21	7.35	0.53
2500			16.74	3.80	10.77	1.30	7.66	0.57
2600			17.41	4.09	11.20	1.40	7.96	0.61
2700			18.08	4.39	11.64	1.50	8.27	0.65
2800			18.75	4.69	12.07	1.61	8.58	0.70
2900			19.42	5.01	12.50	1.72	8.88	0.75
3000					12.93	1.83	9.19	0.80
3100					13.36	1.94	9.50	0.85
3200					13.79	2.06	9.80	0.90
3300					14.22	2.18	10.11	0.95
3400					14.65	2.30	10.42	1.00
3500					15.08	2.43	10.72	1.06
3600					15.52	2.56	11.03	1.12
3700					15.95	2.69	11.33	1.17
3800					16.38	2.83	11.64	1.23
3900					16.81	2.97	11.95	1.29
4000					17.24	3.11	12.25	1.36
4100					17.67	3.26	12.56	1.42
4200					18.10	3.41	12.87	1.48
4300					18.53	3.56	13.17	1.55
4400					18.96	3.71	13.48	1.62
4500					19.40	3.87	13.79	1.69
4600					19.83	4.03	14.09	1.76
4700							14.40	1.83
4800							14.71	1.90
4900							15.01	1.98
5000							15.32	2.05

(Continued)

Note: Shaded areas of chart indicate velocities over 5' per second. **Use with Caution.**

FRICTION LOSS CHARACTERISTICS
PVC SCHEDULE 40 IPS PLASTIC PIPE
(1120, 1220) C = 150
PSI LOSS PER 100 FEET OF TUBE (PSI/100 FT)
Sizes ½" thru 3½"
Flow GPM 1 thru 600

SIZE	0.50	0.75	1.00	1.25	1.50	2.00	2.50	3.00	3.50	SIZE
OD	0.840	1.050	1.315	1.660	1.900	2.375	2.875	3.500	4.000	OD
ID	0.622	0.824	1.049	1.380	1.610	2.067	2.469	3.068	3.548	ID
WALL THK	0.109	0.113	0.133	0.140	0.145	0.154	0.203	0.216	0.226	WALL THK

Flow G.P.M.	Vel F.P.S.	P.S.I. Loss	Vel F.P.S.	P.S.I. Loss	Vel F.P.S.	P.S.I. Loss	Vel F.P.S.	P.S.I. Loss	Vel F.P.S.	P.S.I. Loss	Vel F.P.S.	P.S.I. Loss	Vel F.P.S.	P.S.I. Loss	Vel F.P.S.	P.S.I. Loss	Vel F.P.S.	P.S.I. Loss	Flow G.P.M.
1	0.93	0.32	0.60	0.11	0.37	0.03	0.21	0.01	0.15	0.00									1
2	1.86	1.14	1.20	0.39	0.74	0.12	0.42	0.03	0.31	0.02	0.19	0.00							2
3	2.79	2.42	1.80	0.84	1.11	0.26	0.64	0.07	0.47	0.03	0.28	0.01	0.20	0.00					3
4	3.72	4.13	2.40	1.42	1.48	0.44	0.85	0.12	0.62	0.05	0.38	0.02	0.26	0.01					4
5	4.65	6.24	3.00	2.15	1.85	0.66	1.07	0.18	0.78	0.08	0.47	0.02	0.33	0.01	0.21	0.00			5
6	5.58	8.75	3.60	3.02	2.22	0.93	1.28	0.25	0.94	0.12	0.57	0.03	0.40	0.01	0.26	0.01			6
7	6.51	11.64	4.20	4.01	2.59	1.24	1.49	0.33	1.10	0.15	0.66	0.05	0.46	0.02	0.30	0.01			7
8	7.44	14.90	4.80	5.14	2.96	1.59	1.71	0.42	1.25	0.20	0.76	0.06	0.53	0.02	0.34	0.01	0.25	0.00	8
9	8.37	18.54	5.40	6.39	3.33	1.97	1.92	0.52	1.41	0.25	0.85	0.07	0.60	0.03	0.39	0.01	0.29	0.01	9
10	9.30	22.53	6.00	7.77	3.70	2.40	2.14	0.63	1.57	0.30	0.95	0.09	0.66	0.04	0.43	0.01	0.32	0.01	10
11	10.24	26.88	6.60	9.27	4.07	2.86	2.35	0.75	1.73	0.36	1.05	0.11	0.73	0.04	0.47	0.02	0.35	0.01	11
12	11.17	31.58	7.21	10.89	4.44	3.36	2.57	0.89	1.88	0.42	1.14	0.12	0.80	0.05	0.52	0.02	0.38	0.01	12
14	13.03	42.01	8.41	14.48	5.19	4.47	2.99	1.18	2.20	0.56	1.33	0.17	0.93	0.07	0.60	0.02	0.45	0.01	14
16	14.89	53.80	9.61	18.55	5.93	5.73	3.42	1.51	2.51	0.71	1.52	0.21	1.07	0.09	0.69	0.03	0.51	0.02	16
18	16.75	66.92	10.81	23.07	6.67	7.13	3.85	1.88	2.83	0.89	1.71	0.26	1.20	0.11	0.78	0.04	0.58	0.02	18
20	18.61	81.34	12.01	28.04	7.41	8.66	4.28	2.28	3.14	1.08	1.90	0.32	1.33	0.13	0.86	0.05	0.64	0.02	20
22			13.21	33.45	8.15	10.33	4.71	2.72	3.46	1.29	2.10	0.38	1.47	0.16	0.95	0.06	0.71	0.03	22
24			14.42	39.30	8.89	12.14	5.14	3.20	3.77	1.51	2.29	0.45	1.60	0.19	1.04	0.07	0.77	0.03	24
26			15.62	45.58	9.64	14.08	5.57	3.17	4.09	1.75	2.48	0.52	1.74	0.22	1.12	0.08	0.84	0.04	26
28			16.82	52.28	10.38	16.15	5.99	4.25	4.40	2.01	2.67	0.60	1.87	0.25	1.21	0.09	0.90	0.04	28
30			18.02	59.41	11.12	18.35	6.42	4.83	4.72	2.28	2.86	0.68	2.00	0.29	1.30	0.10	0.97	0.05	30
35					12.97	24.42	7.49	6.43	5.50	3.04	3.34	0.90	2.34	0.38	1.51	0.13	1.13	0.06	35
40					14.83	31.27	8.56	8.23	6.29	3.89	3.81	1.15	2.67	0.49	1.73	0.17	1.29	0.08	40
45					16.68	38.89	9.64	10.24	7.08	4.84	4.29	1.43	3.01	0.60	1.95	0.21	1.45	0.10	45
50					18.53	47.27	10.71	12.45	7.87	5.88	4.77	1.74	3.34	0.73	2.16	0.26	1.62	0.13	50
55							11.78	14.85	8.65	7.01	5.25	2.08	3.68	0.88	2.38	0.30	1.78	0.15	55
60							12.85	17.45	9.44	8.24	5.72	2.44	4.01	1.03	2.60	0.36	1.94	0.18	60
65							13.92	20.23	10.23	9.56	6.20	2.83	4.35	1.19	2.81	0.41	2.10	0.20	65
70							14.99	23.21	11.01	10.96	6.68	3.25	4.68	1.37	3.03	0.48	2.26	0.23	70
75							16.06	26.37	11.80	12.46	7.16	3.69	5.01	1.56	3.25	0.54	2.43	0.27	75
80							17.13	29.72	12.59	14.04	7.63	4.16	5.35	1.75	3.46	0.61	2.59	0.30	80
85							18.21	33.26	13.37	15.71	8.11	4.66	5.68	1.96	3.68	0.68	2.75	0.34	85
90							19.28	36.97	14.16	17.46	8.59	5.18	6.02	2.18	3.90	0.76	2.91	0.37	90
95									14.95	19.30	9.07	5.72	6.35	2.41	4.11	0.84	3.07	0.41	95
100									15.74	21.22	9.54	6.29	6.69	2.65	4.33	0.92	3.24	0.45	100
110									17.31	25.32	10.50	7.51	7.36	3.16	4.76	1.10	3.56	0.54	110
120									18.88	29.75	11.45	8.82	8.03	3.72	5.20	1.29	3.88	0.64	120
130											12.41	10.23	8.70	4.31	5.63	1.50	4.21	0.74	130
140											13.36	11.74	9.37	4.94	6.06	1.72	4.53	0.85	140
150											14.32	13.33	10.03	5.62	6.50	1.95	4.86	0.96	150
160											15.27	15.03	10.70	6.33	6.93	2.20	5.18	1.08	160
170											16.23	16.81	11.37	7.08	7.36	2.46	5.50	1.21	170
180											17.18	18.69	12.04	7.87	7.80	2.74	5.83	1.35	180
190											18.14	20.66	12.71	8.70	8.23	3.02	6.15	1.49	190
200											19.09	22.72	13.38	9.57	8.66	3.33	6.48	1.64	200
225													15.05	11.90	9.75	4.14	7.29	2.04	225
250													16.73	14.47	10.83	5.03	8.10	2.48	250
275													18.40	17.26	11.92	6.00	8.91	2.96	275
300															13.00	7.05	9.72	3.47	300
325															14.08	8.17	10.53	4.03	325
350															15.17	9.38	11.34	4.62	350
375															16.25	10.65	12.15	5.25	375
400															17.33	12.01	12.96	5.92	400
425															18.42	13.43	13.77	6.62	425
450															19.50	14.93	14.58	7.36	450
475																	15.39	8.14	475
500																	16.20	8.95	500
550																	17.82	10.67	550
600																	19.44	12.54	600

(Continued)

Note: Shaded areas of chart indicate velocities over 5' per second. **Use with Caution.**

FRICTION LOSS CHARACTERISTICS
PVC SCHEDULE 40 IPS PLASTIC PIPE
(1120, 1220) C = 150
PSI LOSS PER 100 FEET OF TUBE (PSI/100 FT)

Sizes 4" thru 12"
Flow GPM 1 thru 600

SIZE	4.00	5.00	6.00	8.00	10.00	12.00	SIZE
OD	4.500	5.563	6.625	8.625	10.750	12.750	OD
ID	4.026	5.047	6.065	7.981	10.020	11.814	ID
WALL THK	0.237	0.258	0.280	0.322	0.365	0.406	WALL THK

Flow G.P.M.	4.00 Velocity F.P.S.	P.S.I. Loss	5.00 Velocity F.P.S.	P.S.I. Loss	6.00 Velocity F.P.S.	P.S.I. Loss	8.00 Velocity F.P.S.	P.S.I. Loss	10.00 Velocity F.P.S.	P.S.I. Loss	12.00 Velocity F.P.S.	P.S.I. Loss	Flow G.P.M.
1													1
2													2
3													3
4													4
5													5
6													6
7													7
8													8
9													9
10													10
11													11
12	0.30	0.00											12
14	0.35	0.01											14
16	0.40	0.01											16
18	0.45	0.01											18
20	0.50	0.01											20
22	0.55	0.01	0.35	0.00									22
24	0.60	0.02	0.38	0.01									24
26	0.65	0.02	0.41	0.01									26
28	0.70	0.02	0.44	0.01									28
30	0.75	0.03	0.48	0.01									30
35	0.88	0.04	0.56	0.01	0.38	0.00							35
40	1.00	0.04	0.64	0.01	0.44	0.01							40
45	1.13	0.06	0.72	0.02	0.49	0.01							45
50	1.25	0.07	0.80	0.02	0.55	0.01							50
55	1.38	0.08	0.88	0.03	0.61	0.01							55
60	1.51	0.10	0.96	0.03	0.66	0.01							60
65	1.63	0.11	1.04	0.04	0.72	0.02							65
70	1.76	0.13	1.12	0.04	0.77	0.02	0.44	0.00					70
75	1.88	0.14	1.20	0.05	0.83	0.02	0.48	0.01					75
80	2.01	0.16	1.28	0.05	0.88	0.02	0.51	0.01					80
85	2.13	0.18	1.36	0.06	0.94	0.02	0.54	0.01					85
90	2.26	0.20	1.44	0.07	0.99	0.03	0.57	0.01					90
95	2.39	0.22	1.52	0.07	1.05	0.03	0.60	0.01					95
100	2.51	0.25	1.60	0.08	1.10	0.03	0.64	0.01					100
110	2.76	0.29	1.76	0.10	1.22	0.04	0.70	0.01					110
120	3.02	0.34	1.92	0.11	1.33	0.05	0.76	0.01					120
130	3.27	0.40	2.08	0.13	1.44	0.05	0.83	0.01	0.52	0.00			130
140	3.52	0.46	2.24	0.15	1.55	0.06	0.89	0.02	0.56	0.01			140
150	3.77	0.52	2.40	0.17	1.66	0.07	0.96	0.02	0.60	0.01			150
160	4.02	0.59	2.56	0.20	1.77	0.08	1.02	0.02	0.65	0.01			160
170	4.27	0.66	2.72	0.22	1.88	0.09	1.09	0.02	0.69	0.01			170
180	4.53	0.73	2.88	0.24	1.99	0.10	1.15	0.03	0.73	0.01			180
190	4.78	0.81	3.04	0.27	2.10	0.11	1.21	0.03	0.77	0.01			190
200	5.03	0.89	3.20	0.30	2.21	0.12	1.28	0.03	0.81	0.01	0.58	0.00	200
225	5.66	1.10	3.60	0.37	2.49	0.15	1.44	0.04	0.91	0.01	0.65	0.01	225
250	6.29	1.34	4.00	0.45	2.77	0.18	1.60	0.05	1.01	0.02	0.73	0.01	250
275	6.92	1.60	4.40	0.53	3.05	0.22	1.76	0.06	1.11	0.02	0.80	0.01	275
300	7.55	1.88	4.80	0.63	3.32	0.26	1.92	0.07	1.21	0.02	0.87	0.01	300
325	8.18	2.18	5.20	0.73	3.60	0.30	2.08	0.08	1.32	0.03	0.95	0.01	325
350	8.81	2.50	5.60	0.83	3.88	0.34	2.24	0.09	1.42	0.03	1.02	0.01	350
375	9.43	2.84	6.00	0.95	4.15	0.39	2.40	0.10	1.52	0.03	1.09	0.02	375
400	10.06	3.20	6.40	1.07	4.43	0.44	2.56	0.11	1.62	0.04	1.16	0.02	400
425	10.69	3.58	6.80	1.19	4.71	0.49	2.72	0.13	1.72	0.04	1.24	0.02	425
450	11.32	3.98	7.20	1.33	4.99	0.54	2.88	0.14	1.82	0.05	1.31	0.02	450
475	11.95	4.40	7.60	1.46	5.26	0.60	3.04	0.16	1.93	0.05	1.38	0.02	475
500	12.58	4.84	8.00	1.61	5.54	0.66	3.20	0.17	2.03	0.06	1.46	0.03	500
550	13.84	5.77	8.80	1.92	6.10	0.79	3.52	0.21	2.23	0.07	1.60	0.03	550
600	15.10	6.78	9.61	2.26	6.65	0.92	3.84	0.24	2.43	0.08	1.75	0.04	600

(Continued)

Note: Shaded areas of chart indicate velocities over 5' per second. Use with Caution.

FRICTION LOSS CHARACTERISTICS
PVC SCHEDULE 80 IPS PLASTIC PIPE
(1120, 1220) C = 150
PSI LOSS PER 100 FEET OF TUBE (PSI/100 FT)

Sizes ½" thru 3½"
Flow GPM 1 thru 600

SIZE	0.50	0.75	1.00	1.25	1.50	2.00	2.50	3.00	3.50	SIZE
OD	0.840	1.050	1.315	1.660	1.900	2.375	2.875	3.500	4.000	OD
ID	0.546	0.742	0.957	1.278	1.500	1.939	2.323	2.900	3.364	ID
WALL THK	0.147	0.154	0.179	0.191	0.200	0.218	0.276	0.300	0.318	WALL THK

Flow G.P.M.	0.50 Velocity F.P.S.	0.50 P.S.I. Loss	0.75 Velocity F.P.S.	0.75 P.S.I. Loss	1.00 Velocity F.P.S.	1.00 P.S.I. Loss	1.25 Velocity F.P.S.	1.25 P.S.I. Loss	1.50 Velocity F.P.S.	1.50 P.S.I. Loss	2.00 Velocity F.P.S.	2.00 P.S.I. Loss	2.50 Velocity F.P.S.	2.50 P.S.I. Loss	3.00 Velocity F.P.S.	3.00 P.S.I. Loss	3.50 Velocity F.P.S.	3.50 P.S.I. Loss	Flow G.P.M.
1	1.36	0.81	0.74	0.18	0.44	0.05	0.24	0.01	0.18	0.01	0.10	0.00							1
2	2.73	2.92	1.48	0.66	0.89	0.19	0.49	0.05	0.36	0.02	0.21	0.01							2
3	4.10	6.19	2.22	1.39	1.33	0.40	0.74	0.10	0.54	0.05	0.32	0.01	0.22	0.00					3
4	5.47	10.54	2.96	2.37	1.78	0.69	0.99	0.17	0.72	0.08	0.43	0.02	0.30	0.01					4
5	6.84	15.93	3.70	3.58	2.22	1.04	1.24	0.25	0.90	0.12	0.54	0.03	0.37	0.01	0.24	0.00			5
6	8.21	22.33	4.44	5.02	2.67	1.46	1.49	0.36	1.08	0.16	0.65	0.05	0.45	0.02	0.29	0.01			6
7	9.58	29.71	5.18	6.68	3.11	1.94	1.74	0.47	1.26	0.22	0.75	0.06	0.52	0.03	0.33	0.01	0.25	0.00	7
8	10.94	38.05	5.92	8.56	3.56	2.48	1.99	0.61	1.45	0.28	0.86	0.08	0.60	0.03	0.38	0.01	0.28	0.01	8
9	12.31	47.33	6.66	10.64	4.00	3.09	2.24	0.76	1.63	0.35	0.97	0.10	0.68	0.04	0.43	0.01	0.32	0.01	9
10	13.68	57.52	7.41	12.93	4.45	3.75	2.49	0.92	1.81	0.42	1.08	0.12	0.75	0.05	0.48	0.02	0.36	0.01	10
11	15.05	68.63	8.15	15.43	4.90	4.47	2.74	1.10	1.99	0.50	1.19	0.14	0.83	0.06	0.53	0.02	0.39	0.01	11
12	16.42	80.63	8.89	18.13	5.34	5.26	2.99	1.29	2.17	0.59	1.30	0.17	0.90	0.07	0.58	0.02	0.43	0.01	12
14			10.37	24.12	6.23	6.99	3.49	1.71	2.53	0.79	1.51	0.23	1.05	0.09	0.67	0.03	0.50	0.02	14
16			11.85	30.88	7.12	8.95	3.99	2.19	2.90	1.01	1.73	0.29	1.20	0.12	0.77	0.04	0.57	0.02	16
18			13.33	38.41	8.01	11.14	4.49	2.73	3.26	1.25	1.95	0.36	1.36	0.15	0.87	0.05	0.64	0.02	18
20			14.82	46.69	8.90	13.54	4.99	3.31	3.62	1.52	2.17	0.44	1.51	0.18	0.97	0.06	0.72	0.03	20
22			16.30	55.70	9.80	16.15	5.40	3.95	3.98	1.81	2.38	0.52	1.66	0.22	1.06	0.07	0.79	0.04	22
24			17.78	65.44	10.69	18.97	5.99	4.64	4.35	2.13	2.60	0.61	1.81	0.25	1.16	0.09	0.86	0.04	24
26			19.26	75.90	11.58	22.01	6.49	5.39	4.71	2.47	2.82	0.71	1.96	0.29	1.26	0.10	0.93	0.05	26
28					12.47	25.24	6.99	6.18	5.07	2.83	3.03	0.81	2.11	0.34	1.35	0.11	1.00	0.06	28
30					13.36	28.69	7.49	7.02	5.43	3.22	3.25	0.92	2.26	0.38	1.45	0.13	1.08	0.06	30
35					15.59	38.16	8.74	9.34	6.34	4.29	3.79	1.23	2.64	0.51	1.69	0.17	1.26	0.08	35
40					17.81	48.87	9.99	11.96	7.25	5.49	4.34	1.57	3.02	0.65	1.94	0.22	1.44	0.11	40
45							11.24	14.88	8.16	6.83	4.88	1.96	3.40	0.81	2.18	0.28	1.62	0.13	45
50							12.49	18.09	9.06	8.30	5.42	2.38	3.78	0.99	2.42	0.34	1.80	0.16	50
55							13.73	21.58	9.97	9.90	5.96	2.84	4.15	1.18	2.66	0.40	1.98	0.19	55
60							14.98	25.35	10.87	11.63	6.51	3.33	4.53	1.38	2.91	0.47	2.16	0.23	60
65							16.23	29.40	11.78	13.49	7.05	3.87	4.91	1.61	3.15	0.55	2.34	0.27	65
70							17.48	33.72	12.69	15.47	7.59	4.44	5.29	1.84	3.39	0.63	2.52	0.30	70
75							18.73	38.32	13.59	17.58	8.13	5.04	5.67	2.09	3.63	0.71	2.70	0.35	75
80							19.98	43.19	14.50	19.81	8.68	5.68	6.04	2.36	3.88	0.80	2.88	0.39	80
85									15.41	22.16	9.22	6.36	6.42	2.63	4.12	0.90	3.06	0.44	85
90									16.32	24.64	9.77	7.07	6.80	2.93	4.36	1.00	3.24	0.48	90
95									17.22	27.23	10.30	7.81	7.18	3.24	4.60	1.10	3.42	0.54	95
100									18.13	29.95	10.85	8.59	7.56	3.57	4.85	1.21	3.60	0.59	100
110									19.94	35.73	11.93	10.25	8.31	4.26	5.33	1.45	3.96	0.70	110
120											13.02	12.04	9.07	5.00	5.82	1.70	4.32	0.82	120
130											14.10	13.96	9.82	5.60	6.30	1.97	4.68	0.96	130
140											15.19	16.02	10.58	6.65	6.79	2.27	5.04	1.10	140
150											16.27	18.20	11.34	7.56	7.27	2.57	5.40	1.25	150
160											17.36	20.51	12.09	8.51	7.76	2.89	5.76	1.41	160
170											18.44	22.95	12.85	9.53	8.24	3.24	6.12	1.57	170
180											19.53	25.51	13.60	10.59	8.73	3.60	6.48	1.75	180
190													14.36	11.71	9.21	3.98	6.85	1.93	190
200													15.12	12.87	9.70	4.37	7.21	2.12	200
225													17.01	16.01	10.91	5.44	8.11	2.64	225
250													18.90	19.46	12.12	6.61	9.01	3.21	250
275															13.34	7.89	9.91	3.83	275
300															14.55	9.27	10.81	4.50	300
325															15.76	10.75	11.71	5.22	325
350															16.97	12.33	12.61	5.99	350
375															18.19	14.01	13.52	6.81	375
400															19.40	15.79	14.42	7.67	400
425																	15.32	8.58	425
450																	16.22	9.54	450
475																	17.12	10.54	475
500																	18.02	11.51	500
550																	19.82	13.83	550
600																			600

(Continued)

Note: Shaded areas of chart indicate velocities over 5' per second. Use with Caution.

FRICTION LOSS CHARACTERISTICS
PVC SCHEDULE 80 IPS PLASTIC PIPE
(1120, 1220) C - 150
PSI LOSS PER 100 FEET OF TUBE (PSI/100 FT)

Sizes 4" thru 12"
Flow GPM 1 thru 600

SIZE	4.00		5.00		6.00		8.00		10.00		12.00		SIZE
OD	4.500		5.563		6.625		8.625		10.750		12.750		OD
ID	3.826		4.813		5.761		7.625		9.564		11.376		ID
WALL THK	0.337		0.375		0.432		0.500		0.593		0.687		WALL THK

Flow G.P.M.	Velocity F.P.S.	P.S.I. Loss	Velocity F.P.S.	P.S.I. Loss	Velocity F.P.S.	P.S.I. Loss	Velocity F.P.S.	P.S.I. Loss	Velocity F.P.S.	P.S.I. Loss	Velocity F.P.S.	P.S.I. Loss	Flow G.P.M.
1													1
2													2
3													3
4													4
5													5
6													6
7													7
8													8
9													9
10	0.27	0.00											10
11	0.30	0.01											11
12	0.33	0.01											12
14	0.39	0.01											14
16	0.44	0.01											16
18	0.50	0.01	0.31	0.00									18
20	0.55	0.02	0.35	0.01									20
22	0.61	0.02	0.38	0.01									22
24	0.66	0.02	0.42	0.01									24
26	0.72	0.03	0.45	0.01									26
28	0.78	0.03	0.49	0.01									28
30	0.83	0.03	0.52	0.01	0.36	0.00							30
35	0.97	0.05	0.61	0.01	0.43	0.01							35
40	1.11	0.06	0.70	0.02	0.49	0.01							40
45	1.25	0.07	0.79	0.02	0.55	0.01							45
50	1.39	0.09	0.88	0.03	0.61	0.01							50
55	1.53	0.10	0.96	0.03	0.67	0.01							55
60	1.67	0.12	1.05	0.04	0.73	0.02							60
65	1.81	0.14	1.14	0.05	0.79	0.02	0.45	0.00					65
70	1.95	0.16	1.23	0.05	0.86	0.02	0.49	0.01					70
75	2.09	0.18	1.32	0.06	0.92	0.03	0.52	0.01					75
80	2.22	0.21	1.40	0.07	0.98	0.03	0.56	0.01					80
85	2.36	0.23	1.49	0.08	1.04	0.03	0.59	0.01					85
90	2.50	0.26	1.58	0.08	1.10	0.04	0.63	0.01					90
95	2.64	0.29	1.67	0.09	1.16	0.04	0.66	0.01					95
100	2.78	0.31	1.76	0.10	1.22	0.04	0.70	0.01					100
110	3.06	0.38	1.93	0.12	1.35	0.05	0.77	0.01	0.49	0.00			110
120	3.34	0.44	2.11	0.14	1.47	0.06	0.84	0.02	0.53	0.01			120
130	3.62	0.51	2.28	0.17	1.59	0.07	0.91	0.02	0.57	0.01			130
140	3.90	0.59	2.46	0.19	1.72	0.08	0.98	0.02	0.62	0.01			140
150	4.18	0.67	2.64	0.22	1.84	0.09	1.05	0.02	0.66	0.01			150
160	4.45	0.75	2.81	0.25	1.96	0.10	1.12	0.03	0.71	0.01			160
170	4.73	0.84	2.99	0.28	2.08	0.11	1.19	0.03	0.75	0.01			170
180	5.01	0.93	3.17	0.31	2.21	0.13	1.26	0.03	0.80	0.01	0.56	0.00	180
190	5.29	1.03	3.34	0.34	2.33	0.14	1.33	0.04	0.84	0.01	0.59	0.01	190
200	5.57	1.14	3.52	0.37	2.45	0.16	1.40	0.04	0.89	0.01	0.63	0.01	200
225	6.27	1.41	3.96	0.46	2.76	0.19	1.57	0.05	1.00	0.02	0.70	0.01	225
250	6.96	1.72	4.40	0.56	3.07	0.23	1.75	0.06	1.11	0.02	0.78	0.01	250
275	7.66	2.05	4.84	0.67	3.38	0.28	1.92	0.07	1.22	0.02	0.86	0.01	275
300	8.36	2.41	5.28	0.79	3.68	0.33	2.10	0.08	1.33	0.03	0.94	0.01	300
325	9.05	2.79	5.72	0.91	3.99	0.38	2.28	0.10	1.44	0.03	1.02	0.01	325
350	9.75	3.20	6.16	1.05	4.30	0.44	2.45	0.11	1.56	0.04	1.10	0.02	350
375	10.45	3.64	6.60	1.19	4.60	0.50	2.63	0.13	1.67	0.04	1.18	0.02	375
400	11.14	4.10	7.04	1.34	4.91	0.56	2.80	0.14	1.78	0.05	1.26	0.02	400
425	11.84	4.59	7.48	1.50	5.22	0.63	2.98	0.16	1.89	0.05	1.33	0.02	425
450	12.54	5.10	7.92	1.67	5.53	0.70	3.15	0.18	2.00	0.06	1.41	0.03	450
475	13.23	5.64	8.36	1.85	5.83	0.77	3.33	0.20	2.11	0.07	1.49	0.03	475
500	13.93	6.20	8.80	2.03	6.14	0.85	3.50	0.22	2.23	0.07	1.57	0.03	500
550	15.32	7.40	9.68	2.42	6.76	1.01	3.85	0.26	2.45	0.09	1.73	0.04	550
600	16.72	8.69	10.56	2.84	7.37	1.19	4.21	0.30	2.67	0.10	1.89	0.04	600

(Continued)

Note: Shaded areas of chart indicate velocities over 5' per second. Use with Caution.

FRICTION LOSS CHARACTERISTICS
SCHEDULE 40 STANDARD STEEL PIPE C = 100
PSI LOSS PER 100 FEET OF PIPE (PSI/100 FT)

Sizes ½" thru 3½"
Flow GPM 1 thru 600

SIZE	0.50	0.75	1.00	1.25	1.50	2.00	2.50	3.00	3.50	SIZE
OD	0.840	1.050	1.315	1.660	1.900	2.375	2.875	3.500	4.000	OD
ID	0.622	0.824	1.049	1.380	1.610	2.067	2.469	3.068	3.548	ID
WALL THK	0.109	0.113	0.133	0.140	0.145	0.154	0.203	0.216	0.226	WALL THK

Flow G.P.M.	Velocity F.P.S.	P.S.I. Loss	Velocity F.P.S.	P.S.I. Loss	Velocity F.P.S.	P.S.I. Loss	Velocity F.P.S.	P.S.I. Loss	Velocity F.P.S.	P.S.I. Loss	Velocity F.P.S.	P.S.I. Loss	Velocity F.P.S.	P.S.I. Loss	Velocity F.P.S.	P.S.I. Loss	Velocity F.P.S.	P.S.I. Loss	Flow G.P.M.
1	1.05	0.91	0.60	0.23	0.37	0.07	0.21	0.02	0.15	0.01	0.09	0.00							1
2	2.10	3.28	1.20	0.84	0.74	0.26	0.42	0.07	0.31	0.03	0.19	0.01	0.13	0.00					2
3	3.16	6.95	1.80	1.77	1.11	0.55	0.64	0.14	0.47	0.07	0.28	0.02	0.20	0.01	0.13	0.00			3
4	4.21	11.85	2.40	3.02	1.48	0.93	0.85	0.25	0.62	0.12	0.38	0.03	0.26	0.01	0.17	0.01			4
5	5.27	17.91	3.00	4.56	1.85	1.41	1.07	0.37	0.78	0.18	0.47	0.05	0.33	0.02	0.21	0.01	0.16	0.00	5
6	6.32	25.10	3.60	6.39	2.22	1.97	1.28	0.52	0.94	0.25	0.57	0.07	0.40	0.03	0.26	0.01	0.19	0.01	6
7	7.38	33.40	4.20	8.50	2.59	2.63	1.49	0.69	1.10	0.33	0.66	0.10	0.46	0.04	0.30	0.01	0.22	0.01	7
8	8.43	42.77	4.80	10.89	2.96	3.36	1.71	0.89	1.25	0.42	0.76	0.12	0.53	0.05	0.34	0.02	0.25	0.01	8
9	9.49	53.19	5.40	13.54	3.33	4.18	1.92	1.10	1.41	0.52	0.85	0.15	0.60	0.06	0.39	0.02	0.29	0.01	9
10	10.54	64.65	6.00	16.46	3.70	5.08	2.14	1.34	1.57	0.63	0.95	0.19	0.66	0.08	0.43	0.03	0.32	0.01	10
11	11.60	77.13	6.60	19.63	4.07	6.07	2.35	1.60	1.73	0.75	1.05	0.22	0.73	0.09	0.47	0.03	0.35	0.02	11
12	12.65	90.62	7.21	23.07	4.44	7.13	2.57	1.88	1.88	0.89	1.14	0.26	0.80	0.11	0.52	0.04	0.38	0.02	12
14	14.76	20.56	8.41	30.69	5.19	9.48	2.99	2.50	2.20	1.18	1.33	0.35	0.93	0.15	0.60	0.05	0.45	0.03	14
16	16.87	54.39	9.61	39.30	5.93	12.14	3.42	3.20	2.51	1.51	1.52	0.45	1.07	0.19	0.69	0.07	0.51	0.03	16
18	18.98	92.02	10.81	48.88	6.67	15.10	3.85	3.98	2.83	1.88	1.71	0.56	1.20	0.23	0.78	0.08	0.58	0.04	18
20			12.01	59.41	7.41	18.35	4.28	4.83	3.14	2.28	1.90	0.68	1.33	0.29	0.86	0.10	0.64	0.05	20
22			13.21	70.88	8.15	21.90	4.71	5.77	3.46	2.72	2.10	0.81	1.47	0.34	0.95	0.12	0.71	0.06	22
24			14.42	83.27	8.89	25.72	5.14	6.77	3.77	3.20	2.29	0.95	1.60	0.40	1.04	0.14	0.77	0.07	24
26			15.62	96.57	9.64	29.83	5.57	7.86	4.09	3.71	2.48	1.10	1.74	0.46	1.12	0.16	0.84	0.08	26
28			16.82	10.78	10.38	34.22	5.99	9.01	4.40	4.26	2.67	1.26	1.87	0.53	1.21	0.18	0.90	0.09	28
30			18.02	25.88	11.12	38.89	6.42	10.24	4.72	4.84	2.86	1.43	2.00	0.60	1.30	0.21	0.97	0.10	30
35					12.97	51.74	7.49	13.62	5.50	6.44	3.34	1.91	2.34	0.80	1.51	0.28	1.13	0.14	35
40					14.83	66.25	8.56	17.45	6.29	8.24	3.81	2.44	2.67	1.03	1.73	0.36	1.29	0.18	40
45					16.68	82.40	9.64	21.70	7.08	10.25	4.29	3.04	3.01	1.28	1.95	0.44	1.45	0.22	45
50					18.53	00.16	10.71	26.37	7.87	12.46	4.77	3.69	3.34	1.56	2.16	0.54	1.62	0.27	50
55							11.78	31.47	8.65	14.86	5.25	4.41	3.68	1.86	2.38	0.65	1.78	0.32	55
60							12.85	36.97	9.44	17.46	5.72	5.18	4.01	2.18	2.60	0.76	1.94	0.37	60
65							13.92	42.88	10.23	20.25	6.20	6.00	4.35	2.53	2.81	0.88	2.10	0.43	65
70							14.99	49.18	11.01	23.23	6.68	6.89	4.68	2.90	3.03	1.01	2.26	0.50	70
75							16.06	55.89	11.80	26.40	7.16	7.83	5.01	3.30	3.25	1.15	2.43	0.56	75
80							17.13	62.98	12.59	29.75	7.63	8.82	5.35	3.72	3.46	1.29	2.59	0.64	80
85							18.21	70.47	13.37	33.29	8.11	9.87	5.68	4.16	3.68	1.44	2.75	0.71	85
90							19.28	78.33	14.16	37.00	8.59	10.97	6.02	4.62	3.90	1.61	2.91	0.79	90
95									14.95	40.90	9.07	12.13	6.35	5.11	4.11	1.78	3.07	0.88	95
100									15.74	44.97	9.54	13.33	6.69	5.62	4.33	1.95	3.24	0.96	100
110									17.31	53.66	10.50	15.91	7.36	6.70	4.76	2.33	3.56	1.15	110
120									18.88	63.04	11.45	18.69	8.03	7.87	5.20	2.74	3.88	1.35	120
130											12.41	21.68	8.70	9.13	5.63	3.17	4.21	1.56	130
140											13.36	24.87	9.37	10.47	6.06	3.64	4.53	1.79	140
150											14.32	28.26	10.03	11.90	6.50	4.14	4.86	2.04	150
160											15.27	31.84	10.70	13.41	6.93	4.66	5.18	2.30	160
170											16.23	35.63	11.37	15.01	7.36	5.22	5.50	2.57	170
180											17.18	39.61	12.04	16.68	7.80	5.80	5.83	2.86	180
190											18.14	43.78	12.71	18.44	8.23	6.41	6.15	3.16	190
200											19.09	48.14	13.38	20.28	8.66	7.05	6.48	3.47	200
225													15.08	25.22	9.75	8.76	7.29	4.32	225
250													16.73	30.65	10.83	10.65	8.10	5.25	250
275													18.40	36.57	11.92	12.71	8.91	6.27	275
300															13.00	14.93	9.72	7.36	300
325															14.08	17.32	10.53	8.54	325
350															15.17	19.87	11.34	9.79	350
375															16.25	22.57	12.15	11.13	375
400															17.33	25.44	12.96	12.54	400
425															18.42	28.46	13.77	14.03	425
450															19.50	31.64	14.58	15.60	450
475																	15.39	17.24	475
500																	16.20	18.96	500
550																	17.82	22.62	550
600																	19.44	26.57	600

(Continued)

Note: Shaded areas of chart indicate velocities over 5' per second. **Use with Caution.**

FRICTION LOSS CHARACTERISTICS
SCHEDULE 40 STANDARD STEEL PIPE C = 100
PSI LOSS PER 100 FEET OF PIPE (PSI/100 FT)
Sizes 4" thru 6"
Flow GPM 1 thru 600

SIZE	4.00	5.00	6.00	8.00	10.00	12.00	14.00	16.00	18.00	SIZE
OD	4.500	5.563	6.625	8.625	10.750	12.750	14.000	16.000	18.000	OD
ID	4.026	5.047	6.065	7.981	10.020	11.938	13.126	15.000	16.876	ID
WALL THK	0.237	0.258	0.280	0.322	0.365	0.406	0.438	0.500	0.562	WALL THK

Flow G.P.M.	4.00 Velocity F.P.S.	4.00 P.S.I. Loss	5.00 Velocity F.P.S.	5.00 P.S.I. Loss	6.00 Velocity F.P.S.	6.00 P.S.I. Loss	8.00 Velocity F.P.S.	8.00 P.S.I. Loss	10.00 Velocity F.P.S.	10.00 P.S.I. Loss	12.00 Velocity F.P.S.	12.00 P.S.I. Loss	14.00 Velocity F.P.S.	14.00 P.S.I. Loss	16.00 Velocity F.P.S.	16.00 P.S.I. Loss	18.00 Velocity F.P.S.	18.00 P.S.I. Loss	Flow G.P.M.
1																			1
2																			2
3																			3
4																			4
5																			5
6																			6
7																			7
8	0.20	0.00																	8
9	0.22	0.01																	9
10	0.25	0.01																	10
11	0.27	0.01																	11
12	0.30	0.01																	12
14	0.35	0.01	0.22	0.00															14
16	0.40	0.02	0.25	0.01															16
18	0.45	0.02	0.28	0.01															18
20	0.50	0.03	0.32	0.01															20
22	0.55	0.03	0.35	0.01	0.24	0.00													22
24	0.60	0.04	0.38	0.01	0.26	0.01													24
26	0.65	0.04	0.41	0.01	0.28	0.01													26
28	0.70	0.05	0.44	0.02	0.31	0.01													28
30	0.75	0.06	0.48	0.02	0.33	0.01													30
35	0.88	0.07	0.56	0.02	0.38	0.01													35
40	1.00	0.10	0.64	0.03	0.44	0.01													40
45	1.13	0.12	0.72	0.04	0.49	0.02	0.28	0.00											45
50	1.25	0.14	0.80	0.05	0.55	0.02	0.32	0.01											50
55	1.38	0.17	0.88	0.06	0.61	0.02	0.35	0.01											55
60	1.51	0.20	0.96	0.07	0.66	0.03	0.38	0.01											60
65	1.63	0.23	1.04	0.08	0.72	0.03	0.41	0.01											65
70	1.76	0.27	1.12	0.09	0.77	0.04	0.44	0.01											70
75	1.88	0.31	1.20	0.10	0.83	0.04	0.48	0.01											75
80	2.01	0.34	1.28	0.11	0.88	0.05	0.51	0.00											80
85	2.13	0.39	1.36	0.13	0.94	0.05	0.54	0.01	0.34	0.00									85
90	2.26	0.43	1.44	0.14	0.99	0.06	0.57	0.02	0.36	0.01									90
95	2.39	0.47	1.52	0.16	1.05	0.06	0.60	0.02	0.38	0.01									95
100	2.51	0.52	1.60	0.17	1.10	0.07	0.64	0.02	0.40	0.01									100
110	2.76	0.62	1.76	0.21	1.22	0.08	0.70	0.02	0.44	0.01									110
120	3.02	0.73	1.92	0.24	1.33	0.10	0.76	0.03	0.48	0.01									120
130	3.27	0.85	2.08	0.28	1.44	0.12	0.83	0.03	0.52	0.01									130
140	3.52	0.97	2.24	0.32	1.55	0.13	0.89	0.03	0.56	0.01	0.40	0.00							140
150	3.77	1.10	2.40	0.37	1.66	0.15	0.96	0.04	0.60	0.01	0.42	0.01							150
160	4.02	1.24	2.56	0.41	1.77	0.17	1.02	0.04	0.65	0.01	0.45	0.01							160
170	4.27	1.39	2.72	0.46	1.88	0.19	1.08	0.05	0.69	0.02	0.48	0.01							170
180	4.53	1.55	2.88	0.51	1.99	0.21	1.15	0.06	0.73	0.02	0.51	0.01	0.42	0.00					180
190	4.78	1.71	3.04	0.57	2.10	0.23	1.21	0.06	0.77	0.02	0.54	0.01	0.44	0.01					190
200	5.03	1.88	3.20	0.63	2.21	0.26	1.28	0.07	0.81	0.02	0.57	0.01	0.47	0.01					200
225	5.66	2.34	3.60	0.78	2.49	0.32	1.44	0.08	0.91	0.03	0.64	0.01	0.53	0.01					225
250	6.29	2.84	4.00	0.95	2.77	0.39	1.60	0.10	1.01	0.03	0.71	0.01	0.59	0.01	0.45	0.00			250
275	6.92	3.39	4.40	1.13	3.05	0.46	1.76	0.12	1.11	0.04	0.78	0.02	0.65	0.01	0.49	0.01			275
300	7.55	3.98	4.80	1.33	3.32	0.54	1.92	0.14	1.21	0.05	0.85	0.02	0.71	0.01	0.54	0.01			300
325	8.18	4.62	5.20	1.54	3.60	0.63	2.08	0.17	1.32	0.05	0.93	0.02	0.76	0.01	0.58	0.01			325
350	8.81	5.30	5.60	1.76	3.88	0.72	2.24	0.19	1.42	0.06	1.00	0.03	0.82	0.02	0.63	0.01	0.50	0.00	350
375	9.43	6.02	6.00	2.00	4.15	0.82	2.40	0.22	1.52	0.07	1.07	0.03	0.88	0.02	0.67	0.01	0.53	0.01	375
400	10.06	6.78	6.40	2.26	4.43	0.92	2.56	0.24	1.62	0.08	1.14	0.03	0.94	0.02	0.72	0.01	0.57	0.01	400
425	10.69	7.59	6.80	2.53	4.71	1.03	2.72	0.27	1.72	0.09	1.21	0.04	1.00	0.02	0.77	0.01	0.60	0.01	425
450	11.32	8.43	7.20	2.81	4.99	1.15	2.88	0.30	1.82	0.10	1.28	0.04	1.06	0.03	0.81	0.01	0.64	0.01	450
475	11.95	9.32	7.60	3.10	5.26	1.27	3.04	0.33	1.93	0.11	1.35	0.05	1.12	0.03	0.86	0.02	0.68	0.01	475
500	12.58	10.25	8.00	3.41	5.54	1.40	3.20	0.37	2.03	0.12	1.43	0.05	1.18	0.03	0.90	0.02	0.71	0.01	500
550	13.84	12.23	8.80	4.07	6.10	1.67	3.52	0.44	2.23	0.14	1.57	0.06	1.30	0.04	0.99	0.02	0.78	0.01	550
600	15.10	14.37	9.61	4.78	6.65	1.96	3.84	0.51	2.43	0.17	1.71	0.07	1.42	0.05	1.08	0.02	0.85	0.01	600

(Continued)

Note: Shaded areas of chart indicate velocities over 5' per second. Use with Caution.

LOSS OF PRESSURE DUE TO FRICTION IN UNLINED IRON PIPE

Expressed in Pounds Per Square Inch

Per 100 Feet of Length

PIPE SIZES

G.P.M.	1/2"	3/4"	1"	1¼"	1½"	2"	2½"	3"	4"
1	.9								
2	3.2	.8							
3	6.8	1.8	.6						
4	11.7	3.0	.9	.3					
5	17.8	4.5	1.4	.4	.1				
6	24.9	6.4	1.9	.5	.2				
7		9.1	2.5	.7	.3				
8		10.8	3.4	.8	.4	.1			
10		16.5	5.1	1.3	.6	.2			
12		23.0	7.1	1.8	.8	.3			
14		29.0	9.5	2.4	1.1	.4			
16			12.1	3.1	1.4	.5			
18			15.2	3.9	1.8	.6	.1		
20			18.2	4.8	2.2	.7	.2		
25			27.8	7.2	3.4	1.1	.3	.1	
30				10.2	4.8	1.6	.4	.2	
35				13.5	6.3	2.2	.5	.3	
40				17.3	8.1	2.8	.7	.4	
50				26.0	12.3	4.3	1.4	.6	.1
60					17.2	6.0	1.9	.8	.2
70					23.0	8.0	2.6	1.1	.2
80					29.5	10.3	3.4	1.4	.3
90						12.7	4.2	1.7	.4
100						15.5	5.2	2.1	.5
120						21.7	7.3	3.0	.7
140							9.6	3.9	.9
160							12.6	5.1	1.2
180							15.6	6.4	1.5
200							18.7	7.7	1.9
220							22.6	9.2	2.2
240								10.9	2.6
260								12.5	3.1
280								14.6	3.5
300								16.5	4.0
320								18.4	4.6
340								20.7	5.1
360								23.0	5.7

Per 1000 Feet of Length

PIPE SIZES

G.P.M.	2"	2½"	3"	4"	6"	8"	10"
25	11.7	3.9	1.7	.4			
50	43.0	14.3	6.0	1.3	.2		
100		52.0	21.7	5.2	.7		
150			46.0	11.3	1.6		
200			77.0	19.1	2.6		
250				28.8	4.0	1.0	.3
300				40.7	5.6	1.5	.5
350				53.7	7.5	1.8	.6
400					9.6	2.3	.8
450					11.9	2.9	1.0
500					14.4	3.6	1.2
550					17.3	4.2	1.4
600					20.3	5.0	1.7
650					23.6	5.8	1.9
700					27.1	6.7	2.2
750					30.6	7.6	2.6
800					34.5	8.5	2.9
850					38.8	9.5	3.2
900					43.2	10.6	3.5
950						11.7	3.9
1000						12.9	4.3
1100						15.3	5.2
1200						18.0	6.1
1300						20.8	7.1
1400						23.8	8.1
1500						27.3	9.2
1600							10.3

Figures above shaded area are at velocities of less than 5 feet per second; in shaded area between 5 and 10 feet per second; below shaded area more than 10 feet per second.

Velocities above 5 feet per second should be used with caution. Velocities above 10 feet per second should be avoided.

These friction loss tables are based on the Williams and Hazen Formula, using the coefficient of c equal to 100 as standard. To find friction loss values for other than c = 100 multiply the result in these tables by the following factors:

Type of Pipe	C Factor	Multiplication Factor
Plastic	150	0.45
Asbestos Cement	140	0.53
Smooth and Straight Brass or Aluminum	120	0.714
New Smooth and Straight Iron	110	0.838
Ordinary Iron	100	1.000
Old Iron	90	1.224
Rough Iron	80	1.513

FRICTION LOSS CHARACTERISTICS
POLYETHYLENE (PE) SDR-PRESSURE RATED TUBE
(2306, 3206, 3306) SDR 7, 9, 11.5, 15 C = 140
PSI LOSS PER 100 FEET OF TUBE (PSI/100 FT)

Sizes 1/2" thru 6"
Flow GPM 1 thru 1800

SIZE	0.50	0.75	1.00	1.25	1.50	2.00	2.50	3.00	4.00	6.00	SIZE
OD	0.000	0.000	0.000	0.000	0.000	0.000	0.000	0.000	0.000	0.000	OD
ID	0.622	0.824	1.049	1.380	1.610	2.067	2.469	3.068	4.026	6.065	ID
WALL THK	0.000	0.000	0.000	0.000	0.000	0.000	0.000	0.000	0.000	0.000	WALL THK

Flow GPM	0.50 Vel FPS	0.50 PSI Loss	0.75 Vel FPS	0.75 PSI Loss	1.00 Vel FPS	1.00 PSI Loss	1.25 Vel FPS	1.25 PSI Loss	1.50 Vel FPS	1.50 PSI Loss	2.00 Vel FPS	2.00 PSI Loss	2.50 Vel FPS	2.50 PSI Loss	3.00 Vel FPS	3.00 PSI Loss	4.00 Vel FPS	4.00 PSI Loss	6.00 Vel FPS	6.00 PSI Loss	Flow GPM
1	1.05	0.49	0.60	0.12	0.37	0.04	0.21	0.01	0.15	0.00	0.09	0.00									1
2	2.10	1.76	1.20	0.45	0.74	0.14	0.42	0.04	0.31	0.02	0.19	0.01									2
3	3.16	3.73	1.80	0.95	1.11	0.29	0.64	0.08	0.47	0.04	0.28	0.01	0.20	0.00							3
4	4.21	6.35	2.40	1.62	1.48	0.50	0.85	0.13	0.62	0.06	0.38	0.02	0.26	0.01							4
5	5.27	9.60	3.00	2.44	1.85	0.76	1.07	0.20	0.78	0.09	0.47	0.03	0.33	0.01	0.21	0.00					5
6	6.32	13.46	3.60	3.43	2.22	1.06	1.28	0.28	0.94	0.13	0.57	0.04	0.40	0.02	0.26	0.01					6
7	7.38	17.91	4.20	4.56	2.59	1.41	1.49	0.37	1.10	0.18	0.66	0.05	0.46	0.02	0.30	0.01					7
8	8.43	22.93	4.80	5.84	2.96	1.80	1.71	0.47	1.25	0.22	0.76	0.07	0.53	0.03	0.34	0.01					8
9	9.49	28.52	5.40	7.26	3.33	2.24	1.92	0.59	1.41	0.28	0.85	0.08	0.60	0.03	0.39	0.01					9
10	10.54	34.67	6.00	8.82	3.70	2.73	2.14	0.72	1.57	0.34	0.95	0.10	0.66	0.04	0.43	0.01					10
11	11.60	41.36	6.00	10.53	4.07	3.25	2.35	0.86	1.73	0.40	1.05	0.12	0.73	0.05	0.47	0.02	0.27	0.00			11
12	12.65	48.60	7.21	12.37	4.44	3.82	2.57	1.01	1.88	0.48	1.14	0.14	0.80	0.06	0.52	0.02	0.30	0.01			12
14	14.76	64.65	8.41	16.46	5.19	5.08	2.99	1.34	2.20	0.63	1.33	0.19	0.93	0.08	0.60	0.03	0.35	0.01			14
16	16.87	82.79	9.61	21.07	5.93	6.51	3.42	1.71	2.51	0.81	1.52	0.24	1.07	0.10	0.69	0.04	0.40	0.01			16
18	18.98	102.97	10.81	26.21	6.67	8.10	3.85	2.13	2.83	1.01	1.71	0.30	1.20	0.13	0.78	0.04	0.45	0.01			18
20			12.01	31.86	7.41	9.84	4.28	2.59	3.14	1.22	1.90	0.36	1.33	0.15	0.86	0.05	0.50	0.01			20
22			13.21	38.01	8.15	11.74	4.71	3.09	3.46	1.46	2.10	0.43	1.47	0.18	0.95	0.06	0.55	0.02			22
24			14.42	44.65	8.89	13.79	5.14	3.63	3.77	1.72	2.29	0.51	1.60	0.21	1.04	0.07	0.60	0.02			24
26			15.62	51.79	9.64	16.00	5.57	4.21	4.09	1.99	2.48	0.59	1.74	0.25	1.12	0.09	0.65	0.02			26
28			16.82	59.41	10.38	18.35	5.99	4.83	4.40	2.28	2.67	0.68	1.87	0.29	1.21	0.10	0.70	0.03			28
30			18.02	67.50	11.12	20.85	6.42	5.49	4.72	2.59	2.86	0.77	2.00	0.32	1.30	0.11	0.75	0.03	0.33	0.00	30
35					12.97	27.74	7.49	7.31	5.50	3.45	3.34	1.02	2.34	0.43	1.51	0.15	0.88	0.04	0.38	0.01	35
40					14.83	35.53	8.56	9.36	6.29	4.42	3.81	1.31	2.67	0.55	1.73	0.19	1.00	0.05	0.44	0.01	40
45					16.68	44.19	9.64	11.64	7.08	5.50	4.29	1.63	3.01	0.69	1.95	0.24	1.13	0.06	0.49	0.01	45
50					18.53	53.71	10.71	14.14	7.87	6.68	4.77	1.98	3.34	0.83	2.16	0.29	1.25	0.08	0.55	0.01	50
55							11.78	16.87	8.65	7.97	5.25	2.36	3.68	1.00	2.38	0.35	1.38	0.09	0.61	0.01	55
60							12.85	19.82	9.44	9.36	5.72	2.78	4.01	1.17	2.60	0.41	1.51	0.11	0.66	0.01	60
65							13.92	22.99	10.23	10.86	6.20	3.22	4.35	1.36	2.81	0.47	1.63	0.13	0.72	0.02	65
70							14.99	26.37	11.01	12.46	6.68	3.69	4.68	1.56	3.03	0.54	1.76	0.14	0.77	0.02	70
75							16.06	29.97	11.80	14.16	7.16	4.20	5.01	1.77	3.25	0.61	1.88	0.16	0.83	0.02	75
80							17.13	33.77	12.59	15.95	7.63	4.73	5.35	1.99	3.46	0.69	2.01	0.18	0.88	0.03	80
85							18.21	37.79	13.37	17.85	8.11	5.29	5.68	2.23	3.68	0.77	2.13	0.21	0.94	0.03	85
90							19.28	42.01	14.16	19.84	8.59	5.88	6.02	2.48	3.90	0.86	2.26	0.23	0.99	0.03	90
95									14.95	21.93	9.07	6.50	6.35	2.74	4.11	0.95	2.39	0.25	1.05	0.03	95
100									15.74	24.12	9.54	7.15	6.69	3.01	4.33	1.05	2.51	0.28	1.10	0.04	100
110									17.31	28.77	10.50	8.53	7.36	3.59	4.76	1.25	2.76	0.33	1.22	0.05	110
120									18.88	33.80	11.45	10.02	8.03	4.22	5.20	1.47	3.02	0.39	1.33	0.05	120
130											12.41	11.62	8.70	4.90	5.63	1.70	3.27	0.45	1.44	0.06	130
140											13.36	13.33	9.37	5.62	6.06	1.95	3.52	0.52	1.55	0.07	140
150											14.32	15.15	10.03	6.38	6.50	2.22	3.77	0.59	1.66	0.08	150
160											15.27	17.08	10.70	7.19	6.93	2.50	4.02	0.67	1.77	0.09	160
170											16.23	19.11	11.37	8.05	7.36	2.80	4.27	0.75	1.88	0.10	170
180											17.18	21.24	12.04	8.95	7.79	3.11	4.53	0.83	1.99	0.11	180
190											18.14	23.48	12.71	9.89	8.23	3.44	4.78	0.92	2.10	0.12	190
200											19.09	25.81	13.38	10.87	8.66	3.78	5.03	1.01	2.21	0.14	200
225													15.05	13.52	9.75	4.70	5.66	1.25	2.49	0.17	225
250													16.73	16.44	10.83	5.71	6.29	1.52	2.77	0.21	250
275													18.40	19.61	11.92	6.82	6.92	1.82	3.05	0.25	275
300															13.00	8.01	7.55	2.13	3.32	0.29	300
325															14.08	9.29	8.18	2.48	3.60	0.34	325
350															15.17	10.65	8.81	2.84	3.88	0.39	350
375															16.25	12.10	9.43	3.23	4.15	0.44	375
400															17.33	13.64	10.06	3.64	4.43	0.50	400
425															18.42	15.26	10.69	4.07	4.71	0.55	425
450															19.50	16.97	11.32	4.52	4.99	0.62	450
475																	11.95	5.00	5.26	0.68	475
500																	12.58	5.50	5.54	0.75	500
550																	13.84	6.56	6.10	0.89	550
600																	15.10	7.70	6.65	1.05	600

Flow GPM 650 thru 1800

SIZE	4.00	6.00		
OD	0.000	0.000		
ID	4.026	6.065		
WALL THK	0.000	0.000		

Flow GPM	4.00 Vel FPS	4.00 PSI Loss	6.00 Vel FPS	6.00 PSI Loss
650	16.36	8.94	7.20	1.22
700	17.62	10.25	7.76	1.40
750	18.87	11.65	8.31	1.59
800			8.87	1.79
850			9.42	2.00
900			9.98	2.22
950			10.53	2.46
1000			11.09	2.70
1050			11.64	2.96
1100			12.20	3.22
1150			12.75	3.50
1200			13.31	3.79
1250			13.86	4.09
1300			14.41	4.39
1350			14.97	4.71
1400			15.52	5.04
1450			16.08	5.38
1500			16.63	5.73
1550			17.19	6.09
1600			17.74	6.45
1650			18.30	6.83
1700			18.85	7.22
1750			19.41	7.62
1800			19.96	8.03

Note: Shaded areas of chart indicate velocities over 5' per second. **Use with Caution.**

Appendix 5

Sprinklers

SURE-QUICK® VALVES & KEYS

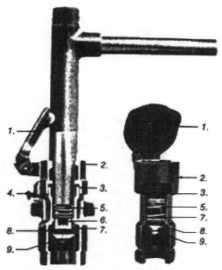

Two Piece Valves Models 33D, 44, 55

One Piece Valve Models 3, 5 and 7

Two Piece Valve Model 44

1. Cover
2. Upper Body
3. Flange Packing
4. Locking Screw
5. Spring
6. Valve Cage
7. Neoprene Valve Disc
8. Disc Retainer
9. Retainer Washer and Screw
10. Locking Valve Cover Key

impact sprinklers/rain guns

51DR rotor: full-circle, double-nozzle configuration.

91DR rotor: full-circle, double-nozzle configuration.

95DR rotor: full-circle or part-circle, double-nozzle configuration.

51DR Performance

psi	Radius'	gpm	Nozzle
50	54	14.3	13×11
	60	15.8	14×11
	61	18.3	16×11
	62	21.3	18×11.5
	62	24.4	20×11.5
60	57	15.8	13×11
	63	17.2	14×11
	65	20.0	16×11
	67	23.4	18×11.5
	68	26.5	20×11.5
70	58	17.0	13×11
	63	18.6	14×11
	66	21.6	16×11
	68	25.3	18×11.5
	70	28.7	20×11.5
	71	33.5	22×13
	73	36.7	24×13
80[a]	60	18.8	13×11
	65	20.7	14×11
	69	24.1	16×11
	72	28.2	18×11.5
	75	31.7	20×11.5
	76	37.2	22×13
	78	40.8	24×13
90	60	19.2	13×11
	65	21.1	14×11
	70	24.6	16×11
	73	28.8	18×11.5
	76	32.4	20×11.5
	77	38.1	22×13
	79	41.7	24×13
100	72	25.9	16×11
	76	30.2	18×11.5
	79	34.3	20×11.5
	80	40.2	22×13
	81	44.0	24×13

91DR Performance

psi	Radius'	gpm	Nozzle
60	66	27.7	18×15
	72	36.3	22×15
	76	39.9	24×15
	80	44.7	26×15
	83	49.2	28×15
	86	53.9	30×15
70	68	29.9	18×15
	76	39.1	22×15
	80	43.0	24×15
	84	48.2	26×15
	87	53.0	28×15
	90	58.3	30×15
80[a]	71	33.6	18×15
	80	43.0	22×15
	85	48.5	24×15
	89	55.0	26×15
	92	60.5	28×15
	95	65.3	30×15
90	72	34.0	18×15
	81	44.5	22×15
	86	49.0	24×15
	90	55.9	26×15
	93	61.5	28×15
	96	66.2	30×15
100	73	35.9	18×15
	82	46.0	22×15
	87	50.6	24×15
	91	58.8	26×15
	95	64.7	28×15
	98	69.6	30×15

95DR Performance

psi	Radius'	gpm	Nozzle
60	66	26.1	18×15
	72	33.8	22×15
	76	37.2	24×15
	80	42.6	26×15
	83	46.9	28×15
	86	51.5	30×15
70	68	28.3	18×15
	76	36.6	22×15
	80	40.3	24×15
	84	46.2	26×15
	87	50.8	28×15
	90	55.9	30×15
80[a]	71	31.7	18×15
	80	41.3	22×15
	85	45.4	24×15
	89	51.7	26×15
	92	56.8	28×15
	95	62.8	30×15
90	72	31.9	18×15
	81	41.8	22×15
	86	46.0	24×15
	90	52.4	26×15
	93	57.6	28×15
	96	63.2	30×15
100	73	33.7	18×15
	82	44.1	22×15
	87	48.5	24×15
	91	55.4	26×15
	95	60.9	28×15
	98	67.2	30×15

All nozzle combinations are given in XX/64 equivalents. The following chart can be used for conversion.

NEW	11	12	13	14	15	16	17	18	20	22	24
OLD	11/64	3/16	13/64	7/32	15/64	1/4	17/64	9/32	5/16	11/32	3/8

[a] Factory pressure setting. Standard performance for 80 psi or higher at the base of rotor.

EAGLE™ Series Rotors

Performance Data
(*Source:* Rain Bird Golf Irrigation)

700/750

Eagle 700 Series Performance Chart.

Nozzles

Base Pressure (psi)	#14/White Radius (ft)	#14/White Flow (gpm)	#16/Blue Radius (ft)	#16/Blue Flow (gpm)	#18/Yellow Radius (ft)	#18/Yellow Flow (gpm)	#20/Orange Radius (ft)	#20/Orange Flow (gpm)	#22/Green Radius (ft)	#22/Green Flow (gpm)	#24/Black Radius (ft)	#24/Black Flow (gpm)
50	56	15.6	59	16.3	62	18.2	66	22.6	66	25.3	68	28.1
60	56	17.1	64	17.9	65	20.1	68	24.9	68	27.9	72	31.2
70	59	18.5	66	19.4	66	21.8	72	27.1	72	30.2	76	33.4
80	60	19.9	70	20.8	68	23.2	74	29.0	78	32.3	80	36.3
90	66	21.0	66	22.1	68	24.8	78	30.8	80	34.5	81	38.7
100	68	22.3	68	23.4	68	25.3	80	32.6	82	36.2	82	40.9

Eagle 750 Series Performance Chart.

Nozzles

Base Pressure (psi)	#14/White Radius (ft)	#14/White Flow (gpm)	#16/Blue Radius (ft)	#16/Blue Flow (gpm)	#18/Yellow Radius (ft)	#18/Yellow Flow (gpm)	#20/Orange Radius (ft)	#20/Orange Flow (gpm)	#22/Green Radius (ft)	#22/Green Flow (gpm)	#24/Black Radius (ft)	#24/Black Flow (gpm)
50	56	15.6	59	16.3	60	18.2	62	23.2	62	26.2	66	28.1
60	56	17.1	62	17.9	63	20.1	64	25.3	66	27.9	70	31.2
70	60	18.5	64	19.4	65	21.8	66	27.7	68	30.2	72	33.4
80	60	19.9	66	20.8	65	23.2	68	29.7	70	32.3	75	36.3
90	62	21.0	66	22.1	66	24.8	70	31.4	72	34.5	76	38.7
100	64	22.3	66	23.4	68	25.3	70	34.0	74	36.2	77	40.9

EAGLE™ Series Rotors

Performance Data
(*Source:* Rain Bird Golf Irrigation)

Eagle 900 Series Performance Chart.

Nozzles

Base Pressure (psi)	#18/White Radius (ft)	#18/White Flow (gpm)	#22/Blue Radius (ft)	#22/Blue Flow (gpm)	#24/Yellow Radius (ft)	#24/Yellow Flow (gpm)	#26/Orange Radius (ft)	#26/Orange Flow (gpm)	#28/Green Radius (ft)	#28/Green Flow (gpm)	#30/Black Radius (ft)	#30/Black Flow (gpm)	#32/Brown Radius (ft)	#32/Brown Flow (gpm)
60	70	20.1	72	28.9	76	31.6	74	36.0	—	—	—	—	—	—
70	72	22.0	74	31.2	78	33.7	80	38.7	87	39.9	90	46.2	87	48.0
80	74	24.1	76	33.0	80	36.2	82	41.5	89	42.6	92	49.3	89	51.1
90	75	25.7	78	35.4	84	38.5	84	43.4	91	45.1	96	52.2	93	53.8
100	76	26.9	78	37.3	84	40.4	86	46.2	95	47.9	98	54.5	95	57.4

Data reflect no pressure reguation. Larger nozzles are recommended for use at pressuresof at least 70 psi.

Eagle 950 Series Performance Chart.

Nozzles

									NewData		New Nozzle		New Nozzle	
Base Pressure (psi)	#18/White Radius (ft)	#18/White Flow (gpm)	#22/Blue Radius (ft)	#22/Blue Flow (gpm)	#24/Yellow Radius (ft)	#24/Yellow Flow (gpm)	#26/Orange Radius (ft)	#26/Orange Flow (gpm)	#28/Green Radius (ft)	#28/Green Flow (gpm)	#30/Black Radius (ft)	#30/Black Flow (gpm)	#32/Brown Radius (ft)	#32/Brown Flow (gpm)
60	69	20.1	72	28.9	77	31.6	77	36.0	—	—	—	—	—	—
70	73	22.0	74	31.2	78	33.7	82	38.7	84	42.9	84	47.3	84	50.4
80	74	24.1	76	33.0	80	36.2	84	41.5	86	47.3	86	50.4	85	53.1
90	76	25.7	78	35.4	82	38.5	85	43.4	89	48.5	90	52.9	88	55.6
100	77	26.9	80	37.3	84	40.4	86	46.2	91	52.2	92	55.8	92	59.4

Data reflect no pressure reguation. Larger nozzles are recommended for use at pressuresof at least 70 psi.

Buckner
10x90 Series
1" Bronze or Plastic Case
Rotary Pop-Up Sprinklers
Impact Drive
Application and Description

This series of turf rotors features all-bronze impact drive sprinklers. Their high value construction offers strength and impact resistance while having a high level of corrosion resistance. These rotary pop-up sprinklers are ideal for industrial, commercial and golf course irrigation systems.

10092

NOZZLE SIZE CODE	NOZZLE SIZE IN's	INLET PRESS. PSI	DIA. FT.	FLOW RATE GPM	RECTANGLE SPACING FT.	RECTANGLE RATE IN/HR	SINGLE ROW SPACING FT.	SINGLE ROW RATE IN/HR
-14-13	7/32 x 13/64	45	113	16.0	62x57	.44	57	.28
		50	115	16.8	63x58	.44	58	.29
		55	117	17.6	64x59	.45	59	.29
		60	119	18.3	65x60	.45	60	.29
		65	121	19.1	67x61	.46	61	.29
		70	123	19.9	68x62	.46	62	.29
		75	124	20.6	68x62	.47	62	.30
		80	125	21.2	69x63	.48	63	.31
		85	126	21.8	69x63	.48	63	.31
-16-13	1/4 x 13/64	45	116	18.4	64x58	.48	58	.31
		50	118	19.3	65x59	.49	59	.31
		55	121	20.2	67x61	.48	61	.31
		60	123	21.1	68x62	.49	62	.32
		65	125	22.0	69x63	.49	63	.32
		70	127	22.9	70x64	.50	64	.32
		75	129	23.8	71x65	.50	65	.32
		80	131	24.6	72x66	.50	66	.32
		85	132	25.4	73x66	.51	66	.33
		90	133	26.1	73x67	.52	67	.33
		95	134	26.8	74x67	.52	67	.34
		100	135	27.6	74x68	.53	68	.34
-18-13	9/32 x 13/64	45	119	20.5	65x60	.51	46	.32
		50	122	21.8	67x61	.51	61	.33
		55	124	22.9	68x62	.52	62	.34
		60	126	23.9	69x63	.53	63	.34
		65	128	25.1	70x64	.54	64	.34
		70	130	26.0	72x65	.54	65	.34
		75	132	27.0	73x66	.54	66	.35
		80	134	27.9	74x67	.54	67	.35
		85	136	28.8	75x68	.55	68	.35
		90	137	29.6	75x69	.55	69	.36
		95	138	30.4	76x69	.56	69	.36
		100	140	31.2	77x70	.56	70	.36
-20-13 ★	5/16 x 13/64	50	126	25.1	69x63	.55	63	.36
		55	128	26.3	70x64	.56	64	.36
		60	130	27.5	72x65	.57	65	.36
		65	133	28.8	73x67	.57	67	.36
		70	135	29.9	74x68	.57	68	.37
		75	138	30.9	76x69	.57	69	.36
		80	140	32.0	77x70	.57	70	.36
		85	142	33.0	78x71	.57	71	.37
		90	144	34.0	79x72	.57	72	.37
		95	145	35.0	80x73	.58	73	.37
		100	147	36.0	81x74	.58	74	.37
-22-13	11/32 x 13/64	50	130	29.7	72x65	.62	65	.39
		55	131	31.1	72x66	.63	66	.41
		60	134	32.6	74x67	.64	67	.41
		65	137	34.1	75x69	.64	69	.41
		70	141	35.4	78x71	.62	71	.40
		75	144	36.7	79x72	.62	72	.40
		80	147	37.9	81x74	.61	74	.39
		85	150	39.2	83x75	.61	75	.39
		90	152	40.3	84x76	.61	76	.39
		95	153	41.4	84x77	.62	77	.40
		100	156	42.6	86x78	.61	78	.39
-24-13	3/8 x 13/64	55	134	35.1	74x67	.68	67	.44
		60	137	36.6	75x69	.68	69	.44
		65	142	38.1	78x71	.66	71	.43
		70	145	39.5	80x73	.66	73	.42
		75	149	41.0	82x75	.65	75	.41
		80	153	42.4	84x77	.63	77	.41
		85	156	43.9	86x78	.63	78	.41
		90	158	45.1	87x79	.63	79	.41
		95	161	46.4	89x81	.63	81	.40
		100	163	47.8	90x82	.63	82	.41

Buckner

10110 Series
1-1/2 in. Cast Iron Case
Rotary Pop-Up Sprinklers
Impact Drive
Application and Description

This series of turf rotors is constructed from cast iron and bronze. These sprinklers are designed for use in large turf grass areas of sportsfields, parks, schools, golf courses and industrial sites. Their large diameter of coverage eliminates/minimizes the number of sprinklers in the playing areas of sportsfields.

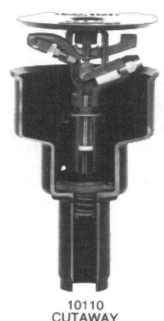

10110
CUTAWAY

NOZZLE SIZE		INLET PRESSURE		DIAMETER		FLOW RATE		RECTANGLE				SINGLE ROW			
								SPACING		PREC. RATE		SPACING		PREC. RATE	
CODE	INCHES (MM)	PSI	ATM	FT.	M	GPM	M³/HR	FT.	M	IN/H	MM/H	FT.	M	IN/H	MM/H
-24	³⁄₈x	60	4.1	167	50.9	41.6	9.4	92x84	28.0x25.5	.52	13	84	25.5	.33	8
-14	⁷⁄₃₂	70	4.8	173	52.7	45.0	10.2	95x87	29.0x26.4	.53	13	87	26.4	.33	8
	(9.5x	80	5.4	178	54.3	47.9	10.9	98x89	29.8x27.1	.53	13	89	27.2	.34	9
	5.6)	90	6.1	183	55.8	50.9	11.6	101x92	30.7x27.9	.53	14	92	27.9	.34	9
		100	6.8	183	55.8	54.1	12.3	101x92	30.7x27.9	.57	14	92	27.9	.36	9
-28	⁷⁄₁₆x	60	4.1	178	54.3	52.7	12.0	98x89	29.8x27.1	.58	15	89	27.2	.37	9
-14	⁷⁄₃₂	70	4.8	184	56.1	57.0	12.9	101x92	30.9x28.0	.59	15	92	28.1	.37	9
	(9.5x	80	5.4	190	57.9	60.9	13.8	105x95	31.9x29.0	.59	15	95	29.0	.37	9
	5.6)	90	6.1	196	59.8	64.7	14.7	108x98	32.9x29.9	.59	15	98	29.9	.37	9
		100	6.8	201	61.3	67.9	15.4	111x101	33.7x30.6	.59	15	101	30.7	.37	9
-32	½x	60	4.1	191	58.2	70.3	16.0	105x96	32.0x29.1	.67	17	96	29.3	.43	11
-16	¼	70	4.8	197	60.1	74.3	16.9	108x99	33.0x30.0	.67	17	98	29.9	.43	11
*	(12.7x	80	5.4	203	61.9	79.4	18.0	110x102	33.5x30.9	.68	17	100	30.5	.43	11
	6.4)	90	6.1	211	64.3	84.1	19.1	112x106	34.1x32.2	.69	17	102	31.1	.43	11
		100	6.8	218	66.5	88.3	20.1	114x109	34.8x33.2	.68	17	104	31.7	.43	11
-34	¹⁷⁄₃₂x	60	4.1	197	60.1	73.7	16.7	108x99	33.0x30.2	.66	17	97	29.6	.43	11
-16	¼	70	4.8	203	61.9	78.8	17.9	110x102	33.5x30.9	.68	17	99	30.2	.44	11
	(13.5x	80	5.4	209	63.7	84.1	19.1	112x105	34.1x31.9	.69	18	100	30.5	.45	11
	6.4)	90	6.1	215	65.5	89.2	20.3	114x108	34.8x32.8	.70	18	103	31.4	.45	11
		100	6.8	221	67.4	94.0	21.3	116x110	35.4x33.5	.71	18	105	32.0	.45	11
-36	⁹⁄₁₆x	60	4.1	204	62.2	85.0	19.3	110x102	33.5x31.1	.73	19	98	29.9	.54	14
-16	¼	70	4.8	210	64.0	90.4	20.5	112x105	34.1x32.0	.74	19	100	30.5	.55	14
	(14.3x	80	5.4	216	65.9	96.5	21.9	114x106	34.8x32.3	.77	20	101	30.8	.55	14
	6.4)	90	6.1	223	68.0	102.3	23.2	116x108	35.4x32.9	.79	20	104	31.7	.56	14
		100	6.8	229	69.8	106.5	24.2	117x110	35.7x33.5	.80	20	106	32.3	.56	14
-40	⅝x	60	4.1	213	67.9	102.8	23.3	114x107	34.8x32.5	.82	21	99	30.2	.54	14
-16	¼	70	4.8	220	67.1	109.2	24.8	116x108	35.4x32.9	.84	21	101	30.8	.55	14
	(15.9x	80	5.4	227	69.2	115.4	26.2	118x109	36.0x33.2	.86	22	102	31.1	.55	14
	6.4)	90	6.1	234	71.3	121.8	27.7	120x111	36.6x33.8	.88	22	104	31.7	.56	14
		100	6.8	240	73.2	128.7	29.2	122x113	37.2x34.5	.90	23	106	32.3	.56	14

Thompson

186/7

186/7 VIH

FEATURES
- Full and part circle rotors
  Full circle (Model 186)
  Part circle (Model 187).

*PSI	P-NOZZLE Radius	GPM	Q-NOZZLE Radius	GPM	R-NOZZLE Radius	GPM	S-NOZZLE Radius	GPM	T-NOZZLE Radius	GPM	U-NOZZLE Radius	GPM	VS-NOZZLE Radius	GPM	V-NOZZLE Radius	GPM	*PSI
10	30	2.0	27	3.3	25	6.1	35	8.2									10
15	33	3.5	32	4.1	38	7.8	40	8.8									15
20	35	3.9	34	4.6	41	8.2	40	9.5	42	13.0	45	17.0	45	17.0			20
25	37	4.3	38	5.2	44	9.2	43	10.5	47	15.0	50	18.5	50	19.0	55	19.8	25
30	38	4.8	40	5.6	46	9.8	50	12.1	53	16.0	55	20.1	55	20.0	56	22.0	30
35	40	5.2	42	6.0	50	10.4	52	13.0	54	17.8	57	22.0	60	22.0	58	24.0	35
40	42	5.3	42	6.4	54	10.9	55	14.0	55	18.9	58	23.5	62	24.0	61	26.0	40
45	43	5.4	39	7.3	57	11.4	56	14.8	56	19.8	60	24.5	64	24.5	65	26.4	45
50	44	5.8	39	7.4	57	12.0	60	16.0	57	20.0	62	26.0	67	26.0	68	28.0	50
55	45	5.9			57	12.3	60	16.2	59	21.5	65	27.0	69	27.0	70	29.5	55
60	44	6.2					54	17.0	61	22.8	66	28.0	71	28.0	72	30.6	60
65							52	17.5	62	23.2	67	29.0	73	28.5	75	31.2	65
70									61	24.2	68	29.5	75	29.0	78	31.8	70
75									55	25.2	70	30.5	80	30.5	80	33.5	75
80									55	25.8	68	31.1	82	31.0	82	34.0	80
85													83	31.5	85	34.9	85
90													81	32.5	86	36.0	90
95													80	33.0	85	37.0	95
100															82	38.0	100

188/9   188/9 VIH

*PSI	W-NOZZLE Radius	GPM	X-NOZZLE Radius	GPM	Y-NOZZLE Radius	GPM	Z-NOZZLE Radius	GPM	PSI*
20									20
25	63	28.0	61	31.0					25
30	64	30.0	63	33.0	67	34.0			30
35	65	31.5	67	35.0	70	37.5			35
40	67	33.0	70	37.0	72	40.0	74	55.0	40
45	69	34.0	71	39.0	74	42.5	76	57.0	45
50	71	36.5	72	42.0	76	45.0	78	59.0	50
55	73	39.0	74	44.0	78	46.5	81	61.0	55
60	73	41.0	78	45.0	81	49.5	85	63.0	60
65	74	42.0	81	47.0	83	51.0	90	65.0	65
70	75	43.0	83	49.0	85	53.0	96	67.0	70
75	77	45.0	85	50.5	87	55.0	98	69.0	75
80	72	46.0	86	52.5	90	56.0	100	71.0	80
85	74	47.0	86	53.0	92	58.0	103	73.0	85
90	76	50.5	88	55.0	95	60.0	104	75.0	90
95	80	53.0	87	57.0	97	62.0	106	78.0	95
100	82	54.0	86	59.0	100	64.0	108	80.0	100

Weather✳matic®

TURF ROTORS

For golf course and large turf area applications.

K70F (FULL CIRCLE) Maximum Operating Pressure 90 psi

Nozzle	Operating Pressure psi	†Spacing (ft.) Δ	□	Flow gpm	Diameter Coverage ft.	Δ ††Precip. in/hr.
E12 x A11 3.16 x 11.64	50	66	61	12.5	102	.31
	60	68	63	13.8	105	.33
	70	69	64	15.1	107	.35
	80	71	66	16.4	110	.36
	90	72	67	17.8	112	.38
E13 x A11 13.64 x 11.64	50	71	66	15.0	110	.33
	60	73	67	16.4	113	.34
	70	76	70	17.8	117	.34
	80	78	72	19.0	120	.34
	90	79	73	20.2	122	.36
..	50	74	68	16.2	114	.32
	60	78	72	17.5	120	.32
	70	79	73	18.8	123	.33
	80	81	75	20.1	126	.34
	90	81	75	21.5	126	.36
E16 x A11 1.4 x 11.64	50	75	69	19.0	116	.37
	60	79	73	20.3	122	.37
	70	81	75	21.7	126	.37
	80	83	77	23.0	129	.37
	90	83	77	24.3	129	.39

K80F (FULL CIRCLE) Maximum Operating Pressure 100 psi

Nozzle	Operating Pressure psi	†Spacing (ft.) Δ	□	Flow gpm	Diameter Coverage ft.	Δ ††Precip. in/hr.
E18 x A12 9.32 x 3.16	60	74	68	23.6	114	.48
	70	78	72	25.3	120	.46
	80	82	76	27.1	126	.45
	90	83	77	28.8	128	.46
	100	84	78	30.7	130	.48
	65	81	75	29.5	125	.50
	70	82	76	30.3	126	.50
	80	82	76	32.0	127	.52
	90	86	79	33.7	133	.51
	100	88	81	35.5	136	.51
E22 x A13 11.32 x 13.64	70	92	85	35.3	142	.46
	80	99	92	37.8	154	.43
	90	100	93	40.2	155	.45
	100	100	93	42.6	155	.47
E24 x A13 3.8 x 13.64	75	99	92	40.3	154	.46
	80	99	92	41.7	154	.47
	90	100	93	44.3	155	.49
	100	100	93	47.1	155	.52

K90F (FULL CIRCLE) Maximum Operating Pressure 100 psi

Nozzle	Operating Pressure psi	†Spacing (ft.) Δ	□	Flow gpm	Diameter Coverage ft.	Δ ††Precip. in/hr.
F24 x A11 3.8 x 11.64	60	110	102	37.0	170	.34
	70	117	108	39.0	180	.32
	80	119	110	41.0	184	.32
	90	121	112	42.0	187	.32
	100	122	113	43.0	188	.32
*	65	119	110	48.0	184	.38
	70	120	111	50.0	186	.38
	80	123	114	52.0	190	.38
	90	126	116	54.0	194	.38
	100	126	116	56.0	194	.39
F32 x A11 1.2 x 11.64	70	126	116	62.0	194	.43
	80	128	118	67.0	198	.45
	90	130	120	71.0	200	.47
	100	131	121	76.0	202	.49
F24 x A14 3.8 x 7.32	60	110	102	40.0	170	.37
	70	117	108	42.8	180	.35
	80	119	110	45.5	184	.36
	90	121	112	48.1	187	.37
	100	122	113	51.0	188	.38
F28 x A14 7.16 x 7.32	65	119	110	51.5	184	.41
	70	120	111	53.3	186	.42
	80	123	114	57.0	190	.42
	90	126	116	60.5	194	.43
	100	126	116	64.1	194	.46
F32 x A14 1.2 x 7.32	70	126	116	66.7	194	.48
	80	128	118	70.5	198	.49
	90	130	120	74.5	200	.50
	100	131	121	78.3	202	.51

K90P (PART CIRCLE) Maximum Operating Pressure 100 psi

Nozzle	Operating Pressure psi	†Spacing (ft.) Δ	□	Flow gpm	Radius Coverage ft.	Δ ††Precip. in/hr.
F24 x A11 3.8 x 11.64	60	110	102	37.0	85	.68
	70	117	108	39.0	90	.64
	80	119	110	41.0	92	.64
	90	121	112	42.0	93	.64
	100	122	113	43.0	94	.64
*	65	119	110	48.0	92	.76
	70	120	111	50.0	93	.76
	80	123	114	52.0	95	.76
	90	126	116	54.0	97	.76
	100	126	116	56.0	97	.78
F32 x A11 1.2 x 11.64	70	126	116	62.0	97	.86
	80	128	118	67.0	99	.90
	90	130	120	71.0	100	.94
	100	131	121	76.0	101	.98
F24 x A14 3.8 x 7.32	60	110	102	40.0	85	.74
	70	117	108	42.8	90	.70
	80	119	110	45.5	92	.72
	90	121	112	48.1	93	.74
	100	122	113	51.0	94	.76
F28 x A14 7.16 x 7.32	65	119	110	51.5	92	.82
	70	120	111	53.3	93	.84
	80	123	114	57.0	95	.84
	90	126	116	60.5	97	.86
	100	126	116	64.1	97	.92
F32 x A14 1.2 x 7.32	70	126	116	66.7	97	.96
	80	128	118	70.5	99	.98
	90	130	120	74.5	100	1.00
	100	131	121	78.3	101	1.02

†Spacings are for still air. Derate for wind. See Rotary Head Design Data Page.

††Precipitation rates indicated are for overlapping sprinklers on triangular spacing. Part Circle precipitation rates listed are based on 180° arc.

Hunter

HUNTER I-44 SPRINKLER

The Hunter I-44 sod cup sprinkler holds a plug of living sod on top of the riser. When the riser is retracted, the sod cup makes the sprinkler totally invisible and presents a completely consistent playing surface which is highly desired on golf greens, bowling greens and grass tennis courts. The I-44 was developed from Hunter's I-40 sprinkler and has exactly the same performance characteristics as that sprinkler. The I-44 is ideal for golf greens and sport fields where short cut grasses are essential.

Nozzle Number	PSI	Radius	GPM	Nozzle Number	PSI	Radius	GPM
1	40	40'	3.8	**5**	50	53'	9.6
	50	41'	4.3		60	54'	10.8
	60	42'	4.7		70	55'	11.5
	70	43'	5.1		80	56'	12.1
2	40	42'	4.4	**6**	50	56'	10.5
	50	43'	4.8		60	57'	11.5
	60	45'	5.3		70	58'	12.4
	70	46'	5.6		80	59'	13.3
3	40	48'	6.6	**7**	50	60'	13.2
	50	49'	7.0		60	62'	14.1
	60	49'	7.5		70	62'	15.0
	70	50'	7.9		80	63'	15.9
4	40	48'	6.7	**8**	50	62'	16.6
	50	49'	7.8		60	64'	18.1
	60	51'	8.7		70	64'	19.3
	70	52'	9.2		80	65'	20.2

Data represents test results in zero wind. Adjust for local conditions. Radius may be reduced up to 30% with nozzle retaining screw.

Nozzle Number	PSI	Radius	GPM
41	50	44'	10.2
	60	44'	11.5
	70	45'	12.6
	80	46'	13.5
42	50	46'	11.0
	60	47'	12.3
	70	49'	13.5
	80	50'	14.4
43	50	51'	14.2
	60	52'	15.5
	70	52'	16.3
	80	53'	18.1
44	60	58'	20.0
	70	58'	21.8
	80	60'	23.8
	90	60'	24.9
45	60	60'	22.0
	70	62'	24.3
	80	64'	25.9
	90	65'	27.5

Data represents test results in zero wind. Adjust for local conditions. Radius may be reduced up to 30% with nozzle retaining screw.

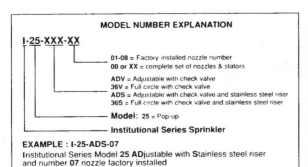

MODEL NUMBER EXPLANATION

I-25-XXX-XX

01-08 = Factory installed nozzle number
00 or XX = complete set of nozzles & stators

ADV = Adjustable with check valve
36V = Full circle with check valve
ADS = Adjustable with check valve and stainless steel riser
36S = Full circle with check valve and stainless steel riser

Model: 25 = Pop-up

Institutional Series Sprinkler

EXAMPLE : I-25-ADS-07
Institutional Series Model 25 ADjustable with Stainless steel riser and number 07 nozzle factory installed

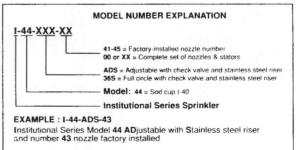

MODEL NUMBER EXPLANATION

I-44-XXX-XX

41-45 = Factory-installed nozzle number
00 or XX = Complete set of nozzles & stators

ADS = Adjustable with check valve and stainless steel riser
36S = Full circle with check valve and stainless steel riser

Model: 44 = Sod cup I-40

Institutional Series Sprinkler

EXAMPLE : I-44-ADS-43
Institutional Series Model 44 ADjustable with Stainless steel riser and number 43 nozzle factory installed

Hunter

MATCHED PRECIPITATION RATES

.4"/hr. at 50 PSI

Square Spacing	90° Nozzle #	90° GPM	180° Nozzle #	180° GPM	360° Nozzle #	360° GPM
25'	1	.7	3	1.2	6	2.7
30'	2	0.9	5	2.0	8	4.2
35'	3	1.2	6	2.7	9	5.5
40'	4	1.6	7	3.4	10	6.8
45'	5	2.0	8	4.2	11	8.9

When the arc/nozzle/pressure combinations are spaced as indicated, the precipitation rate will be approximately .4"/hr. at 50 PSI.

.4"/hr. at 60 PSI

Square Spacing	90° Nozzle #	90° GPM	180° Nozzle #	180° GPM	360° Nozzle #	360° GPM
25'						
30'						
35'	2	1.0	5	2.2	8	4.6
40'	3	1.3	6	2.9	9	6.0
45'	4	1.8	7	3.7	10	7.6

When the arc/nozzle/pressure combinations are spaced as indicated, the precipitation rate will be approximately .4"/hr. at 60 PSI.

.5"/hr. at 40 PSI

Square Spacing	90° Nozzle #	90° GPM	180° Nozzle #	180° GPM	360° Nozzle #	360° GPM
25'	2	0.8	5	1.8	8	3.7
30'	3	1.0	6	2.4	9	4.9
35'	4	1.4	7	3.0	10	6.0
40'	5	1.8	8	3.7	11	8.0
45'	6	2.4	9	4.9	12	11.4

When the arc/nozzle/pressure combinations are spaced as indicated, the precipitation rate will be approximately .5"/hr. at 40 PSI.

.5"/hr. at 50 PSI

Square Spacing	90° Nozzle #	90° GPM	180° Nozzle #	180° GPM	360° Nozzle #	360° GPM
25'	2	0.9	4	1.6	7	3.4
30'	3	1.2	6	2.7	9	5.5
35'	4	1.6	7	3.4	10	6.8
40'	5	2.0	8	4.2	11	8.9
45'	6	2.7	9	5.5	12	12.2

When the arc/nozzle/pressure combinations are spaced as indicated, the precipitation rate will be approximately .5"/hr. at 50 PSI.

.5"/hr. at 60 PSI

Square Spacing	90° Nozzle #	90° GPM	180° Nozzle #	180° GPM	360° Nozzle #	360° GPM
25'						
30'						
35'	4	1.8	7	3.7	10	7.6
40'	5	2.2	8	4.6	11	9.8
45'	6	2.9	9	6.0	12	13.2

When the arc/nozzle/pressure combinations are spaced as indicated, the precipitation rate will be approximately .5"/hr. at 60 PSI.

1802
1804
1806
1812

RAIN✦BIRD

MPR PLASTIC NOZZLES

Features

- Matched precipitation rates across sets
- For use on all 1800 Series sprinklers and the PA-8S Plastic Shrub Adapter
- Adjustable flow and radius
- Stainless steel adjustment screw
- New 1800 Screens shipped with nozzles

Operating Range

Precipitation Rate:
1.52 to 2.63 inches per hour
Spacing: 7 to 15 feet
Pressure: 15 to 30 psi

NOTE: Nozzle GPM has been rounded off to nearest hundredth — any deviation from MPR is due to rounding off only. All nozzles are designed to provide matched precipitation rate (MPR).

■ Square spacing based on 50% diameter of throw
△ Triangular spacing based on 50% diameter of throw

15 Series				Standard 30° Trajectory	
Nozzle	PSI	Radius	GPM	Precip ◆	Precip △
15F	15	11	2.60	2.07	2.39
	20	12	3.00	2.01	2.32
	25	14	3.30	1.62	1.87
	30	15	3.70	1.58	1.83
15TQ	15	11	1.95	2.07	2.39
	20	12	2.25	2.01	2.32
	25	14	2.48	1.62	1.87
	30	15	2.78	1.58	1.83
15TT	15	11	1.74	2.07	2.39
	20	12	2.01	2.01	2.32
	25	14	2.21	1.62	1.87
	30	15	2.48	1.58	1.83
15H	15	11	1.30	2.07	2.39
	20	12	1.50	2.01	2.32
	25	14	1.65	1.62	1.87
	30	15	1.85	1.58	1.83
15T	15	11	0.87	2.07	2.39
	20	12	1.00	2.01	2.32
	25	14	1.10	1.62	1.87
	30	15	1.23	1.58	1.83
15Q	15	11	0.65	2.07	2.39
	20	12	0.75	2.01	2.32
	25	14	0.83	1.62	1.87
	30	15	0.93	1.58	1.83

Toro 300 Series
Steam Rotor
Sprinklers
(15 ft – 30 ft radius)

Features:

• Multiple rotating steam pattern;
• Matched precipitation rate nozzles and arc discs;
• Choice of 6 nozzles and 9 interchangeable arc discs;
• Gear-driven design;
• Large basket filter screen;
•Durable cycolac and stainless steel construction.

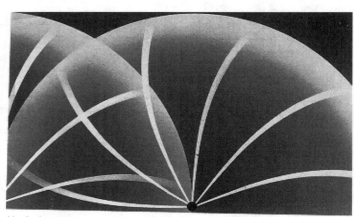

Matched precipitation rate refers to uniform delivery of water across each square foot of an irrigated area. Each sprinkler's coverage provides every blade with no more — and no less — water than the next. The result is high-precision application — and consistently green healthy turf.

					360°	202.5°	180°	112°	90°
Nozzle	PSI	Radius	Prec. Rates* △	□	Gallons Per Minute				
	35	15'	1.69	1.46	3.41	1.91	1.70	1.06	0.85
	35	18'	1.37	1.19	4.00	2.25	2.00	1.24	1.00
	35	21'	1.15	1.00	4.58	2.58	2.29	1.42	1.15
	35	24'	0.99 *	0.86	5.17	2.91	2.58	1.60	1.29
300-15-ADJ	35	26'	0.95	0.82	5.76	3.24	2.88	1.79	1.44
300-25-ADJ	50	18'	1.60	1.38	4.65	2.62	2.33	1.44	1.16
	50	21'	1.55	1.17	5.36	3.02	2.68	1.66	1.34
	50	24'	1.17	1.02	6.08	3.42	3.04	1.88	1.52
	50	27'	1.04	0.90	6.79	3.82	3.40	2.10	1.70
	50	30'	0.93	0.80	7.51	4.23	3.75	2.33	1.88

300-15 and 300-25 Omni Adjustable Radius Nozzle Performance Chart

Greens or Tees—Part- or Full-Circle—Single or 2-Speed

650 Series

NOZZLE PERFORMANCE — U.S.

BASE PRES.	NOZZLE SET 54		NOZZLE SET 55		NOZZLE SET 56		NOZZLE SET 57		NOZZLE SET 58		NOZZLE SET 59	
PSI	Rad	GPM	Rad	GPM	Rad	GPM	Rad	GPM	Rad	GPM	Rad	GPM
50	56	13.0	61	16.4	64	17.8	65	22.4	66	24.7	67	28.7
55	56	13.6	62	17.2	65	18.7	67	23.6	68	25.8	71	30.7
60	57	14.2	62	18.0	67	19.6	70	25.1	71	26.9	76	33.7
65	57	14.8	63	18.7	68	20.4	71	26.1	73	28.1	78	34.9
70	58	15.4	64	19.4	69	21.1	72	27.0	75	29.7	80	36.5
75	58	15.9	64	20.0	70	21.9	74	27.9	76	30.5	81	37.5
80	59	16.4	65	20.7	71	22.6	75	28.8	78	31.7	82	38.8
85	60	17.0	66	21.3	72	23.6	76	29.6	79	32.6	83	40.0
90	60	17.5	68	21.9	74	24.5	77	30.6	80	33.5	84	41.7
95	61	18.1	69	22.5	75	25.5	78	31.3	81	34.4	85	42.5
100	61	18.6	70	23.1	76	26.4	79	32.1	82	35.3	86	43.7

Rad = feet   GPM = gallons per minute   ▨ = Pressure regulation   ▨ = Nozzles not recommended at this pressure

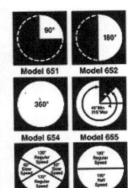

Model 651   Model 652
Model 654   Model 655
Model 656   Model 658

Fairways—Single-Row—Part- or Full-Circle—Single or 2-Speed

690 Series

NOZZLE PERFORMANCE — U.S.

BASE PRES.	NOZZLE SET 90		NOZZLE SET 91		NOZZLE SET 92	
PSI	Rad	GPM	Rad	GPM	Rad	GPM
50	73	40.5	84	50.7	88	62.3
55	75	42.8	86	52.2	90	63.0
60	78	45.1	87	53.7	92	63.6
65	80	47.4	89	56.3	94	64.3
70	82	48.6	91	57.2	96	67.5
75	85	49.8	94	59.2	98	70.7
80	87	51.0	96	61.2	100	74.0
85	88	52.5	97	64.3	102	76.1
90	89	54.1	98	67.3	105	78.0
95	89	55.6	99	70.4	106	80.2
100	90	57.1	100	73.4	108	82.2

Rad = feet   GPM = gallons per minute
▨ = Pressure regulation models
▨ = Nozzles not recommended at this pressure

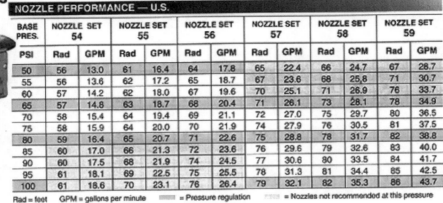

Model 691   Model 692
Model 694   Model 696
Model 698

ORDERING INFORMATION

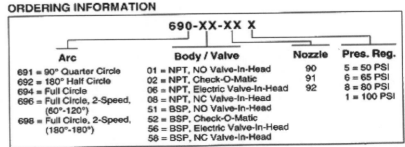

690-XX-XX X

Arc	Body / Valve	Nozzle	Pres. Reg.
691 = 90° Quarter Circle	01 = NPT, NO Valve-In-Head	90	5 = 50 PSI
692 = 180° Half Circle	02 = NPT, Check-O-Matic	91	6 = 65 PSI
694 = Full Circle	06 = NPT, Electric Valve-In-Head	92	8 = 80 PSI
696 = Full Circle, 2-Speed, (60°-120°)	08 = NPT, NC Valve-In-Head		1 = 100 PSI
698 = Full Circle, 2-Speed, (180°-180°)	51 = BSP, NO Valve-In-Head		
	52 = BSP, Check-O-Matic		
	56 = BSP, Electric Valve-In-Head		
	58 = BSP, NC Valve-In-Head		

Note: Pressure regulation available on Electric and Normally Closed Valve-In-Head models only.

Appendix 6

Resistance Method of Sizing Wire

Formulas

Current (amps) multiplied by the total wire resistance, or voltage loss, equals:

$$V = \frac{2L \times I \times R}{1000}$$

$$R = \frac{1000\,(V)}{2L \times I}$$

where

- L = one-way wire length in ft
- I = current (amps) flowing through the wire section (length L)
- R = resistance of wire in ohms/1,000 ft. For underground installations, use the 77°F charts; for aboveground installations, use the 149°F charts.
- V = Allowable voltage loss, in volts, where for controller power wire, the allowable voltage loss equals minimum available voltage at the power source less the minimum operating voltage required at the controller.
  For valve wire sizing, the allowable voltage loss equals controller output voltage less minimum solenoid operating voltage.

The following calculations use the Buckner Recommended Wire Sizing Procedure and Charts for power and valve wires.

Example 1

Find the wire size required to power a Buckner M-12DD controller as shown in the sketch. The clock is operating a single 20930 valve per station. Refer to Buckner M-12DD specification sheet to find the current (amps) required and minimum operating voltage.

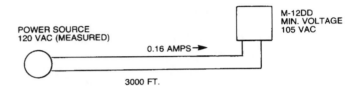

POWER SOURCE 120 VAC (MEASURED)

0.16 AMPS →

3000 FT.

M-12DD MIN. VOLTAGE 105 VAC

Solution

From the formula

$$
\begin{aligned}
R &= \frac{1000\,(V)}{2 \times L \times I} \\
&= \frac{1000\,(120 - 105)}{2 \times 3000 \times 0.16} \\
&= \frac{15000}{960} \\
&= 15.625 \text{ ohms per 1000 ft} \\
&= \text{allowable wire resistance}
\end{aligned}
$$

Looking at the chart for wire resistance, at 77° you will look for a value equal to or less than 15.625 ohms/1000 ft. The closest smaller number is 6.510, which corresponds to 18 AWG wire. However, 18 AWG wire lacks enough mechanical strength. A larger wire or a multi-conductor cable should be used. The minimum single wire size should be *14 AWG* and its resistance appears as 2.580 ohms/100 ft or wire.

Find the voltage loss between the power source and the controller in example 1. Find the available voltage at the controller.

To find the voltage loss in the two-wire circuit, use the formula:

$$V = \frac{2L \times I \times R}{1000 \text{ ft}}$$

$$= \frac{2\,(3000 \text{ ft}) \times (0.16 \text{ amps}) \times (2.58)}{1000}$$

$$= \frac{2476.8}{1000}$$

$$= 2.48 \text{ Volts}$$

To find the available voltage at the controller, subtract 2.48 volts from the source voltage: (i.e., 120 VAC – 2.48 VAC = 117.52 VAC or approximately 117 VAC.

Wire Size Chart

(These charts are to be used with Buckner's recommended wire sizing procedures.)

Wire Gauge (AWG)	Wire Type (UF)	Insulation Thickness (in.)	Resistance (ohms/1000 ft)	
			77°F (25°C)	149°F (65°C)
00	Stranded	5/64	0.079	0.092
0	Stranded	5/64	0.100	0.116
2	Stranded	5/64	0.159	0.184
4	Stranded	5/64	0.253	0.292
6	Stranded	5/64	0.403	0.465
8	Solid	5/64	0.641	0.739
10	Solid	4/64	1.020	1.180
12	Solid	4/64	1.620	1.870
14	Solid	4/64	2.580	2.970
16	Solid	4.64	4.090	4.730
18	Solid	4/64	6.510	7.510

Note: A circular mil is the unit of area applied to electrical wire and equals the area of a circle one mil (0.001 in.) in diameter. The area of a circle in circular mils equals the square of its diameter in mils.

Valve Model vs. Controller Matrix

	Valve Series		
Controller Series	20040, 20500	20930, 26000	20010, 20120, 20000
Mechatonic Controller	Chart I	Chart II	Chart III
VC24-P 0VC24-WFC VC24FC	Chart IV	Chart V	Chart VI

Wire Sizing Charts for Voltage Wire Between Controller and Valve

(Maximum Equivalent Wire Length in Feet)

Chart 1. Mechatronic and 20040/20500

Wire Size (AWG) Annealed Copper		Controller Input Voltage (VAC)				
Control	Common	105	110	115	120	125
18	18	960	1280	1600	1920	2240
18	16	1197	1572	1965	2358	2752
16	16	1528	2037	2547	3056	3566
16	14	1874	2499	3123	3748	4373
16	12	2189	2919	3649	4378	5108
14	14	2422	3230	4037	4845	5652
14	12	2976	3968	4960	5952	6944
14	10	3472	4630	5787	6944	8102
12	12	3858	5144	6430	7716	9002
12	10	4735	6313	7891	9470	11048
10	10	6127	8170	10212	12255	14297

Chart 2. Mechatronic and 20930/26000

Wire Size (AWG) Annealed Copper		Controller Input Voltage (VAC)				
Control	Common	105	110	115	120	125
18	18	219	439	658	878	1097
18	16	270	539	809	1078	1348
16	16	349	699	1048	1397	1746
16	14	428	857	1285	1713	2142
16	12	500	1001	1501	2002	2502
14	14	554	1107	1661	2215	2769
14	12	680	1361	2041	2721	3401
14	10	794	1587	2381	3175	3968
12	12	882	1764	2646	3527	4409
12	10	1082	2165	3247	4329	5411
10	10	1401	2801	4202	5602	7003

Chart 3. Mechatronic and 20000/20120/20010/20130

Wire Size (AWG) Annealed Copper		Controller Input Voltage (VAC)				
Control	Common	105	110	115	120	125
18	18	183	366	549	731	914
18	16	225	449	674	898	1123
16	16	291	582	873	1164	1455
16	14	357	714	1071	1428	1785
16	12	417	834	1251	1668	2085
14	14	461	923	1384	1846	2307
14	12	567	1134	1701	2268	2834
14	10	661	1323	1984	2646	3307
12	12	735	1470	2205	2939	3674
12	10	902	1804	2706	3608	4509
10	10	1167	2334	3501	4669	5836

Chart 4. Series 700 ("COPS") and 20040/20500

Wire Size (AWG) Annealed Copper		Controller Input Voltage (VAC)
Control	Common	120
18	18	2720
18	16	3341
16	16	4330
16	14	5310
16	12	6203
14	14	6864
14	12	8433
14	10	9838
12	12	10931
12	10	13415
10	10	17361

Chart 5. Series 700 ("COPS") and 20930/26000

Wire Size (AWG) Annealed Copper		Controller Input Voltage (VAC)
Control	Common	120
18	18	1426
18	16	1752
16	16	2270
16	14	2784
16	12	3252
14	14	3599
14	12	4422
14	10	5159
12	12	5732
12	10	7035
10	10	9104

Chart 6. Series 700 ("COPS") and 20000/20120/20010/20130

Wire Size (AWG) Annealed Copper		Controller Input Voltage (VAC)
Control	Common	120
18	18	1189
18	16	1460
16	16	1892
16	14	2320
16	12	2710
14	14	2999
14	12	3685
14	10	4299
12	12	4777
12	10	5862
10	10	7586

Note: Shaded areas represent applications where only multi-conductor cables may be used because of the tensile strength of the wire.

Valve Wire Sizing Procedure

Step 1. Determine actual distance, along wire run, from controller out to the first valve on a circuit and between each succeeding valve on a multiple valve circuit (as shown in example below).

Example

(Solenoid electrical control valves 2 W and 26.5-V transformer controllers, with 150 psi water pressure at valves.)

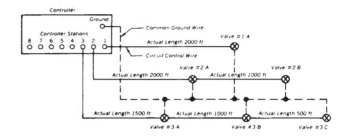

Equivalent Length Calculations:

Station #1—Equiv. Length = 1 valve × 2000 ft = 2000 ft
Station #2—Equiv. Length = 1 valve × 1000 ft) + (2 valves × 2000 ft) = 5000 ft
Station #3—Equiv. Length = (1 valve × 500 ft) + (2 valves × 1000 ft) + (3 valves 1500 ft) = 7000 ft

Step 2. Calculate the equivalent circuit length for each valve circuit on the controller.

Step 3. Selecting common ground wire size—Using longest equivalent length calculated above, go to the appropriate valve chart (based on transformer voltage, 26.5 V, and water pressure at valve) and select ground wire and control wore combination that are as near the same size as possible (ground wire size should always be equal to—or one size larger than—the control wire size).

In the example above, circuit for Station 3 has longest equivalent length of 7000 ft. In chart (for this example use high pressure chart for 150 psi water pressure at valve and 26.5-V transformer) select a wire size combination of size 14 and 12 wire. Select common ground wire as size 12 wire. Since one common ground wire shall be used for all valves on the controller, you have now established the common ground wire size for that controller as size 12 wire.

Step 4. Sizing circuit control wires—Using the common ground wire size selected in Step 3 (size #12)—proceed to select each control wire size, from the chart, using the calculated equivalent length for each circuit.

Station #1—Equivalent Length = 2000 ft
Select size #18 control wire.
Station #2—Equivalent Length = 5000 ft
Select size #16 control wire.
Station #3—Equivalent Length = 7000 ft
Select size #14 control wire.

Wire Sizing Charts for Rain Bird 24VAC Solenoid Valves

Wire Sizing Chart—Equivalent Circuit Length. Rain Bird 5.5 VA Solenoid Electric Valves with 26.5 Volt Transformers

Common Wire Size	80 psi Water Pressure at Valve Control Wire Size							
	18	16	14	12	10	8	6	4
18	3000	3700	4300	4800	5200	5500	5200	5800
16	3700	4800	5900	6900	7700	8300	8800	9100
14	4300	5900	7700	9400	11000	12300	13300	14000
12	4800	6900	9400	12200	15000	17500	19600	21100
10	5200	7700	11000	15000	19400	23900	27800	31100
8	5500	8300	12300	17500	23900	30900	38000	44300
6	5700	8800	13300	19600	27800	38000	49200	60400
4	5800	9100	14000	21100	31100	44300	60400	78200

Common Wire Size	100 psi Water Pressure at Valve Control Wire Size							
	18	16	14	12	10	8	6	4
18	2800	3500	4100	4500	4900	5200	5400	5500
16	3500	4500	5500	6500	7300	7800	8300	8500
14	4100	5500	7200	8900	10300	11600	12500	13200
12	4500	6500	8900	11500	14100	16500	18400	19900
10	4900	7300	10300	14100	18300	22500	26200	29300
8	5200	7800	11600	16500	22500	29100	35700	41700
6	5400	8300	12500	18400	26200	35700	46300	56900
4	5500	8500	13200	19900	29300	41700	56900	73600

Common Wire Size	125 psi Water Pressure Valve Control Wire Size							
	18	16	14	12	10	8	6	4
18	2600	3200	3800	4200	4600	4800	5000	5100
16	3200	4200	5200	6000	6700	7300	7700	7900
14	3800	5200	6700	8200	9600	10800	11600	12200
12	4200	6000	8200	10700	13100	15300	17100	18500
10	4600	6700	9600	13100	17000	20900	24400	27300
8	4800	7300	10800	15300	20900	27100	33200	38800
6	5000	7700	11600	17100	24400	33200	43100	52900
4	5100	7900	12200	18500	27300	38800	52900	68500

Common Wire Size	150 psi Water Pressure at Valve Control Wire Size							
	18	16	14	12	10	8	6	4
18	2400	3000	3500	3900	4300	4500	4600	4700
16	3000	3900	4800	5600	6300	6800	7200	7400
14	3500	4800	6200	7700	9000	10000	10800	11400
12	3900	5600	7700	10000	12200	14300	16000	17300
10	4300	6300	9000	12200	15900	19500	22800	25400
8	4500	6800	10000	14300	19500	25300	31000	36200
6	4600	7200	10800	16000	22800	31000	40200	49400
4	4700	7400	11400	17300	25400	36200	49400	63900

Wire Sizing Chart for Rain Bird Valves

Common Wire Size	Wire Sizing Chart—Equivalent Circuit Length TH Thermal Hydraulic Valves with 26.5 Volt Transformer					
	Control Wire Size					
	14	12	10	8	6	4
14	800	1000	1200	1250	1500	1560
12	1000	1300	1700	2000	2190	2380
10	1200	1700	2200	2700	3200	3410
8	1250	2000	2700	3500	4300	5000
6	1500	2190	3200	4300	5600	6720
4	1560	2380	3410	5000	6720	8650

Maximum and minimum equivalent circuit lengths for wire runs to Rain Bird TH Thermal Hydraulic control valves from Controller terminals.

Note: TH Valve must have from 22 volts to 26.5 volts at the valve. If any equivalent circuit length falls below the minimum—DO NOT use a 30 volt transformer.

Controller Power Wire Sizing Procedure

Chart #1. Electrical Current Requirements of Rain Bird Controllers and Valves for Power Wire Sizing

Type of Controller or Valve	120 Volt Primary Current Requirements (Amps)
Controllers Only	With Power ON— Not in a Cycle
RC-7i/1260i	0.13
RC-7A/1230/1260	0.13
RC-1260AB-MT	0.13
RC-1860AB/2360AB	0.26
MC-3S	0.29
SC-1230/1260	0.27
RC-12B/18B/23B	0.27
EZ-1	0.03
CRC-4/6/8	0.03
RCM-4/68/12	0.03
CIC-8/12	0.03
ISC-4/8/12/16/24/32	0.03
MIC-4/8	0.03
Valves Only	Current Draw When Energized
Solenoid Valve	0.07

Chart #2

Wire Size	"F" Factor
#18	13.02
#16	8.18
#14	5.16
#12	3.24
#10	2.04
#8	1.28
#6	0.81
#4	0.51

120 Volt Primary Wire Sizing Procedure for Rain Bird Controllers

Procedure

1. Determining current requirements of Controller Unit, Master Valve and Station Valves from Chart #1.
2. Determine maximum allowable voltage drop.
   a. Determine voltage at power source.
   b. Determine voltage desired at Controller (Limits: 110 volts minimum and 125 volts maximum).
3. Calculate equivalent circuit length.
4. By using formula, calculate "F" factor:

$$F = \frac{\text{Allowable Voltage Drop}}{\text{Amps (per unit)} \times \text{Equiv. Length (in 1000' s of ft)}}$$

5. Select power wire size from Chart #2—(select wire size with an "F" factor *equal* to or *less* than the calculated "F" factor).

Example

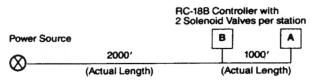

1. Controller Primary Current Requirements (from Chart #1):
   RC-18B Controller Only   0.27
   2 Solenoid Valves (2 × 0.07)   0.14
                     Total = 0.41 Amps
2. Maximum Allowable Voltage Drop:
   a.   Power Source = 119 Volts

b.   Desired Voltage at Controller (min.) = 112 Volts
Volts Drop Allowable = 119V − 112V = 7 Volts

3. Equivalent Circuit Length:
1 (Controller A)   1000 ft   = 1000 ft
2 (Controllers A&B)   2000 ft = 4000 ft
                                         5000 ft = 5.0
                                         (in 1000's of feet)

4. $F = \dfrac{\text{Allowable Voltage Drop}}{\text{Amps (per unit)} \times \text{Equiv. Length (in 1000's of ft)}}$

5. Wire Size Required:
From Chart #2 select Size #12 wire (Chart Value = 3.24)

Determining Maximum Number of Valves per Station of a Controller

Chart #3. Secondary Current Requirements of Rain Bird Controllers and Valves for Determining Maximum Number and Combinations of Valves per Station

Type of Controller or Valve	Max. Secondary Current (Amps) Allowable for Controller[a]	24 Volt Secondary Current (Amps) Requirement for Controllers and Valves
Controllers		
RC-7i/1260i[b]	1.25	0.35
RC-7A/1230/1260	1.5	0.35
RC-1260AB-MT	2.5	0.35
RC-1860AB/2360AB	2.5	0.35
MC-3S	2.5[c]	0.47
SC-1230/1260[d]	1.5	0.35
RC-12B/18B/23B	2.25	0.34
EZ-1	0.5	0.03
CRC-4/6/8[c]	0.75	0.03
RCM-4/6/8/12[c]	1.5	0.03
CIC-8/12[c]	1.5	0.03
ISC-4/8/12/16/24/32	1.5	0.03
MIC-4	1.5	0.03
MIC-8	2.5	0.03
Valves		
Solenoid Valve	—	0.27

[a]   Including Controller, Master Valve and maximum number of valves on a station.
[b]   Maximum of 2 valves per station.
[c]   Maximum number of satellite panels that maybe connected to a master is 75. (25 satellites on each of 3 irrigation clocks.)
[d]   For SC-1230/SC-1260 Satellite Unit it is necessary to count the maximum number of valves on a station of each panel for total secondary or primary load.

Procedure

1. From Chart #3 determine the maximum secondary current (amps) allowable for the model controller selected.
2. Determine the secondary current requirement of the controller only and the Master Valve (if a Master Valve is being used).
3. Calculate the remaining current available for valve operation. Max. Secondary Current allowable *minus* the current for Controller and Master Valve.
4. Number of valves per station equals:

$$\frac{\text{Remaining Available Current}}{\text{Current Required for 1 Valve}}$$

Example

Determine maximum number of Solenoid Valves that may be operated per station of an RC-12B Controller using a Master Valve.

1. From Chart #3 Max. Secondary current allowable:
RC-12B = 2.25 Amps

2. Secondary current requirement for Controller and Master Valve:

RC-12B Controller Only   0.34
Solenoid Master Valve   0.27
                    Total =   0.61

3. Remaining current available:
2.25 − 0.61 = 1.64 Amps

4. Maximum number of Solenoid Valves per station:

$$\frac{1.64 \text{ Amps (available)}}{0.27 \text{ Amps / Solenoid}} = 6.07$$

Max. No. of Solenoids per Station = 6

VALVE WIRE SIZING TABLES

Table A Base Table. *Caution! Do Not* use table lengths without applying conversion factors for applicable controller in Tables A-1, A-2, or A-3.

AWG[b] Size No.	Maximum Allowable Length for Each of Two Wires per Valve[a]				
	Maximum Static Water Pressure Not Exceeding				
	50 lb/in.2 (ft)	75 lb/in.2 (ft)	100 lb/in.2 (ft)	125 lb/in.2 (ft)	150 lb/in.2 (ft)
18	2750	2070	1480	970	530
16	4380	3290	2360	1550	840
14	6990	5250	3760	2470	1340
12	11080	8330	5970	3930	2140
10	17640	13260	9500	6250	3400
8	28050	21090	15120	9940	5420
6	44590	33520	24030	15810	8610
4	70920	53310	38220	25140	13700

[a] The two wires (control and common of same gauge) may be of unequal length if total is not more than twice the table length.
[b] American Wire Gauge: annealed copper.

Example

A MK12A Controller will be used in the following figure; minimum controller electrical input that will be available has been determined to be 120 volts; maximum static water pressure will not exceed 75 lb/in.[2]

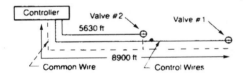

Step 1: The required wire size to the most distant valve, in terms of wire length, is determined first:

Referring to (base) Table A, nearest base wire length to the 8900 ft of wire length to valve #1 is 8330 ft for No. 12. But, Table A-1 indicates that with 120 volts to he MK12A Controller, wire length can be 15% more. Applying the conversion factor for 75 lb/in.[2] to the base footage shows that 9596 ft of No. 12 wire can be used: 8330 ft × 1.152 = 9596 ft. Therefore 8900 ft of control wire and 8900 ft of common wire is easily allowable.

Step 2: The length of control wire to valve #2 is 5630 ft. It is then determined that No. 14 wire can be used for the control wire: 5250 ft × 1.152 = 6048 ft—common wire remains No. 12.

Note: 1. Quantity of valves per station refers to Weather-matic valves only.
2. Master valve must be included in quantity of valves per station when master valve circuit is utilized.

3. Normal table use: If minimum job voltage is not listed, use next lower listed voltage; if maximum job static water pressure is not listed use next higher listed pressure. However, multiplying factors may be interpolated between nearest listed voltages; do not interpolate between water pressures.
4. Do not operate more valves than those for which multiplying factors are listed. More may be operated only with certain conditions; consult factory before attempting.
5. Input to controllers must not exceed 125 V, either operating or not operating, or be less than 105 V when operating, otherwise controller may malfunction or be damaged. Recommendation: Provide voltage of between 110 V and 120 V to controllers. (Input voltage is measured at terminals L1 and L2 in the controller.)
6. Metric information on valve wiring is contained in Weather-matic's technical booklet TD103.
7. Use of Weather-matic pump start accessory package no. 952 does not affect valve wiring data shown on chart. *Caution:* Chart calculations may not apply if relay other than Weather-matic no. 952 use.

Installation

Wiring

Type UF is recommended for valve circuit wiring. Type UF, with its heavier insulation, offers the advantage of longer trouble-free service. However, wire type and method of installation should be in accordance with local codes for NEC Class II circuits of 30 V a.c. or less.

Table A-1. One Valve Per Station

Minimum Input: 60 Hz. a.c. Volts	Maximum Static Water Pressure Not Exceeding:				
	50 lb/in.2	75 lb/in.2	100 lb/in.2	125 lb/in.2	150 lb/in.2
RM7/11 Controllers (see note 7)					
110	0.882	0.859	0.823	0.758	0.597
115	1.026	1.036	1.053	1.083	1.157
120	1.170	1.213	1.282	1.408	1.717
125	1.314	1.390	1.511	1.734	2.276
RM18/22 Controllers (see note 7)					
110	0.910	0.896	0.874	0.834	0.735
115	1.058	1.078	1.109	1.167	1.308
120	1.206	1.259	1.344	1.501	1.882
125	1.354	1.441	1.579	1.834	2.455
MK6A/8A/12A Controllers (see note 7)					
110	0.832	0.798	0.744	0.646	0.405
115	0.976	0.975	0.974	0.971	0.964
120	1.120	1.152	1.203	1.296	1.524
125	1.265	1.329	1.432	1.622	2.084
MK16A/MK24A Controllers (see note 7)					
110	0.832	0.798	0.744	0.646	0.405
115	0.976	0975	0.974	0.971	0.964
120	1.120	1.152	1.203	1.296	1.524
125	1.265	1.329	1.432	1.622	2.084
MK200/MK400/MK12S Controllers					
110	0.907	0.892	0.868	0.825	0.718
115	1.049	1.066	1094	1.145	1.268
120	1.191	1.241	1.320	1.465	1.819
125	1.333	1.415	1.545	1.785	2.369

A single wire connecting each solenoid to the controller and a common neutral wire to all solenoids from the controller serves as the power supply.

Valve Hook-Up

All wire splices should be soldered, or joined by positive mechanical connectors. Splices must be properly insulated and waterproofed. Dri-splice connectors are recommended. See Controller Accessories.

An expansion curl should be provided within three (3) feet of each wire connection to a solenoid, and at least every 100 ft of wire length on runs more than 100 ft in length. Expansion curls are easily formed by wrapping at least five (5) turns of wire around a rod or pipe 1 inch or more in diameter, then withdrawing rod.

RECOMMENDED WIRE SIZING PROCEDURE AND CHARTS

Controller Power Wire (120 V-AC High-Voltage)

To size the 120 V-AC power wires between the power source and the Buckner controller you will need to do the following:

Step 1. Determine the actual minimum available voltage at the power source. This may be measured or obtained from the local power company.

Step 2. Determine the controller's current (amps) draw and minimum input (operating) voltage. This information is available from the controller's specification sheet. Determine the permissible voltage loss between power source and the controller.

Table A-2. Two Valves Per Station

Minimum Input: 60 Hz. a.c. Volts	Maximum Static Water Pressure Not Exceeding:				
	50 lb/in.2	75 lb/in.2	100 lb/in.2	125 lb/in.2	150 lb/in.2
RM7/11 Controllers (see note 7)					
110	0.412	0.392	0.359	0.299	0.152
115	0.484	0.480	0.474	0.462	0.432
120	0.557	0.569	0.588	0.624	0.712
125	0.629	0.657	0.703	0.787	0.992
RM18/22 Controllers (see note 7)					
110	0.437	0.424	0.404	0.366	0.274
115	0.511	0.515	0.521	0.533	0.561
120	0.585	0.605	0.638	0.699	0.847
125	0.659	0.696	0.756	0.866	1.134
MK6A/8A/12A Controllers (see note 7)					
110	0.387	0.361	0.320	0.243	0.056
115	0.460	0.450	0.434	0.406	0.336
120	0.532	0.538	0.549	0.568	0.616
125	0.604	0.627	0.664	0.731	0.895
MK16A/MK24A Controllers (see note 7)					
110	0.387	0.361	0.320	0.243	0.056
115	0.460	0.450	0.434	0.406	0.336
120	0.532	0.538	0.549	0.568	0.616
125	0.604	0.627	0.664	0.731	0.895
MK200/MK400/MK12S Controllers					
110	0.433	0.419	0.397	0.355	0.254
115	0.504	0.506	0.509	0.515	0.530
120	0.575	0.593	0.622	0.675	0.805
125	0.646	0.680	0.735	0.835	1.080

Step 3. Calculate the *allowable wire resistance.*

Step 4. Select the proper wire size (AWG) from the wire sizing chart for annealed copper wire.

Step 5. Calculate the actual voltage loss between the power source and the controller. Determine actual available voltage at controller.

Example 1

Determine the wire sizes required to operate a Buckner MT-12DP controller as shown in the sketch. The controller operates a single model 20040 valve per station. Refer to the MT-12DP product specification sheet for the required current and minimum input (operating) voltage.

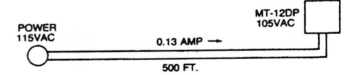

Solution

Step 1. Determine actual minimum voltage available at power source—From the diagram, the measured voltage at the source is 115 V-AC minimum.

Step 2. Determine controller amperage draw and minimum input voltage—From the product specification sheet we find the primary current required for a MT-12DP operating one

Table A-3. Three Valves Per Station

| Minimum Input: 60 Hz. a.c. Volts | Maximum Static Water Pressure Not Exceeding: | | | | |
	50 lb/in.2	75 lb/in.2	100 lb/in.2	125 lb/in.2	150 lb/in.2
	RM7/11 Controllers (see note 7)				
110	0.256	0.236	0.204	0.146	N/A
115	0.304	0.295	0.281	0.255	0.191
120	0.352	0.354	0.357	0.363	0.377
125	0.400	0.413	0.434	0.471	0.564
	RM18/22 Controllers (see note 7)				
110	0.279	0.267	0.247	0.210	0.120
115	0.328	0.327	0.325	0.321	0.312
120	0.378	0.387	0.403	0.432	0.503
125	0.427	0.448	0.482	0.543	0.694
	MK6A/8A/12A Controllers (see note 7)				
110	0.239	0.216	0.178	0.109	N/A
115	0.287	0.275	0.254	0.217	0.126
120	0.335	0.334	0.331	0.326	0.313
125	0.384	0.393	0.407	0.434	0.499
	MK16A/MK24A Controllers (see note 7)				
110	0.239	0.216	0.178	0.109	N/A
115	0.287	0.275	0.254	0.217	0.126
120	0.335	0.334	0.331	0.326	0.313
125	0.384	0.393	0.407	0.434	0.499
	MK200/MK400/MK12S Controllers				
110	0.275	0.261	0.239	0.199	0.100
115	0.322	0.319	0.314	0.305	0.283
120	0.370	0.377	0.390	0.412	0.467
125	0.417	0.436	0.465	0.519	0.650

model 20040 RCV equals 0.13 amps. The minimum input voltage for all Buckner controllers is 105 V-AC. The permissible voltage loss between power source and the controller is (115 – 105 V-AC) = 10 V-AC.

Step 3. Calculate the allowable wire resistance—Using the formula:

$$R = \frac{1000\,(V)}{2\,LI}$$

where:

R = allowable wire resistance
L = one way wire distance (ft)
I = primary current (amps)
V = permissible voltage loss

We get the following for this example:

$$R = \frac{1000(115-105)}{2(500\ \text{ft})(0.13)} = \frac{10,000}{130}$$

$$= 76.92\ \frac{\text{ohms}}{1000\ \text{ft}}$$

Step 4. Select proper wire sizes—From the wire sizing chart for annealed copper, we look in the 77°F column for a resistance value equal to or smaller than our calculated value. In this case, it is 6.510 ohms/1000 ft. This value equates to 18 AWG wire. However, as a matter of mechanical strength, the smallest wire we want to use is 14 AWG. Therefore, both power wires will be 14 AWG with an R value of 2.58 ohms/1000 ft.

Note: Local codes require a third wire to be used as earth ground. This should be a 14 AWG green wire.

Step 5. Calculate actual voltage loss between power source and controller—Using the formula

$$(V) = \frac{2\,LIR}{1000\ ft}$$

where:

R = actual resistance of wire size selected
(V) = actual voltage loss (drop)

We get the following for this example:

$$(V) = \frac{2(500)(0.13)(2.58)}{1000\ ft} = 0.34\ V$$

This available voltage at the controller is:

$$(115 - 0.34) = 114.6\ V\text{-}AC$$

Example 2

Determine the wire sizes for each leg of the following. The voltage at the source is measured at 117 V-AC. Each controller operates at the same time and provides power to a single model 20930 valve per station. The controllers are model M-12DD and require 105 V-AC minimum to operate.

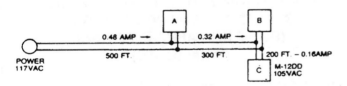

Step 1. From the diagram the actual voltage at the source is 117 V-AC minimum.

Step 2. From the product specification sheet, we find 0.16 amps primary current required for each M-12DD operating one 20930 valve per station. The wires from A to B carry 0.32 amps. The wire between A and the power source carry (3 × 0.16 = 0.48 amps).

The minimum input (operating) voltage to each controller is 105 V-AC. However, adding some factor for safety, let us use 107 V-AC. The allowable voltage loss is 117 V-AC – 107 V-AC = 10 V.

The total wire length involved is (500 ft + 300 ft + 200 ft) = 1000 ft. The proportional voltage losses in each wire section are:

Controller B to C

$$\frac{200\ ft}{1000\ ft} = 0.2 \times 10\ V = 2\ V$$

Controller A to B

$$\frac{300\ ft}{1000\ ft} = 0.3 \times 10\ V = 3\ V$$

Controller A to Source

$$\frac{500\ ft}{1000\ ft} = 0.5 \times 10\ V = 5\ V$$

Step 3. The allowable wire resistance for each wire section is:

Controller B to C

$$R = \frac{1000(2\ V)}{2(200\ ft)(0.16)} = \frac{2000}{64}$$
$$= 31.25\ ohms\,/\,1000\ ft$$

Controller A to B

$$R = \frac{1000(3\ V)}{2(300\ ft)(0.32)} = \frac{3000}{192}$$
$$= 15.63\ ohms\,/\,1000\ ft$$

Controller A to Source

$$R = \frac{1000(5\ V)}{2(500\ ft)(0.48)} = \frac{5000}{480}$$
$$= 10.42\ ohms\,/\,1000\ ft$$

Step 4. Select proper wire size.

From the wire sizing charts for annealed copper, we see that all of our R values are greater than that for 18 AWG wire (6.510 ohms/1000 ft). As before, we will use 14 AWG because of its better mechanical characteristics. *Note:* A 14 AWG green wire is required by code to be used as an earth ground.

Step 5. The actual voltage losses for each section will be:

Controller B to C

$$\frac{2(200\ ft)(0.16)(2.58)}{1000\ ft} = \frac{165}{1000}\ 0.17\ V$$

Controller A to B

$$\frac{2(300\ ft)(0.32)(2.58)}{1000\ ft} = \frac{495}{1000} = 0.50\ V$$

Controller A to Source

$$\frac{2(500\ ft)(0.48)(2.58)}{1000\ ft} = \frac{1239}{1000} = 1.24\ V$$

The total actual voltage loss = (1.24 + 0.50 + 0.17) = 1.91 V. Therefore, the actual voltage available in the worst case to the controller C is (117 − 1.91) = 115.09 V-AC. The 115 V-AC is greater than the required 107 V-AC, so there should be no problem having enough voltage to guarantee operation.

Wire Gauge (AWG)	Wire Type (UF)	Insulation Thickness (in.)	Resistance (ohms/1000 ft) 77°F (25°C)	Resistance (ohms/1000 ft) 149°F (65°C)
00	Stranded	5/64	0.079	0.092
0	Stranded	5/64	0.100	0.116
2	Stranded	5/64	0.159	0.184
4	Stranded	5/64	0.253	0.292
6	Stranded	5/64	0.403	0.465
8	Solid	5/64	0.641	0.739
10	Solid	4/64	1.020	1.180
12	Solid	4/64	1.620	1.870
14	Solid	4/64	2.580	2.970
16	Solid	4.64	4.090	4.730
18	Solid	4/64	6.510	7.510

Controller to Valve Wires (24 V-AC Low Voltage)

To size the low voltage wire between Buckner controllers and Buckner remove control valves, you will need to do the following:

Step 1. Determine the actual wire run length between the controller and the first valve on the circuit. This may be obtained by measurement from the irrigation plan.

Step 2. Calculate the equivalent circuit length values for any circuit having more than one valve on it.

Step 3. Determine the minimum input voltage to the controller by measuring the available voltage or by calculation using the technique described in the power wire sizing section of this procedure.

Step 4. Select the proper wire size chart from the following table.

	Valve Series		
Controller Series	20040, 20500	20930, 26000	20010, 20120, 20000
Mechatonic Controller	Chart I	Chart II	Chart III
VC24-P VC24-WFC VC24FC	Chart IV	Chart V	Chart VI

Step 5. Check the actual or equivalent circuit length against the appropriate chart. Beneath the minimum input find a charted wire length that is equal to or greater than the measured or calculated circuit length.

Step 6. Read horizontally to the left to find the wire sizes required. If two different wires are indicated, the larger sizes should be used as the common ground.

EVERYTHING YOU WANTED TO KNOW ABOUT POWER WIRE SIZING... (BUT WERE AFRAID TO ASK) (Release 2.0)

120-V-AC Controller Power Wiring

Remember: Each leg of the wiring (from controller to controller) must be calculated and then added up to determine total loss at the furthest, or worst, location.

Example

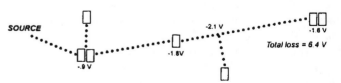

Step 1. Determine total amperage draw on each leg of wire to be sized.

Step 2. Determine existing wire size or size you want to use on each leg.

Step 3. Find constant on sizing chart for total amps and size. Multiply this number by the total feet of the wire leg to get voltage loss.

Step 4. Add all losses to get total loss at worst location. If final voltage is less than 105, at the worst point, up-size wire on certain legs and recalculate.

Example

If 120 V at source, then 15 V total loss is acceptable.

120-V-AC Line Conditioners

Step 1. Determine total amperage draw on the entire power source.

Step 2. Multiply total amps by 120 (line voltage)

Step 3. Add appropriate 30% to total.

Step 4. The total represents Watts or VA (voltamps). Divide by 1000 to determine kW or kVA.

Example

(20) Network LTC Satellites
20×0.51 (amp draw) = 10.2 (total amps)
10.2×120 (line voltage) = 1224 (VA)
$1224 \times 30\%$ = 1591 (VA)
so
1.5 kVA Line Conditioner

Note: Line Conditioners are available in standard kVA sizes. 0.50, 0.75, 1.0, 1.5, 2.0, 2.5, etc.

Toro Controller Amperage Draws

Solenoids Operating	Network 800 Satellite Amps	Network LTC Satellite Amps	OSMAC Satellite Amps	Vari-Time Satellite Amps
0	0.35	0.15	0.05	0.21
1	0.41	0.21	0.11	0.11
2	0.47	0.27	0.17	0.17
3	0.53	0.33	0.23	0.23
4	0.59	0.39	0.29	0.29
5	0.65	0.45	0.35	
6	0.71	0.51	0.41	
7	0.77	0.57	0.47	
8	0.83	0.63	0.53	
9	0.89	0.69	0.59	
10	0.95	0.75	0.65	
11	1.01	0.81	0.71	
12	1.07	0.87	0.77	
13	1.13		0.83	
14	1.19		0.89	
15	1.25		0.95	
16	1.31			
17	1.37			
18	1.43			

Voltage Loss per Foot of Wire

Amps	#14	#12	#10	#8	#6	#4	#2
2	0.001013	0.000637	0.000401	0.000243	0.000153	0.000096	0.000060
4	0.002026	0.001274	0.000801	0.000485	0.000305	0.000192	0.000121
6	0.003038	0.001911	0.001202	0.000728	0.000458	0.000289	0.000181
8	0.004051	0.002548	0.001602	0.000970	0.000610	0.000385	0.000242
1.0	0.005064	0.003185	0.002003	0.001213	0.000763	0.000481	0.000302
1.5	0.007596	0.004778	0.003005	0.001820	0.001145	0.000722	0.000453
2.0	0.010128	0.006370	0.004006	0.002426	0.001526	0.000962	0.000604
2.5	0.012660	0.007963	0.005008	0.003033	0.001908	0.001203	0.000755
3.0	0.015192	0.009555	0.006009	0.003639	0.002289	0.001443	0.000906
3.5	0.017724	0.011148	0.007011	0.004246	0.002671	0.001684	0.001057
4.0	0.020256	0.012740	0.008012	0.004852	0.003052	0.001924	0.001208
4.5	0.022788	0.014333	0.009014	0.005459	0.003434	0.002165	0.001359
5.0	0.025320	0.015925	0.010015	0.006065	0.003815	0.002405	0.001510
5.5	0.027852	0.017518	0.011017	0.006672	0.004197	0.002646	0.001661
6.0	0.030384	0.019110	0.012018	0.007278	0.004578	0.002886	0.001812
6.5	0.032916	0.020703	0.013020	0.007885	0.004960	0.003127	0.001963
7.0	0.035448	0.022295	0.014021	0.008491	0.005341	0.003367	0.002114

Appendix 7

Pumping Station Control Valving

Pressure Reducing Valve

MODEL 90-01 690-01

Schematic Diagram

Item	Description
1	Hytrol (Main Valve)
2	X58 Restriction Fitting
3	CRD Pressure Reducing Control

Optional Features

Item	Description
A	X46A Flow Clean Strainer
B	CK2 Cock (Isolation Valve)
C	CV Speed Control (Closing)
D	Check Valves with Cock
S	CV Speed Control (Opening)
Y	X43 "Y" Strainer

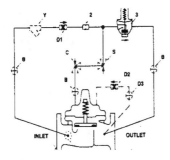

Valves 6" and larger with a "D" feature must be installed with the main valve stem in a vertical position.

Typical Applications

Typical pressure reducing valve station using Model 90-01AB/690-01AB and Model 90-01AS/690-01AS in parallel to handle wide range of flow rates. Larger Model 90-01AB/690-01AB valve takes care of peak loads and smaller Model 90-01AS/690-01AS handles low flows.

Pressure Relief, Pressure Sustaining Valve

MODEL 50-01 650-01

Schematic Diagram

Item	Description
1	Hytrol (Main Valve)
2	X42N-2 Strainer and Needle Valve
3	CRL Pressure Relief Control

Optional Features

Item	Description
B	CK2 Cock (Isolation Valve)
D	Check Valves with Cock
F	Remote Pilot Sensing
H	Drain to Atmosphere
S	CV Speed Control Opening

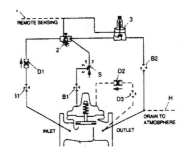

Valves 6" and larger with a "D" feature must be installed with the main valve stem in a vertical position.

Typical Applications

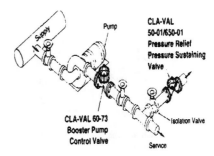

Pressure Relief Service

To provide protection for the system against high pressure surges when pumps are shut down, this fast opening—slow closing relief valve dissipates the excess pressure.

Combination Pressure Reducing and Pressure Sustaining Valve

— MODEL — **92-01**
692-01

Schematic Diagram

Item	Description
1	Hytrol (Main Valve)
2	X44A Strainer and Orifice
3	CRD Pressure Reducing Control
4	CRL Pressure Relief Control
5	CV Flow Control (Opening)

Optional Features

Item	Description
B	CK2 Cock (Isolation Valve)
C	CV Flow Control (Closing)
D	Check Valves with Cock
F	Remote Pilot Sensing

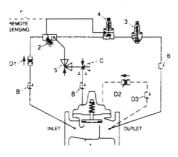

Valves 6" and larger with a "D" feature must be installed with the main valve stem in a vertical position.

Typical Applications

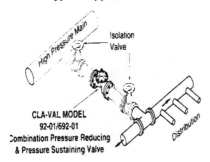

CLA-VAL MODEL
92-01/692-01
Combination Pressure Reducing
& Pressure Sustaining Valve

A typical application for a Combination Pressure Reducing and Pressure Sustaining Valve is to automatically reduce pressure for the downstream distribution network and sustain a minimum pressure in the high pressure main regardless of distribution demand.

Check Valve

— MODEL — **81-02**
681-02

Schematic Diagram

Item	Diagram
1	Hytrol (Reverse Flow Main Valve)
2	CGA Angle Valve (Closing)
3	CNA Needle Valve (Opening)
4	CSC Swing Check Valve

Valve to the system

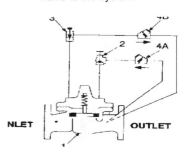

Typical Applications

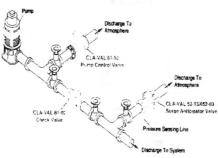

Deep Well Pump

This valve should be an integral part of any well designed pumping system. It is used to prevent damaging and sometimes expensive flow reversals.

Appendix 8

Automatic Controllers

LEVELS 1 and 2

VT 4000 Central

Level 1 control system for golf course applications.

Features
- Up to six central control modules or 3 central and 3 syringe
- Each module controls up to 20 satellites
- Standard module features 24-hour programming with start time resolution of 15 minutes and syringe timing of 0 to 5 minutes
- Manual start and cancel available on each module

Electrical Specifications
- 117 VAC, 60 Hz
- 0.57 amps (70 W) typical (4 centrals plus 1 syringe module)

Central/Satellite Preference Table. Central/Satellite configurations which operate together.

Central	Satellite	#of Stations
VT 4000	186-xx-02	11
VT II	186-xx-04	11
	116-xx-x5	16
	LTC-xx-x5	16
OSMAC	J16YSD0040	8
	J16YSB0040	8–48
Network LTC	186-xx-04	11
	116-xx-x4	16
	LTC-xx-x4	16
Network 8000	132-x6-x8	32

VT Satellite 186-56-04

VT Satellite

Features
- 11 stations
- Electric and Normally Open Hydraulic versions available
- Station run times: 0 to 30 minutes
- Can be run manually or under central control
- Locking, weather-resistant pedestal mount

Electrical Specifications
- Input power
  - 117 VAC, 60 Hz
  - 0.03 amps (electric, idle)
  - 0.21 amps (electric, operating)
  - 0.03 amps (hydraulic, operating)
  - 0.12 amps (hydraulic, operating)
- Output power (electric models)
  - 24 VAC
  - 1.50 amps (36 VA) (Toro VIH sprinklers per station maximum)

Specifying Information

191-09-XX Mounting	
11—1 Central/Syringe	44—4 Central/Syringe
22—2 Central/Syringe	55—5 Central/Syringe
33—3 Central/Syringe	66—6 Central/Syringe

For example: When specifying a VT II 4 Central/Syringe, you would specify: 191-09-44

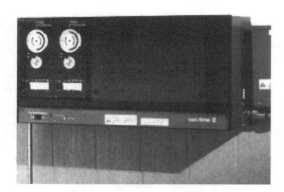

VT II Central

VT II Central

Level 2 control system for golf course applications.

Features
- Up to six central control modules
- Controls up to 72 satellites with a single pair of signal wires
- Dual-program control module features 14-day programming with start time resolution of 15 minutes
- Syringe control integrated into each module with 0- to 9-minute timing
- Manual start, syringe and cancel available on each module
- Rain gauge input for automatic cancellation of irrigation cycle

Electrical Specifications
- 117 VAC, 60 Hz
- 0.46 maps (54 W) (6 central panels)

Command/Satellite System

An affordable, versatile and reliable irrigation control system designed specifically for golf courses and other large turf/landscape projects.

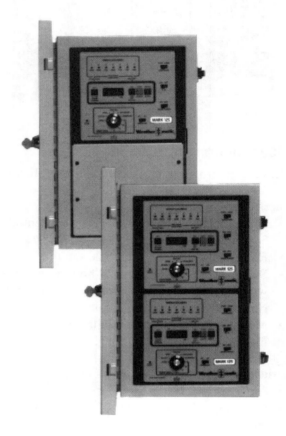

MK12S/24S Satellite Control

Features

Simple and Flexible. Microcomputer accuracy and flexibility. The exclusive function selector switch and four large, clearly labeled programming buttons minimize user training and installer callbacks.

12 or 24 Stations. MK12S model has 12 stations. MK24S model has 2 MK12S controls in the same housing.

Day-Bright™ LED Display. Large digital display of time, date and program values is filtered and recessed for superior outdoor daylight visibility.

Easy Installation. Quick-disconnect for chassis installation/removal from housing with no disturbance to field wiring. All wiring is 24 v except line input to transformer.

Independent Power Supply. Satellites may be wired to independent power supply without affecting operation.

Valve Short Indicator/Lamp. Lamp on face panel flashes to indicate short. Controller skips shorted station and continues watering all other programmed stations. Shorted stations easily identified.

Solid-State Timing. Variable from 0 to 99 minutes on each station in accurate 1-minute increments.

System Shutdown Alert Circuit. The MK12S Satellite alerts the MK12C Command prior to the last scheduled valve completing its watering. The MK12C signals the pumping system, allowing for a gradual shutdown of flow prior to valve closure.

Remote/Satellite Switch (R/S). Allows the Satellite to be used for Remote (independent) operation or for operation as a Satellite.

6 or 7 Day Calendar. Allows programmed watering for specific days of the week, or even-day, odd-day, third day watering. LEDs show days selected for current week's watering, allowing for "Week-at-a-Glance" review. (Effective only when Satellite is in Remote [independent] mode.)

Dual Watering Programs. Two independent programs with four start times each. Start times may be set on-the-minute. (Effective only when Satellite is in Remote [independent] mode.)

Pump Protection Circuit. Overrides back-up program time on unused stations.

Valve Power Switch. Turns off stations without disturbing program; may be used for dry indexing.

Semi-Automatic and Manual Operation. Operator can initiate watering on a single station only in sequence; controller will shut down automatically when watering is completed.

Master Valve Circuit. May be used to operate a local master valve.

Battery Back-Up. Retains program and clock time for 60 minutes if power fails; flashing display indicates program loss and battery depletion. (Requires two NiCad rechargeable batteries, which are not included.) Built-in circuit keeps batteries charged.

Input Surge Protection. Built-in protection from surges on incoming power line.

Output Surge Protection. Shields controller from power surges incoming on field wiring.

Industrial-Grade Housing. Urethane coated, heavy gauge steel, for outdoor use; has hinged cover with two side toggle latches and cylinder lock.

Housing Dimensions
MK12S/MK24S Satellite (Outdoor/Indoor)
11 in. W, 17 in H, 8.5 in D
28 cm, 43.2 cm, 21.6 cm

Electrical Requirements
Input: Each MK12S requires 115 V a.c. 60 Hz @ 0.38 A for three valves. Input voltage 110 V a.c. minimum, 125 V a.c. maximum.

Command Satellite Communications. Satellites may be connected to the Command in parallel (allowing simultaneous starts of up to twenty MK12S Satellites or 10 MK24S Satellites), series (each Satellite starts a successive Satellite at the completion of its watering cycle), or parallel/series (a combination of parallel and series). Requires only two wires from the MK12C Command wired common to all associated Satellites.

Communication Output to Command: 24 V a.c.
Valve Output: 24 V a.c.

Operates: Operates up to 3 Weathermatic valves simultaneously for a maximum current draw of 1.5 A. For additional multiple valve operation, consult factory.

Modular Series Electronic Controllers

Application and Description

The electronic modular controller series is designed for use in all applications including residential, commercial, industrial, agricultural and golf. These controllers incorporate the accuracy of crystal and solid state timing with a simple, functional keyboard for programming.

Features and Benefits: (All)
- Available in many station combinations from 8 to 51.
- Audible programming tone for errors.
- Seven (7) independent programs that can run in series, parallel or independently (5 automatically). Up to six stations or valves can run simultaneously.
- Variable schedule lengths from 1–32 days.
- Four (4) start times for each automatic program.
- Station run times in hours/minutes or minutes/seconds modes.
- Semi-automatic operation by program and start time.
- Input and output lightning protection. (Heavy duty on C & D chassis).
- Manual program for operating stations individually, and up to 6 stations sequentially or simultaneously.
- Sensing overload protection, then skips the station rather than blowing fuses and reports failure when program is completed.
- The 10 year lithium battery provides complete onboard program and time retention in event of power failure.

- Incorporates provisions for radio communication options.
- Percent scaling multiplier (1–512%) of station run times for water management and seasonal adjustment.
- Independent loop program for (1–9 or continuous) repeats with optional delays between each repeat.
- Watering days for each program can be programmed to skip all starts from 1–999 days.
- Rain shut down of the complete controller without losing time or programs.
- Pump/master vale circuit activated independently.
- Single key copying of run times, within a program, from start time to start time.
- Heavy duty, exterior grade, weather resistant, stainless steel cabinet. (Powder coated on C & D chassis.)

Specifications

The written specifications should include all features and benefits of each chassis specified. Particular attention should be made in specifying grounding. Regional standards or local codes should be adhered to and a minimum method of grounding shall be one (1) 5/8 in ´ 8 ft copper clad ground rod per controller driven into the ground it's full length and connected via welded/soldered joints to a minimum #6 solid copper wire direct to all green leads within the Buckner controller. The resulting resistance reading from the controller to ground shall be 15 ohms or less.

The controllers shall be installed outside the branch canopy of any tree or shrub.

Special Requirements

Power Wires: Three wires needed to provide electrical power to any components of the Modular System. Two of these wires (black and white) should be of sufficient size to ensure a minimum voltage of 105 VAC (210 VAC for 230 VAC models). The third wire shall be (green) with a minimum gauge of 14 (or per local codes) which will be connected to the earth ground.

Valve Wires: Information on wire sizes is available in Buckner's "Recommended Wire Sizing Procedures" and charts.

Note: No splices shall be allowed on any of the wires in the installation except at the controller or valve locations. In the event splices inadvertently occur or are required they must be waterproofed and placed in a valve box.

Mechatronic Central Controller System

Application and Description

The Buckner 36300 series Central Field Controller system is designed for large turf and ag applications. The system combines the reliability and accuracy of proven solid state devices with the simplicity of electromechanical programming. The system is easy to install and operate, requiring no special tools or training. It is especially suited for golf courses and parks where central control is required for synchronization of many Field Controllers' operation.

Features and Benefits

Central Controller

- Separate 14-day schedules for normal irrigation and syringe cycling.
- Syringe timing programmable from 1–12 minutes in one minute increments.
- Up to 23 irrigation starts per day provides maximum flexibility.
- 12 programmable syringe starts per day.

Field Controller

- 12 or 24 station (outputs) available.
- Solid state station time settings selectable as follows: 0, 2, 4, 6, 8, 10, 12, 14, 17, 20, 25 and 30 minutes.
- STATION TIME MULTIPLIER for water budgeting or seasonal adjustments (50%, 100%, or 200% of set times).
- Indicator lights show which output is active.
- Unique "CHECK" switch providing 1, 2, and 4 minute timing on all stations for fast system check without changing programmed times.
- AUTOMATIC, SEMI-AUTOMATIC, RAIN OMIT (OFF) or MANUAL capabilities.
- Pump/Master valve output (24 VAC).
- Isolation switch allows Field Controller to ignore Central commands.
- Lightning protection on both primary (input) and secondary (output) circuits of central and field components provides protection against most surges.
- Standard stainless steel case (36712) only

Model Identification Matrix

Cat. No.	Model No.	Description	Ht.	Width	Depth
36304	M-CP4	Central Controller-4 Panel	17-1/4"	14-1/2"	6"
36312	M-12FC	Field Controller-12 Stations	9-1/2"	12-7/8"	5"
36324	M-24FC	Field Controller-24 Stations	17-1/4"	14-1/2"	6"
36712	M-12FCP	Field Controller-12 Station in Stainless Steel Pedestal	3/4"	13-1/4"	10-1/2"

Options

- 115 VAC 50 Hz input (/85)
- 230 VAC 50 Hz input (/87)
- Pedestal mount PED-1 (37100)
- Stainless steel case (/91) 36312 and 36324 only

Electrical Data

Primary (Input)—105–125 VAC 60 Hz, current requirements (amps) as follows:

Valve Series	Primary Current Required at 115 VAC 60 Hz (in amps)				
	Controller Only	Number of Valves Per Station[a]			
		1	2	3	4
20400, 20500	0.13	0.15	0.18	0.24	0.30
20930, 26000	0.13	0.16	0.24	0.30	—
20000, 20120, 20010, 20130	0.13	0.20	0.28	—	—

[a] When "Pump" output is used, it must be considered as one of the valves. The pump relay coil shall be 24 VAC and 0.5 amps (maximum). Relay contacts to be rated for load. Secondary (outputs): 24 VAC 60 Hz, 1.0 ampere maximum for valves and pump.

Solid-State Controller Matrix

Standard Features

This information is intended to be used as a general reference for controller selection. For the complete description of a controller, please refer to the catalog.

Model Number	ISC-4, ISC-8, ISC-12, ISC-16, ISC-24, ISC-32
Number of Stations	4, 8, 12, 16, 24, 32
Independent Station Programming	X
Independent Dual Programming	
Station Timing	0–99 Minutes/ 0–9.9 Hours
Plastic Cabinet—Indoor	
Plastic Cabinet—Outdoor	
Steel Cabinet—Outdoor	X
Built-In Key Lock Cabinet	X
Automatic Starts:	
Up to 3 per day	
Up to 4 per day (per station)	x
Up to 6 per day	
Up to 15 per day (per station)	
Programming Schedule:	
14-day fixed	
Variable 1–8 day cycle	X
Variable 1–7 day cycle	
0–60 days (per station)	
Water Budgeting	X
Sequential Station Operation	X
Simultaneous Station Operation	
12-Hour A.M./P.M. Clock	X
24-Hour Time Clock	X
Manual Operation	X
Single Station	X
Multiple Station	
Cycle Start	X
Manual Station Advance	X
Built-In Stand-By Watering Schedule—	
Each station waters for:	
10 minutes (once per day)	
5 minutes (twice per day)	X
Auto/Off Key	X
UL Listed and Tested	X
Uses Rechargeable NiCad Battery	X
Uses Alkaline Battery	
Battery Back-up maintains	
program for up to:	
24 hours	X
20 hours	
4 hours	
Terminal Strip	ISC-4,8,12,16 (in cabinet) ISC-24,32 (in pedestal)

Quick Disconnect Cables	X
Quick Disconnect Pig Tails	ISC-24,32
Internal Transformer	X
External Transformer	
220/240/260 VAC Transformer 50 Hz	♦
Station Capacity:	
1 valve per station	
Up to 2 valves per station	
Up to 4 valves per station	X*
Master Valve Pump Start Circuit	X
Micro-Circuitry protected by MOV	X
Optical Couplers	X
Secondary Electrical Surge Protection	♦
Remote Control Interface	X
Pedestal Mount	♦
Wall Mount	X

♦ Optional Feature
** Connect only 3 valves per station if master valve output is used.

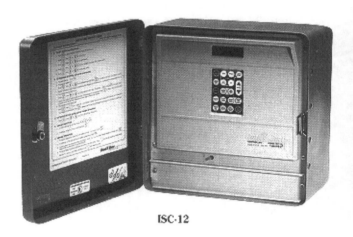

ISC-12

ISC-8-B/ISC-12-B/ISC-16-B/ISC-24-B/ ISC-32-B Independent Station Computer Controllers

Operating Specifications
- ISC-8-B: 8 stations
  ISC-12-B: 12 stations
  ISC-16-B: 16 stations
  ISC-24-B: 24 stations
  ISC-32-B: 32 stations
- Station timing: 0–99 minutes (in 1 minute increments) or 0–99 hours (in 0.1 hour increments)
- Automatic starts: On any quarter hour, up to 8 per day for each station

- Programming schedule: Based on a variable day cycle for every day, every other day, every third day starts, etc.

Water Management Features
- 8 start times
- 16-day variable cycle
- Variable test program 2–9 minutes
- Nonvolatile memory
- Time keeping during power outage
- Ditto programming key

Features
- User friendly
- Independent station programming
- System water budgeting from 25% to 200% of programmed time, in increments of 25%
- 12-hour am/pm or 24-hour clock
- Manual operation
- Station advance key
- Single station operation
- Monitor capability
- Master valve/remote pump start circuit capability
- Auto/off key
- Sequential station operation
- Primary electrical surge protection
- Valve wire terminal strip in all models

Electrical Characteristics
- Input required: 117 volt AV, 60 Hz
- Output: 26.5 volt, 60 Hz, 1.5A

- Circuit breaker: 1.5A holding, 2.5A break
- UL listed and tested
- Sequential operation: When more than one station is programmed to start at the same time, those stations will water in sequence starting from the station with the lowest number.
- Multi-valve station capacity: Three 7VA valves per station if master valve/pump start circuit is used; four 7VA valves per station if master valve/pump start circuit is not used.

Specifications for Pedestals
    Material: Epoxy painted steel
    Dimensions: Height: 23-1/2"; Width: 10-1/2"; Depth: 5"
    Field Wiring Connection: PED-16, none (terminal strip is in the ISC-8-B, ISC-12-B, ISC-16-B); PED-24/32, two screw terminal blocks connectable by means of the pre-installed cable harness, as an alternate to terminal strip in ISC-24-B/32-B.
    How to specify:

This specifies an ISC-B controller; 8 stations indoor and outdoor mountable (see catalog data).

RC-1860AB/RC-2360AB Dual Program Electromechanical Controllers

Operating Specifications
- RC-1860AB: 18 stations
  RC-2360AB: 23 stations
- Station timing:
  RC-1860AB: 11 stations with 6–60 minutes (in 2 minute increments) and 7 stations with 3–60 minutes (in 1 minute increments)
  RC-2360AB: 6–60 minutes (in 2 minute increments)
- Automatic starts: 1 to 23 per day
- Programming schedule: 14-day calendar dial for every day, every other day starts, etc., for a two-week period

Features
- A/AB dual program
- Weatherproof, heavy-duty, locking, steel cabinet
- Independent station timing knob
- 24-hour clock (automatic starts on any hour except midnight)
- Master valve/remote pump start circuit capability
- Master on/off switch

Dimensions
- Controller: Width: 16-1/4"; Height: 18-1/2"; Depth: 7-5/8". *Note:* Add 1-3/4" to depth for door handle.

- Pedestal: Width: 11"; Height: 23-7/8"; Depth: 5-1/2"

Electrical Characteristics
- Input required: 117 volt AC, 60 Hz
- Output: 26.5 volt, 60 Hz, 2.25A
- Reset circuit breaker: 30A holding, 4.5A break
- Master valve circuit: 26.5V, 60 Hz
- Multivalve station capacity: maximum is seven 7VA solenoid valves

Option and Accessories
- WB Turf Wall Mount Bracket
- PED Turf pedestal
- LPP-K primary surge protection kit
- LPV-K valve output surge protection kit
- MSR Multistation Relay
- 50 Hz 220/240/260 volt input transformer

RC-1260S "Add On" Controller
An Economical Way to Add Stations

A new, effective and affordable way to expand systems controlled by RC-C Series controllers. The RC-1260S, a 12-station add-on controller, simply hooks up to the last station of an RC-C Series controller, expanding the system by 11 stations. The last station of the RC-C controller is used exclusively for starting the RC-1260S.

Features
- 12 stations (increases total system stations by 11)
- Individual time control knobs
- Station timing of 6–60 minutes
- Circuit breaker reset button
- Key-lock, weatherproof plastic cabinet
- Economical price
- Powered and controlled by RC-C Series controller utilizing four low-voltage wires
- Master valve/pump start circuit

Expanded Control
The RC-1260S controller essentially transforms the RC-4C, -7C, and -1260C models into 15-, 18-, and 23-station controllers (respectively). Because the RC-1260S is not a stand-alone controller and takes its power and start command from the master controller, there's no danger of overlapping schedules and no need for hour or day dials on the unit. Programming the controller couldn't

be simpler, just set the station runtimes and set the mode switch to "AUTO."

Operating Specifications
- Number of stations: 12
- Station timing: 6–60 minutes (in 2-minute increments)
- Station capacity: 1 valve per station plus a master valve

Electrical Specifications
- Power requirement: 26.5 volts (supplied by RC-C Series controller's 26.5 volt output)

- Circuit breaker: protects controller from damage due to current overload or electrical short circuit (reset button restores power after problem is remedied)
- Pump start circuit: Uses power from master's pump start circuitry

Dimensions
Width: 11-1/2"; Height: 9-3/4"; Depth: 6-3/4"

Level 3

OSMAC Central

Level 3 control system for golf course applications.

Features
- Manages up to 150 RDR satellite units—each RDR is capable of supporting up to 48 stations
- Flexible software with flow analysis allows you to schedule irrigation that maximizes system efficiency, minimizes the water time window and, ultimately, saves you money
- Up to 100 programs may operate at any time
  —Schedules can be given absolute start or stop time, or be assigned relative to sunset or sunrise

- Schedules may be saved and recalled allowing unlimited operational flexibility
- Stations can be assigned to groups, and stations and groups can be combined into sectors for additional programming flexibility
- Manual adjustment factors allow scheduling refinement by group, sector and all satellites
- Uses "bullet-proof" paging technology for absolute security of transmitted data
- Paging terminal also serves as a voice-radio base station
- Paging terminal has alarm inputs allowing automated response to external events
- Manual control of RDR satellites from central software or directly from paging terminal
- Central paging terminal supports local and remote (via telephone) paging and telephone line extension—allows you to receive phone calls through your walkie-talkie while working in the field
- Includes one-year subscription to the exclusive Toro National Support Network—the most advanced computer hardware/software support system in the irrigation industry
- Built for Toro with Six-Sigma quality by Motorola—the world leader in communications

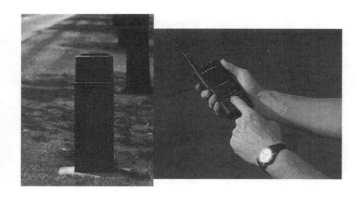

OSMAC Satellite

Features

- Modular solid-state design for expandability from 8 to 48 stations
- Standard RDR great for retrofitting existing installations. Pedestal versions available for new golf courses
- Hand-held remote control puts satellite control into the palm of your hand
- Multi-function radio allows control and voice transmissions from the same unit

- Runs up to 10 stations simultaneously—from the central or from a hand-held remote
- Programmable syringe time from 30 seconds to 128 minutes in 30-second increments
- Optional relay card available
- Build for Toro with Six-Sigma quality by Motorola—the world leader in communications

Electrical Specifications
- Input power
  — 120/240 VAC, 60 Hz
  — 0.05 amps @ 120 VAC (power consumption does not include electric valve loads)
- Output power
  — 24 VAC
  — 0.60 amps (14 VA) per station
  — 3 amps (72 VA) total

Mechanical Specifications
Dimensions:
- RDR: 12″ W × 6-1/2″ H × 4-3/4″ D
- Small pedestal: 12-5/8″ × 41-1/2″ H × 8-1/4″ D
- Large pedestal: 13″ W × 45-1/2″ H × 13″ D

Level 4

Level 4 control system for golf course applications.

Features
- Manages up to 500 Network LTC satellites
- Multi-lingual (English, French, German, Italian, Spanish and Japanese are built in)
- Manages up to 50 satellite groups
- Up to 8 irrigation program types per group
  — Up to 12 start times per program
  — Up to 3 repeats with 0- to 59-minute soak time per program
  — Independent program scheduling: calendar, interval or off
- Automated multi-source flow management
- Manual adjustment factors allow scheduling refinement by station, satellite, program, group and for all satellites
- Employs true two-way data communications allowing the central to transmit program data and receive status data from the satellite
- Transmits fully flow-managed irrigation schedule to satellites for proper system operation—even if the central is taken off-line
- Can send start and syringe commands to VT II compatible satellites with optional VT II Surge Box
- Includes one-year subscription to the exclusive Toro National Support Network—the most advanced computer hardware/software support system in the irrigation industry

Network LTC Satellite in stainless steel cabinet.

Network LTC Satellite in slope top cabinet.

Network LTC Satellite

Features
- Advanced solid-state design
- User-friendly interface simplifies satellite operation
- May be run as stand-alone controller (great for installation) or under full management of the Network LTC central
- 16 stations
- Electric and Hydraulic versions available
- 8 independent programs—up to 4 may be run simultaneously
  — 14-day calendar or 1- to 30-day interval scheduling by program
  — 0 to 3 repeats per program
  — 0- to 59-minute soak time per program
  — Up to 12 starts per program
- Station run times from 1 minute to 8 hours and 59 minutes
- Percent adjust by program (1 to 999%)
- Up to 2 non-irrigation (switch) programs with central
- Non-volatile memory saves program data for up to 10 years without power

- Manual operation by program (normal or syringe), independent station or multiple stations (up to 6)

Electrical Specifications
- Input power
  — 100, 120, 200, 220, 240 VAC, 50/60 Hz
  — 0.10 amps (12 W) @ 120 VAC (power consumption does not include electric valve loads), 60 Hz

- Output power
  — 24 VAC
  — 0.75 amps (18 VA) per station
  — 3 amps (72 VA) total

Level 5

Network 8000 Central

Level 5 control system for golf course applications.
- Manages up to 800 Network 8000 satellites
- Fully automatic irrigation scheduling by station based on environment, irrigation system design, plant materials, soil conditions, terrain and operator requirements
- Manages up to 16 satellite groups
- Up to 16 irrigation program types per group
  — Up to 8 start times per program
  — Up to 3 repeats per program
  — 12- or 14-day program scheduling
- Automated flow management
- Manual adjustment factors allow scheduling refinement by station, satellite, program, group and for all satellites

- Can use on-site or dial-up (CIMIS) weather station
- Sequential multi-manual and multi-manual master programming for advanced non-irrigation watering (overseeding, frost and dew removal)
- Programmable satellite alarms handle emergency situations automatically—with up to 10 responses for any alarm
- Control Codes feature allows the user to activate central-based function from any field satellite
- Switch control programs for automation of non-irrigation functions (lights, fountains, aerators, etc.)
- Growing Degree Days database tracks pest development
- Employs true two-way data communications allowing the central to transmit program data as well as receive status and sensor data from the satellite
- Transmits fully flow-managed irrigation schedule to satellites for proper system operation—even if the central is taken off-line
- Database export utility for using system data in other applications
- Includes one year subscription to the exclusive Toro National Support Network—the most advanced computer hardware/software support system in the irrigation industry

Network 8000 Satellite

- Advanced solid-state design that allows in-field upgrades (no charge with the TORO NSN program)
- User-friendly displays/menu system (English or Metric) simplifies satellite operation
- May be run as stand-alone controller (great for installation) or under full management of the Network 8000 Central
- 32 stations; up to 6 may run simultaneously
- Faulty stations (shorted or open) automatically detected
- 12 independent programs
  — 12- or 14-day scheduling
  — 0 to 3 repeats per program
  — 0- to 59-minute soak time per program
- Up to 8 starts per program with pre-programmed syringe cycles
- Station run-times from 1 minute to 4 hours and 15 minutes

- Built-in Lithium battery maintains program data for up to 90 continuous days without AC power
- Percent adjust by independent station (0 to 900%)—adjust factors may be automatically uploaded to central for permanent use
- Manual operation by program (normal or syringe), independent station or multiple stations (up to 6)
- Pause control for temporary suspension of irrigation cycle
- Control Request function allows in-field actuation of central-based features
- Built-in temperature and pressure sensor ports (requires sensor kits)
- Can monitor and store data for up to 6 digital channels (flow, wind, rain, status). Automatically uploads collected data to central as required (requires Sensor Input Board)
- Can automatically generate alarms (message to central) on sensor data

Electrical Specifications
- Input power
  — 117 VAC, 50/60 Hz
  — 0.35 amps (41 W) @ 120 VAC (power consumption does not include electric valve loads), 60 Hz
- Output power
  — 24 VAC
  — 0.75 amps (18 VA) per station
- 4.50 amps (108 VA) total

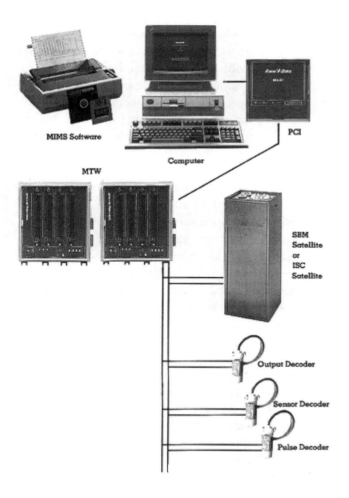

MIMS Software

PCI

MTW

Computer

SBM
Satellite
or
ISC
Satellite

Output Decoder

Sensor Decoder

Pulse Decoder

MAXI® System V
Computer Control System

Features
- Controls up to 224 field satellites or decoders
- Exclusive **Flo-Director**™ software manages the demand on a multiple pump station pumping system (up to five stations)
- True central control with constant communication between the central and the field satellites providing interaction between satellites
- Unique, Cycle + Soak feature optimizes the watering of poor drainage sites, slopes, and heavy soil areas. Applies water at or below soil intake rate.
- Capable of communicating with an on-site weather station and automatically adjusting station times.
- Automatic ET-sensitive response capability— puts back only amount of water actually used by the plant.

- Not limited to irrigation; can control lighting systems, security gates systems, fountains, etc.

Field Satellite Controllers

Features
- Choice of electromechanical or solid-state satellites for controlling electric or hydraulic systems
- Random access of all stations by MAXI Central
- May be operated manually in the field
- 1 to 99 minute or 0.1 to 9.9 hour station timing in the standby program or manual mode
- 26.5 VAC output for operating up to four 7VA electric solenoid valves
- Solid-state design
- Computer Command Center—Operates with IBM-AT's, -XT's and PS/2 models
- MIMS™ Software—Heart of the MAXI-ET System.
- PCI (Personal Computer Interface)—Runs up to 224 field satellites or sensors.
- MTW (Multiple Two-Wire Field Interface)— Constant communication link to the field satellites and sensors.
- Two-Wire Path—Communication
- Satellites—The link between the computer and the field valves.
- SBM Satellites—12-station electromechanical satellite in both electric and hydraulic configurations.
- ISC Satellites—100% solid-state 12 and 24-station models.
- Decoders—Operate and/or sense any switch opening or closing device. Used to sense field conditions.

Wireless...the new dimension

"Wireless" joins "telephone" and "two-wire" as communications options for operating Rain Bird's powerful MAXI® System V computer control system. The introduction of the new "wireless" capability simplifies the installation of MAXI for a variety of applications that range from retrofitting an established course to adding another nine holes to an existing layout or developing an entire new course.

The "wireless" communication option requires no trenching, saving time and money while eliminating any prolonged disruption to the existing course.

Plus, the "wireless" option, utilizing proven Motorola R-Net™ Series Telemetry Radios, provides superintendents with a two-way communication system capable of accessing the full selection of the MAXI system's vast number of special features. Flo-Manager™, Cycle& Soak™, ET-based operation, multiple pump station management, branch and flo-zone management and fail-safe field back-up protection are all available using MAXI with the "wireless" communications option.

LM-1230-SS/LM-1230
Field Satellite Controllers

Features

- 12 station controllers
- 1 to 23 automatic starts per day
- May be operated independent of the Master Controller
- Choice of two electromechanical models
  — LM-1230-SS—Housed in locking, brushed stainless steel pedestal
  — LM-1230—Housed in locking, metal cabinet for wall mounting
- See Brochure D31228

Stainless Steel
Field Satellite
Controllers

Decoders

Features

- Capable of operating and/or sensing any switch opening or closing device
- Used to sense field conditions, activate fountains, control security lights, gates, etc.
- MAXI can utilize up to 224 decoders
- Choice of three models
  — F69199 Output Decoder
  — F69200 Sensor Decoder
  — F69300 Pulse Decoder
- See Brochure D31215

WS-100 Weather Stations

Features

- Communicates directly with the MAXI Central Computer via direct wire interface.
- Powerful internal micro-logger collects and logs essential ET parameters
- Weather software calculates ET values and stores daily and historic ET values
- Simple to service

Links Master™
Master Satellite Control System

Features
- Controls up to 75 field satellites
- Expandable to up to six groups of satellites for greater scheduling flexibility
- Linking permits uninterrupted watering between groups and eliminates the need for cycle time calculations
- Stacking up to four irrigation starts and four syringe starts eliminates the need to calculate cycle lengths
- Easily upgradeable to a MAXI Central control system
- See Brochure D31228

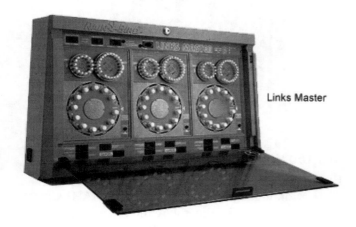

Links Master

Appendix 9

COPS 4 Computer Oriented Programming System

Application and Description

The Buckner COPS System is designed to provide a single source water management tool for monitoring and programming local and remote irrigation systems. The system is comprised of a computer interface, software, field controllers and includes on-site system check-out and training by factory personnel. The customer is free to use or purchase a personal computer of their choice from the best local source.

In order to handle the varying needs and future growth of a golf course, university, municipality, highway project, or park system, the COPS has TRUE two-way communication which can be accomplished four different ways: Direct bury cable that does not loop back to the central; telephone modem; radio direct to the field controllers and weather stations; or telephone modem to remote interface and then radio to the field controllers and weather stations.

Features and benefits (interface):

- True two-way communications/programming between the central computer and all field controllers, either by direct hard wire or telephone modems. This provides for complete interaction from any access location.
- The ability to operate up to 999 individual remote field controllers each within 99 different designated systems.
- Incorporates the most advanced primary, secondary and communication lightning and surge protection in the industry.
- Utilizes small 18 gauge communication cable between the interface and field controllers with up to 10,000 allowable feet or run between each field controller before signal boosting is needed.
- User friendly menu driven programming with on-screen footnote prompting and multi-screen windows for ease of operation.
- Up to 96 standard module programs to provide maximum flexibility in an irrigation program.
- Field controller can be completely reprogrammed in seconds from a database of pre-developed programs.
- Signal transmission lights provide visual verification that the interface is in operation which aids in troubleshooting.
- Automatic "System Function" diagnostics which provides for complete component/communication test at the central computer. Automatic time of day synchronization of all controllers within the system every time the central is turned on.
- Modem communication capability at the central computer and the field controller locations for use

- on large geographical systems where hard wiring is prohibited (/90 option).
- Modem interface is integrated via full duplex operation. Modem interface utilizes a standard RJ-11 connection to a dedicated or non-dedicated telephone line.
- Modem interface incorporates a series of LED status displays for visual verification of operation.
- The modem interface has the capability of identifying itself by telephone number allowing for communication of up to 99 systems.
- Central interface and Modem interface are enclosed in a weather resistant, lockable, steel cabinet. Both central interface and modem interface incorporate a 3-way splitter for ease in communication cable routing.
- Radio has 3 mile range covering most projects, large cities are handled by modem/radio systems or repeater stations.
- Radio interface is easily serviceable or frequency changed by plug-in transmitter and receiver.
- "Emergency Off" uses a selectable shutdown delay between field controllers to protect the piping and pumping plant from excessive pressure surges (from 1–30 second delay, defaulting to 3 seconds).
- Selectable system program "Coupling" provides for the automatic adjustment of start times which eliminates open gaps in operation for smooth, efficient pump operation.
- "Duplicate Program Entry" which gives the user the ability to enter just a program number and all schedules, start times, and run times are automatically duplicated from previous entries.
- Report generation providing complete data reporting of all system operating activities to either the screen, printer or on disk.
- The time the day changes, within a controller, may be selected to best suit the project needs. (This function defaults to noon if no selection is made.)
- Water budgeting by way of run time multiplier from 1% to 200% in 1% increments providing seasonal adjustments to any standard module program or station.
- Software program troubleshooting guides, parts breakdown and pictoral identification of systems components for ease in service and maintenance.
- System security code allowing only authorized personnel to the programs.

- Free factory technical advice and on-site system checkout with a report verifying proper or improper installation according to Buckner's specifications.
- Free software updates, as they are developed, providing new and innovative use of the system.
- Monitor screen automatically turns off the menu displayed after the keyboard has not had an entry for 6 minutes to save the monitor from image burn in. (Except in report generator mode.)
- Any station can be manually operated in any field controller at any time from the central computer.
- Hydraulic flow analysis through a hydrograph that graphically displays the scheduled system flow. The system flow may be displayed showing a 24 hour period or down to a 30 minute segment for precise control.
- Data base entry for plant type, spacing, gpm, plant use factor, elevation, solar radiation, wind, rain, humidity, and temperature for automatic precipitation rate. Automatic ET calculations are estimated providing for ease in establishing suggested run times. The weather data can be entered manually or taken from Buckner's weather station.

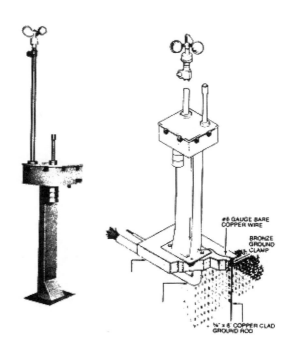

Field Controller

- (User-Friendly): Easy-to-use, easy-to-ad key pad programming with large digit display.
- 24 stations: 4 modules of six stations each. Provides any combination of 6, 12, 18, or 24 stations simply by assigning group numbers. All four modules can be operated at one time or individually or at different times for scheduling flexibility.
- Each module is capable of eight (8) start or repeat times per day.
- Multi-manual feature permits all six station of any one module to be operated simultaneously.
- Each module is capable of independent, daily irrigation scheduling selectable from one to fourteen days.
- Broad range of "STATION ON TIMES," 1 to 59 minutes in one minute increments and 1 to 9 hours in one hour increments.
- Each module is capable of operating three (3) independent programs with two (2) starts each.
- Automatic or manual syringe cycle of 1 to 59 minutes on any selected module or group of modules.
- Built-in memory retention in the event of power failure. Battery backup retains program for several years.
- Lightning and electrical surge protection on each primary and secondary circuit (input and outputs).

Weather Station

- Identical communication connections as the field controllers utilizing the same COPS cable.
- User friendly menu driven programming for ease in operation.
- Up to four (4) weather stations can be operated per system
- Records the following information needed to calculate ETo:
    1. Average wind run in miles per hour (1–65 mph) and total day wind in 24 hours.
    2. Mean relative humidity (1 min.–99% max.)
    3. Total daily solar radiation (1–800 Langleys)
    4. Mean temperature in degrees Fahrenheit (1–130)
    5. Rainfall (0.01–9.99 inches)
- Electronic processor and circuitry is powered through a 120 volt constant voltage transformer for a clean supply.
- Operation and data is maintained by a gel cell rechargeable battery in the event power is interrupted.
- Readings are taken once every minute, averaged and recorded every hour and compiled in a daily log within the memory of the weather station.
- The last 30 days of daily data is stored in the non-volatile memory for referencing.
- Daily, monthly, or current data can be retrieved by the PC at any time.

Appendix 10

Automatic Irrigation System Bid Document and Bidding Specifications

INVITATION TO BID

AUTOMATIC IRRIGATION SYSTEM

Sealed bids will be received on an Irrigation Contract for furnishing all labor, tools, equipment and materials required to perform irrigation improvements designated as the _____, and all work incidental or necessary thereto as detailed in the specifications and plan(s).

Bids will be received until _____ p.m. on _____, 19___, and then, at that office, will be publicly opened and read.

The Instructions to Bidders, Bid Forms, Contract, Plans, Specifications and other Contract Documents are attached.

The successful Bidder shall furnish a performance and payment bond equal to 100% of the bid price. The premium for the bonds shall be included in the bid price.

The Bidder is to submit bids on the attached forms and in the manner requested. Bids must be in a sealed envelope with the Bidder's name marked on the front.

No Bidder may withdraw their bid within thirty (30) days after the actual date of the opening thereof.

Award of the Contract for this project will be based on cost, Contractor's experience on this type of installation, and financial stability of the Contractor. The Owner reserves the right to accept any bid or reject any or all bids and waive any formalities in the bidding when such action is deemed in the best interest of the Owner.

SECTION 1

BID INSTRUCTIONS

1.01 BIDS

A. Bids shall be made upon the Bid Form which is provided, in the manner requested, and in strict accordance with these plan(s) and specifications.

B. Included in these Documents are a complete set of Bidding and Agreement forms which are for the convenience of the Bidders. They are to be filled out, executed and detached from the Contract Documents for bidding purposes.

C. Any deviations or special conditions shall be included on the Bid form and quoted as voluntary alternates only.

D. Bid forms must be filled in by typewriter, computer or ink. Pencil entries will not be accepted.

E. Bid Documents shall be enclosed in an envelope which shall be sealed and clearly labeled with the words "BID DOCUMENTS" and the name of the Bidder.

F. The Contract will be awarded by _____ to a responsible Bidder of its choice. The Contract will require the completion of the work in accordance with the Contract Documents.

1.02 CORRECTIONS

A. Erasures or other changes in the bids must be explained or noted over the signature of the Bidder.

1.03 QUALIFICATIONS OF BIDDERS

A. The Bidder shall submit, with his Bid, a list of three (3) previously completed projects of a similar nature to this project. The Owner reserves the right to request additional data information necessary to qualify the Bidder. The Owner also reserves the right to reject any bid, if in the opinion of the Owner, the Contractor is not qualified to properly perform the work described by the plan and specifications. A list of previous references with telephone numbers and contact names shall be included.

1.04 SITE INSPECTIONS

A. Each Bidder should visit the site of the proposed work and be fully acquainted with the conditions there relating to construction and labor, and should be fully informed as to the facilities involved, the difficulties, and restrictions attending the performance of the Contract. The Bidder shall thoroughly examine and be familiarized with the drawing, specifications, and all other Bid and Contract documents. The Contractor, by execution of the Contract, shall not be relieved of any obligation under such Contract due to failure of receiving or examining any forms or legal documents, or failure to visit the site and be acquainted with the existing conditions. The Owner shall be justified in rejecting any claims thereof.

B. The Owner shall make available to all prospective Bidders, at any pre-bid meetings and prior to the submittal of Bids, all information that the Owner may have as to subsoil conditions in the vicinity of the work, drainage and irrigation lines, and other information that might assist the Bidder in properly evaluating the amount of work that will be required. Owner's information shall be the best factual information available without the assumption of responsibility for its accuracy. The Contractor shall satisfy himself as to the nature and location of the work, the general conditions, and all other matters which can affect work under this Contract.

C. A pre-bid meeting is scheduled for _____ at _____. All Bidders are requested to attend this meeting.

1.05 BID WITHDRAWAL

A. Bids may be withdrawn by written or fax request dispatched by the Bidder and received by the Owner prior to their decision to award the bid.

1.06 AWARD OF CONTRACT

A. The Contract will be awarded to the responsible Bidder of the Owner's choice, complying with the conditions of the Specifications, and upon receipt by the Owner of all Environmental Permits. The awarded Bidder will be notified at the earliest possible date.

B. It is the intent of _____ to notify the successful Bidder of their decision by _____ with Contract signing no later than _____. The Owner, however, reserves the right in its sole and absolute discretion, with or without cause, to reject any and all bids and to waive any formality in bids received whenever it deems such rejection of waiver to be in its best interest.

1.07 PERFORMANCE BOND

A. The successful Bidder shall furnish a surety bond in a penal sum not less than the amount of the Contract and a labor and material bond for payment of all persons, firms or corporations to whom the Contractor may become legally indebted for labor, materials, tools, equipment and services of any nature including transportation in performing the work. Such bonds shall bear the same date, or a date subsequent to that of the agreement. These bonds shall be signed and issued by a guarantee or surety company authorized and qualified to do business in the state. The current power of attorney for the person who signs for any surety company shall be attached to such bonds.

B. The failure of the successful Bidder to supply the required bonds or submit required insurance policies within ten (10) days of the Agreement signing shall constitute a default. The Owner may either award the Contract to the next responsible Bidder of its choice or re-advertise for bids. If a more favorable bid is received the defaulting Bidder shall have no claim against the Owner.

1.08 TAXES

A. _____ is not exempt from payment of State Sales Tax. The Contractor shall include in his bid any and all Sales Tax due to the State under this proposal.

SECTION 2

GENERAL SPECIFICATIONS

2.01 PURPOSE

A. The objective of these specifications is to provide an assembled, installed and fully functioning automatic irrigation system which will efficiently irrigate all areas to be covered and shall be acceptable in all aspects to the Owner. These specifications, design details, and irrigation plan are to be considered part of the Contract, and the Contractor shall follow the specifications with due perseverance.

2.02 OWNER'S AUTHORIZED REPRESENTATIVE(S)

A. The only authority to approve work performed by the Contractor, make field changes that are deemed necessary, and approve estimates submitted by the Contractor for payment, is the Owner or his authorized representative(s). The Owner will notify the Contractor in writing if he designates any other authorized representative. If the Owner is not in direct contact with the

Contractor, that person who represents the Owner in awarding the Contract is deemed an authorized representative.

B. For the scope of these plans and specifications, the term "Owner" refers to _____ with the authorized representative as _____.

2.03 SCOPE OF WORK

A. The work is located on the property of the _____.

B. The Contract contemplated by these specifications consists of the Contractor furnishing all supervision, labor, equipment and materials required for all work described herein to install a fully automatic irrigation system as further defined in the plan and specifications.

C. Unless otherwise specified, the plan and specifications are intended to include everything obviously requisite and necessary for the proper installation and completion of the work whether each necessary item is mentioned herein or not.

D. The plan(s) and specifications are intended to be cooperative and any item called for in one and not the other shall be as binding as if called for in both.

E. All work herein specified or called for on the drawing shall be executed in compliance with all governing ordinances, laws and regulations of the State or any other authority having jurisdiction over the work. Additionally, any changes and/or additions in the work necessary to meet these ordinances, laws, regulations, and/or conditions will be made without additional cost to the Owner.

2.04 DESCRIPTION OF WORK

A. The work to be done under this Contract includes, but is not limited, to the following:

1. Furnish and install new pump station including pumps, fittings, valves, suction lines with foot valves, and discharge piping at new pond. The Contractor will also provide concrete pad and pumphouse building. The Owner will be responsible for providing adequate power to the pump location.

2. Furnish and install all new Class 200 PVC piping, fittings, valves, and necessary pipe line appurtenances for irrigation system.

3. Furnish and install new green, tee, fairway, and rough valve-in-head sprinklers with pre-assembled PVC swing joint assemblies and all necessary fittings.

4. Furnish and install new ____ quick-coupling valves at every green and tee shut-off valve location as indicated on the plan. Quick-couplers shall be installed on ____ galvanized swing joint assemblies next to shut-off valve box.

5. Furnish and install radio-activated satellite controllers with proper grounding equipment, antennas, necessary electrical supplies, and 120-volt power wiring. The Contractor will be responsible for arranging a Radio Frequency Survey and obtaining the proper FCC license.

6. Furnish and install all new 24 volt control and common wiring from the field controllers to all valve-in-head sprinklers.

7. Furnish and install central computer system including necessary electrical supplies, antenna, and wiring at maintenance building.

8. Provide one winterization and spring stat-up of the irrigation system after the entire installation has been completed and approved by the Owner.

2.05 TIME OF COMPLETION

A. Delays caused by the Owner, others who are not party to the Contract, but whose act or action must precede work performed under this Contract, Acts of God, organized labor disputes, and fire are considered legitimate delays to the time of completion.

B. Severe inclement weather conditions which may prohibit the Contractor from performing quality installation techniques may be considered legitimate delays after mutual agreement by the Owner and Contractor.

2.06 UTILITIES AND PROTECTION

A. The Owner shall make available to the Contractor all necessary information from existing utility drawings regarding the known locations of existing utilities or drainage lines within the site's property lines. Before beginning any work, the Contractor shall mark locations of such utilities and/or underground obstructions at the site. The Contractor shall be fully liable for the damages to and the cost of repairing or replacing any buried conduit, cables, or piping encountered during the installation, unless it was not previously informed of such underground utilities. If the Contractor is aware of such buried lines, he shall immediately have the incurred damages repaired at his own expense. Conversely, the Owner shall be liable for the cost of replacing or repairing damages to any of those existing utilities of which the Contractor had not been previously informed.

B. Components of the existing irrigation system must be maintained as long as they are needed to irrigate portions of the golf course. Such determination of need shall be made by the Owner.

2.07 CONDUCT OF WORK

A. The Contractor shall not close more than one hole (tee, green and fairway) at any one time. The Contractor shall not open more trenches that can be backfilled in a full workday period.

2.08 SUBLETTING OF THE CONTRACT

A. The Contractor shall not assign or sublet in whole or any part of this Contract without obtaining the prior written consent of the Owner approving the specific Party to whom it is proposed to sublet the same. Acceptance by the Owner of the Subcontractor does not relieve the Contractor of any responsibilities as outlined in the specifications.

2.09 LIABILITY INSURANCE

A. The Contractor shall protect the Owner against all liabilities, claims, or demands for injuries or damages to any person or property growing out of the performance of the work under this Contract. The Contractor shall assume all liability for any injuries or damages occasioned by his agents or employees acting within the scope of his employment on the premises of the Owner.

B. The Contractor shall protect the Owner against all claims arising from the use of passenger automobiles, motor trucks, and other motor vehicles owned and operated by the Contractor and/or his employees in connection with the work herein specified.

2.10 WORKER'S COMPENSATION INSURANCE

A. The Contractor shall accept the provisions of Worker's Compensation Insurance Act of the State and shall procure the same in full force and effect until the work covered by these plans and specifications has been fully completed. The Contractor shall file with the Owner certificates of insurance complying with the provisions of this paragraph, prior to the commencement of any work.

2.11 SOCIAL SECURITY

A. The Contractor shall pay the contributions measured by the wages of his employees and the employees of any Sub-Contractors required by the Social Security Act and for the Public Laws of the State and assume exclusive liability for said contributions. The Contractor shall further agree to hold harmless the Owner on account of any contributions measured by the wages as above stated,

of employees of the Contractor/Subcontractor, assessed against the Owner under the authority of said Act and the Public Laws of the State.

2.12 WAGE LAWS

A. While working on the premises of the Owner, Contractor agrees to comply with all requirements of the Wage and Hour Act and shall be held responsible for compliance.

2.13 HANDLING OF CONTRACTOR'S EQUIPMENT

A. The Contractor shall provide and pay for all transportation required to deliver and remove from the site all parts and equipment, as required for all the work shown and specified. It is the Contractor's responsibility to provide and pay for any forklifts or other equipment necessary for unloading of materials delivered to the job site.

B. The Contractor shall be responsible for providing any trailers or containers for storage of parts and equipment necessary for this Contract. Any security and/or insurance for protection of parts and equipment will be arranged by the Contractor.

2.14 EQUIPMENT, TOOLS AND LABOR

A. The Contractor shall furnish all such equipment, tools, and labor necessary to pursue the work in an acceptable manner, toward rapid completion. This Contract is based upon the Contractor furnishing equipment, tools and labor which are suitable to carry out this Contract in a professional and thorough manner, unless otherwise herein specified.

2.15 CODES AND INSPECTIONS

A. The entire installation shall fully comply with all governing laws, regulations, and ordinances of the State or any other authority having jurisdiction over the work. Compliance includes any necessary licensing for installers and service personnel.

B. The Contractor shall take out all required permits, arrange for all necessary inspections, and shall pay any fees and expenses in conjunction with the same as part of the work under this Contract.

2.16 OWNER'S SUPERVISION

A. The Owner assumes no responsibility in the supervision and inspection of the work involved in the execution of this Contract beyond ensuring, to the Owner's satisfaction, that the plan, general conditions, and specifications are being properly interpreted and implemented. This supervision and inspection will not relieve the Contractor of any responsibility for the performance of this work in accordance with the plan, general conditions, and these specifications.

B. A representative of the irrigation distributor will be available to perform weekly onsite inspections with the Owner to assess progress of the work.

2.17 CHANGES IN THE WORK

A. The Owner shall have the right to require alterations of, additions to, and deductions from, the work shown on the drawing or described in the specifications without rendering void the Contract. All such changes shall be in the form of a Change Order prepared by the Contractor. The Contractor will compute a value of the work and submit in proposal form, but will not proceed with the changes until signed authorization has been given by the Owner. In each case the price agreed to be paid for the work under the Contract shall be increased or decreased for the work added or omitted. In the event the value of the work or cost adjustment furnished by the Contractor is unacceptable to the Owner, the Contract shall be performed without reference to said Change Order.

2.18 TERMINATION OF CONTRACT

A. If the Contractor refuses or fails to execute the work with such diligence as will ensure its completion within the time specified in these Contract Documents, the Owner, by written notice to the Contractor, may terminate the Contractor's right to proceed with the work.

2.19 ERRORS IN THE PLAN(S) OR SPECIFICATIONS

A. The Contractor shall immediately notify the Owner should he find any errors or conflicts in the drawings and/or specifications. The Owner will render interpretations or instructions on the item(s), in writing, as soon as possible.

B. Any work performed by the Contractor with regard to errors will be done so at his own risk unless he has received written prior approval from the Owner.

SECTION 3

MATERIAL SPECIFICATIONS

3.01 MATERIALS

A. The materials chosen for the design of the irrigation system have been specifically referred to by the manufacturer so as to enable the owner to establish the level of quality and performance required by the system design. Equipment by other manufacturers may be considered only after written application, at least seven (7) days prior to bid opening including five (5) copies of specifications, is made by the Bidder and written approval is received from the Owner.

B. The Contractor shall supply the materials necessary for a complete automatic irrigation system. The Contractor shall arrange for the materials to be onsite prior the the arrival to start the work.

C. The following is a list of the major components to be installed for the complete irrigation system. This is an estimated material list. It is the Bidder's responsibility to verify all quantities.

DESCRIPTION	QUANTITY
Toro 674-06-74 Single-Row Fairway Sprinkler	109
Toro 754-06-54 Two-Row Fairway Sprinkler	104
Toro 658-06-56 Two-Speed Green Sprinkler	87
Toro 754-06-53 Full-Circle Tee Sprinkler	63
Toro 655-06 Series Part-Circle Sprinkler	15
1-1/2" Swing Joint Assembly w/Gate Valve	109
1-1/2" Swing Joint Assembly w/o Gate Valve	269
Toro 474-00 Quick-Coupling Valve	38
1" Galvanized Swing Joint	38
Toro OSMAC Central Computer	
Toro OSMAC RDR50 Field Controller	8
8" C1200 SDR21 PVC Pipe	6202
6" C1200 SDR21 PVC Pipe	

3.02 FULL-CIRCLE V.I.H. SINGLE-ROW FAIRWAY SPRINKLER

A. The full-circle sprinklers shall be Toro model 67406-74 gear-driven rotary, in-ground type and designed with an integral hydraulic valve for remote electrical actuation. The sprinkler shall be capable of covering a 102-foot radius at 100 pounds per square inch pressure with a discharge rate of 66.8 gallons per minute. Water distribution shall be via three (3) nozzles. Rotation shall be accomplished by a sealed oil packed gear assembly isolated from the water supply.

B. The sprinkler shall incorporate a solenoid for actuation of the integral control valve. The solenoid operator shall be suitable for 24-V.A.C., 60-cycle service with an inrush of 0.42 amps and holding of 0.21 amps at 24 V.A.C.

C. An on-off-auto manual selector shall be provided and accessible from the surface with a special key. Pressure regulation shall be preset by the factory according to nozzle set but adjustable to any other pressure desired.

D. The sprinkler shall have positive spring retraction.

E. The sprinkler housing shall be of high-impact molded plastic with 1-1/2 inch I.P.S. connection. The sprinkler shall have a large strainer so as to prevent nozzle clogging. The sprinkler shall be so constructed that the drive assembly, screen and valve are accessible through the top of the sprinkler without disturbing case installation.

F. The sprinkler shall be manufactured by The Toro Company, Irrigation Division, Riverside, California.

3.03 FULL-CIRCLE V.I.H. TWO-ROW FAIRWAY SPRINKLER

A. The full-circle sprinklers shall be Toro model 754-06-54, single-speed series gear-driven rotary, in-ground type and designed with an integral hydraulic valve for remote electrical actuation. The sprinkler shall be capable of covering a 74-foot radius at 80 pounds per square inch pressure with a discharge rate of 25.3 gallons per minute. Water distribution shall be via two (2) color nozzles and shall have a variable stator which does not require changing when switching to different size nozzles. Rotation shall be accomplished by a sealed oil packed gear assembly isolated from the water supply.

B. The sprinkler shall incorporate a solenoid for actuation of the integral control valve. The solenoid operator shall be suitable for 24-V.A.C., 60-cycle service with an inrush of 0.42 amps and holding of 0.21 amps at 24 V.A.C.

C. An on-off-auto manual selector shall be provided and accessible from the surface with a special key. Pressure regulation shall be preset by the factory according to nozzle set but adjustable to any other pressure desired. A bowl-vented discharge shall be incorporated to minimize the differential pressure required for regulation and ensure positive valve closure.

D. The sprinkler shall have a full three (3) inch pop-up height and have positive spring retraction.

E. The sprinkler housing shall be of high-impact molded plastic with 1-1/2 inch I.P.S connection. The sprinkler shall have a large strainer so as to prevent nozzle clogging. The sprinkler shall be so constructed that the drive assembly, screen and valve are accessible through the top of the sprinkler without disturbing case installation. The cap shall also serve as a yardage marker.

F. The sprinkler shall be manufactured by The Toro Company, Irrigation Division, Riverside, California.

3.03 2-SPEED (180-180) V.I.H. FULL-CIRCLE GREEN SPRINKLER

A. The full-circle sprinklers shall be Toro model 658-06-56, 2-speed series gear-driven rotary, in-ground type and designed with an integral hydraulic valve for remote electrical operation. The sprinkler shall be capable of covering a 71-foot radius at 80 pounds per square inch pressure with a discharge rate of 22.6 gallons per minute. Water distribution shall be via three (3) nozzles. Rotation shall be accomplished by a sealed oil packed gear assembly isolated from the water supply.

B. The full-circle sprinkler shall be 2-speed having one-half (180 degrees) rotating at one-half the speed of the other half (180 degrees).

C. The sprinkler shall incorporate a solenoid for actuation of the integral control valve. The solenoid operator shall be suitable for 24-V.A.C., 60-cycle service with an inrush of 0.42 amps and holding of 0.21 amps at 24 V.A.C.

D. An on-off-auto manual selector shall be provided and accessible from the surface with a special key. Pressure regulation shall be preset by the factory according to nozzle set but may be adjusted to any other pressure.

E. The sprinkler shall have positive spring retraction.

F. The sprinkler housing shall be of high-impact molded plastic with a 1-1/2 inch I.P.S. connection. The sprinkler shall have a large strainer so as to prevent nozzle clogging. The sprinkler shall be so constructed that the drive assembly, screen and valve are accessible through the top of the sprinkler without disturbing case installation.

G. The sprinkler shall be manufactured by The Toro Company, Irrigation Division, Riverside, California.

3.05 FULL-CIRCLE V.I.H. TEE SPRINKLER

A. The full-circle sprinklers shall be Toro model 754-06-53, single-speed series gear-driven rotary, in-ground type and designed with an integral hydraulic valve for remote electrical actuation. The sprinkler shall be capable of covering a 68-foot radius at 80 pounds per square inch pressure with a discharge rate of 21.1 gallons per minute. Water distribution shall be via two (2) color-coded nozzles and shall have a variable stator which does not require changing when switching to different size nozzles. Rotation shall be accomplished by a sealed oil packed gear assembly isolated from the water supply.

B. The sprinkler shall incorporate a solenoid for actuation of the integral control valve. The solenoid operator shall be suitable for 24-V.A.C., 60-cycle service with an inrush of 0.42 amps and holding of 0.21 amps at 24 V.A.C.

C. An on-off-auto manual selector shall be provided and accessible from the surface with a special key. Pressure regulation shall be preset by the factory according to nozzle set but adjustable to any other pressure desired. A bowl-vented discharge shall be incorporated to minimize the differential pressure required for regulation and ensure positive valve closure.

D. The sprinkler shall have a full three (3) inch pop-up height and have positive spring retraction.

E. The sprinkler housing shall be of high-impact molded plastic with 1-1/2 inch I.P.S. connection. The sprinkler shall have a large strainer so as to prevent nozzle clogging. The sprinkler shall be so constructed that the drive assembly, screen and valve are accessible through the top of the sprinkler without disturbing case installation. The cap shall also serve as a yardage marker.

F. The sprinkler shall be manufactured by The Toro Company, Irrigation Division, Riverside, California.

3.06 ADJUSTABLE PART-CIRCLE V.I.H GREEN AND TEE SPRINKLER

A. The adjustable part-circle sprinklers shall be Toro model 655-06 series, gear-driven rotary, in-ground type and designed with an integral hydraulic valve for remote electrical actuation. The sprinkler shall be capable of covering a 65- or 71-foot radius at 80 pounds per square inch pressure with a discharge rate of 20.7 or 22.6 gallons per minute depending on size of main nozzle used (#55 or #56). Water distribution shall be via three (3) nozzles. Rotation shall be accomplished by a sealed oil packed gear assembly isolated from the water supply.

B. The sprinkler arc of throw shall be fully adjustable from 45 degrees to 315 degrees in 1-degree increments.

C. The sprinkler shall incorporate a solenoid for actuation of the integral control valve. The solenoid operator shall be suitable for 24-V.A.C., 60-cycle service with an inrush at 0.42 amps and holding of 0.21 amps at 24 V.A.C.

D. An on-off-auto manual selector shall be provided and accessible from the surface with a special key. Pressure

regulation shall be preset by the factory according to nozzle set but adjustable to any other pressure desired.

E. The sprinkler shall have positive spring retraction.

F. The sprinkler housing shall be of high-impact molded plastic with 1-1/2 inch I.P.S. connection. The sprinkler shall have a large strainer so as to prevent nozzle clogging. The sprinkler shall be so constructed that the drive assembly, screen and valve are accessible through the top of the sprinkler without disturbing case installation.

G. The sprinkler shall be manufactured by The Toro Company, Irrigation Division, Riverside, California.

3.07 IRRIGATION CONTROL SYSTEM

A. The irrigation control system shall be the Toro/Motorola OSMAC model and shall utilize a computerized central control station which controls up to 150 field controllers (satellites) with up to 48 stations each. The control system shall not utilize communications cable or phone lines between the central station and the field controllers in order to keep installation and operating costs as low as possible. Because of its proven reliability, high sensitivity, falsing protection and ability to perform in difficult RF environments, the central control station shall utilize standard digital radio paging techniques to communicate with and control the field controllers.

B. The central station shall be comprised of an IBM compatible computer running a menu-driven, flexible and user-friendly irrigation control software program, which will control a desktop control paging station which includes either an internal transceiver or provisions to control an external transceiver capable of sending digital paging messages. (Golay Sequential Code, non-return to zero, FSK modulation.)

C. The central paging station shall in turn send digital paging radio frequency transmissions to the field controllers, causing field units to activate or deactivate irrigation solenoids.

D. The central control paging station shall be capable of being controlled remotely from two-way portable radios equipped with a DTMF signaling option or from a touch-tone telephone. This remote control feature will allow activation and deactivation of individual irrigation solenoids, groups of solenoids simultaneously and multiple irrigation valves sequentially to be accomplished from anywhere within radio range or telephone range of the central paging system.

E. Individual specification requirements for each component of the irrigation control system as describe in the following sections.

F. The irrigation control system shall be manufactured by Motorola Incorporated, Boynton Beach, Florida and The Toro Company, Riverside, California.

3.08 FIELD CONTROLLER

A. The field controller shall be Toro/Motorola OSMAC model RDR50 and be capable of controlling up to 48 independent outputs (stations), expandable in groups of (8) eight, by using up to six modular eight-output triac control printed circuit cards.

B. The field controller shall contain a radio frequency receiver and decoder capable of receiving and decoding standard digital paging signals (Golay Sequential Code, non-return to zero, FSK modulation).

C. The controller shall have means to decode the numeric message received into various actions to activate, deactivate, enable, and disable numerous combinations of stations within the field controller either simultaneously or sequentially.

D. The field controller shall only respond to a command upon proper receipt of its paging address.

E. The controller shall have two paging addresses, the first being an individual address settable by the user, and the second being an all unit (group call) address set to be the same on all units in the factory but reprogrammable in the field.

F. The field controller shall have 255 user-selectable individual addresses to choose from, settable by an 8-position DIP switch.

G. The field controller shall be housed in a highly durable, non-metallic enclosure, which is weather-resistant and meets the stringent requirements of ML spec. 810D for rain. The enclosure shall use a vandal-resistant lockable door which is hinged for easy service access. The enclosure shall have two access holes for connecting to AC power and the irrigation valve field wires. The opening for the AC power shall be for 1/2" conduit. The opening for the field wires shall be for either 3/4" or 1-1/4" conduit. All electrical connections to the field unit shall be according to the National Electric Code and local code requirements. The enclosure shall contain "slots" into which receiver/decoder cards, triac output control cards and the power supply may be easily inserted or removed. The field unit shall be mountable in either a wall mount or top mount configuration.

H. The field controller and enclosure shall be installed inside a steel pedestal mount cabinet. The cabinet shall be locking, weatherproof type, constructed of heavy-gauge steel with corrosion-resistant forest green finish inside and out. Access to all wiring shall be through a lockable access door, and four (4) bolts shall secure the pedestal to a concrete pad.

I. The power supply within the field unit shall be capable of connecting to either 110/120 or 220/240 VAC, 50/60 Hz. It shall provide all power necessary to run the field unit and all of its expansion output control cards as well as providing 24 VAC at 3 amps maximum for the control of standard 24-VAC electric irrigation solenoids.

J. The field unit shall be microprocessor-controlled and capable of responding to the following radio data paging commands and OTA programming commands:

   1. Turn a specified list of stations on.
   2. Turn a specified list of stations off.
   3. Syringe a specified list of individual stations for a specified number of hours, minutes, and seconds concurrently.
   4. Disable a specified list of individual stations. (After receipt of this command the stations disabled will not respond to any command until re-enabled.)
   5. Enable a specified list of individual stations.
   6. Disable all stations in the field unit.
   7. Enable all stations in the field unit.
   8. Sequentially syringe a specified list of individual stations in the order that they are entered for a predetermined number of 30-second time intervals already defined in the field unit by a previously received over-the-air (OTA) page.
   9. Sequentially syringe a range of individual stations from the first station specified to the second station specified for a predetermined number of 30-second time intervals already defined in the field unit by a previously received OTA page.
   10. Sequentially syringe all stations in the field unit for a predetermined number of 30-second time intervals already defined in the field unit by a previously received OTA page.
   11. OTA command to define a station as a pump start station (slaved output).
   12. OTA command to change the password for OTA commands.
   13. OTA commands to define the number of 30-second intervals in a syringe.
   14. OTA command to change the individual or group call address.
   15. OTA command to enable/disable the sequential shutdown function for total system disable situations (such as rain shutdown).

K. The field unit shall conform to the following specifications:

   1. Maximum number of stations = up to 48 in blocks of 8

2. Maximum power per station = 0.6 amps @ 24 VAC holding and 1.4 amps @ 24 VAC surge

3. Maximum total secondary power = 3.0 amps @ 24 VAC

4. Input Voltage = 110/120 220/240 VAC, 50/60 Hz

5. Standby current drain = 50 mA @ 110 VAC, 25 mA @ 220V

6. Operating temperature = –20 F to 130°F

7. Environmental Capabilities = Applicable sec. MIL 810D

8. Address codes per unit = Two (1 individual, 1 group)

9. Addresses settable by DIP switch = 255

10. Range of OTA syringe times = 30 sec to 128 min

11. "All Off" field unit delay time = 0 to 45 sec

12. Agency approvals = UL: UL 508 (Fifteenth edition) CSA: C22.2 No. 142 M1987

13. Enclosure dimensions = 12" × 6.75" × 4.75"

14. Weight = 9 pounds

15. Receiver frequencies = 406 to 512 MHZ

16. Rx. selectivity = 65 db. @ ±25khz

17. Rx. sensitivity = 15 uV/M

L. The OSMAC field controller shall be manufactured by Motorola Incorporated, Boynton Beach, Florida and The Toro Company, Riverside, California.

3.09 CENTRAL CONTROLLER

A. The central controller shall be Toro/Motorola OSMAC model with control software program which shall be a menu-driven application program for use on IBM-compatible computers.

B. The irrigation control program shall permit true random access of all stations in the system and allow programs to be constructed with any combination of stations regardless of wiring sequence or field satellite designation.

C. The program shall allow precision tinting of all station run times to at least 5-second accuracy.

D. The program shall allow up to 100 simultaneous programs (irrigation sequences) to be constructed and run concurrently.

E. The central control program shall allow each program constructed to contain up to 100 events. The program shall allow the construction of up to 200 irrigation groups with up to 20 (station) members in each group which can be called up and acted upon within an irrigation program simultaneously.

F. The irrigation program shall allow "water budgeting" to be easily accommodated through the use of "Scaling factors" which can be applied globally or by individual field units.

G. The program shall allow the scaling of programs (or sequences) to be dynamic with relation to events. (Program events will "Link" without gaps or overlap as the total program run time increases or decreases due to use of scaling factors.)

H. The program shall contain a built-in astronomical clock which will automatically calculate sunrise and sunset based upon longitude, latitude and date, and allow programs to be constructed which can be started or stopped in relation to sunrise or sunset.

I. The irrigation control program shall provide the user a means to view a graph of the predicted (calculated) flow vs. time for any single irrigation program or group of simultaneous irrigation programs. This feature will allow the user to level load his overall irrigation schedule to ensure a balanced load on the pumping system in a "dry run" mode before the program is activated.

J. The control program shall keep a daily record of the total water consumption (calculated) on a daily basis and store it in a history file for up to 3 years. The program shall allow the user to view in graphical form his historical water consumption for up to a 3-year period.

K. The program shall contain a screen which allows manual control of the irrigation system allowing individual or group control of stations to start, stop, enable, disable, and syringe for a selectable time as well as to send standard "Pages" to paging receivers within the range of the central control paging station.

L. The central control computer shall be an IBM-compatible machine with the following minimum characteristics:

1. Machine Type = 486DX/66 MHZ.

2. Monitor = Color SVGA

3. Floppy Drive = One 3.5 or 5.25 inch

4. Hard Disk Drive = 540 MB

5. Modem = 9600 Baud

6. DOS Version = 6.0 or higher

7. Uninterruptible Power Source = 300 W to 500 w

8. MS or Compatible Bus Mouse = Yes

9. Real Time Clock = Yes

10. RAM = 16 MB

M. The central control paging station shall be a desktop paging encoder that includes a built-in RF, FM transceiver to provide local area coverage for paging control and two-way communications. The paging encoder shall also contain provisions for controlling an external, higher- power radio frequency Base Station to provide wide area coverage for paging, control signaling, and two-way communications.

N. The paging station shall contain a built-in telephone interface module to allow remote paging and control signaling from any ordinary telephone with DTMF capability.

O. The paging station shall contain a built-in remote paging and radio-telephone interconnect capability to provide:

1. Remote paging from a DTMF portable radio.
2. Remote control signaling from a DTMF portable radio to activate or deactivate irrigation solenoids.
3. Ability to receive telephone calls at the portable radio.
4. Ability to dial out phone calls from the portable radio through single-line telephone systems.

P. The paging station shall be capable of monitoring up to 4 dual, dry-contact, external sensors such as rain gauges and wind speed sensors. Upon detection of a contact closure, the paging station shall automatically send a pre-defined alphanumeric message or control signal. This capability allows automatic action such as system shutdown due to rain or high winds.

Q. The central control paging station shall contain connectors to accommodate:

1. 12-VAC power supply connection.
2. Telephone line connection (RJ11 jack).
3. RS-232C connection to a PC
4. Provisions for external alarm inputs.
5. Provisions for external antenna.

R. The paging station shall utilize an encoder capable of signaling tone-alert, tone and voice, and display pagers. This includes two-tone, 5/6, Golay and Pocsag types.

S. The station shall have the ability to signal up to 500 individual or group call paging addresses.

T. The paging station shall have a 16-digit display with 1/2" high characters to prompt the operator for data entry. The prompting function shall cause the operator to follow valid character entry, format and message length requirements through interactive processing of keypad characters.

U. The central pager shall be controlled by a MC6809 microprocessor whose general operating characteristics reside in a replaceable (pluggable) read-only memory device.

V. The paging station shall contain self-test diagnostics which check all major circuits in the encoder every time power is applied to the unit. In the event of failure, an error message shall be displayed on its 16-character display indicating the area of malfunction. If the test is successful, a message, "Self-Test OK," shall be displayed.

X. The paging station shall utilize a synthesized voice to prompt a caller though paging, control signaling, or telephone interface process.

Y. The paging station shall meet the following specifications:

1. Frequencies = UHF (440–512 MHz)
2. Data Rate (RF Transmission) = 600 BPS
3. Speaker Audio Level = 500 mW with < 5% distortion
4. Power Supply = External 12.5-volt supply
5. Weight = 3.5 pounds
6. Dimensions = 11.75" × 8.25" × 3.42"
7. Pager Types = Tone Alert, Voice, Numeric, and Alphanumeric
8. Operating Temperature = 0°C to +40°C
9. Stability = ±7 ppm
10. Channel Spacing = 25 KHz
11. Sensitivity = 0.5 uV max. (12 db Sinad)
12. Modulation Acceptance = ±5.0 kHz min.
13. Intermodulation = 65 db (EIA Sinad)
14. Spurious Rejection = 60 db
15. Adjacent Channel Selectivity = 70 db
16. RF Power Output = 2 watts
17. Modulation = ±3.5 kHz (data page) ±5.0 KHz (voice page)
18. Frequency Stability = 2.5 ppm
19. FM Noise = 40 db
20. Spurious Emissions = –50 dbc

Z. The OSMAC central computer shall be manufactured by Motorola Incorporated, Boynton Beach, Florida and The Toro Company, Riverside, California.

3.10 ONE-INCH QUICK-COUPLING VALVE

A. The quick-coupling valve shall be one (1) inch, one piece single-lug type.

B. The valve shall be constructed of brass with a wall thickness guaranteed to withstand a normal working pressure of 150 psi without leakage. The quick-coupler valve shall accept a Toro model 464-01 quick-coupler key with a top connection of three-quarter (3/4) inch female pipe thread and one (1) inch male pipe thread.

C. The quick-coupler valve shall be Toro 474-00 manufactured by The Toro Company, Irrigation Division, Riverside, California.

3.11 SWING JOINT ASSEMBLY

A. All 1-1/2" swing joint assemblies for sprinkler heads shall be pre-assembled units manufactured of Schedule

80 PVC material. Swing shall consist of (4) 90-degree elbows and (1) twelve-inch-long riser nipple with 90-degree bend on one end.

B. All connections shall consist of Buttress Threads and double "O" Rings providing leak-free 360-degree adjustment. Wall construction shall be Schedule 80+ with special emphasis at inside cornrs on change of direction fittings.

C. Swing joint assemblies shall be made from virgin PVC Type 1, Cell Classification 12454-B material listed for potable water conveyance by NSF. Working pressure shall be 200 psi combined static and surge.

D. All PVC swing joints shall be factory assembled.

E. All PVC swing joint assemblies shall be Unibody Ultra Swing Joint manufactured by Dura Plastic Products Inc., Beaumont, California or approved equal.

F. The 1-inch swing joint assembly for the quick-coupler valves shall be comprised of: one (1) 1-inch by 12-inch galvanized nipple and three (3) 1-inch galvanized street elbows.

G. Galvanized swing joints shall be assembled with Teflon® tape or sealant completely covering all threads of all components used. Assemblies shall be tightened to the point at which no leakage shall occur and swing joint is still able to pivot and absorb any shock or pressure which it is designed for.

H. If backfill material removed from sprinkler location contains rocks or heavy clay, all PVC and galvanized swing joint assemblies and sprinklers attached shall be completely backfilled with sand to within 3 inches of final grade.

I. In applications where a shut-off valve is used in-line before the swing joint assembly, 4-inch ADS perforated drainage pipe capped off with a Toro model 850-00 valve cover shall be installed as a reach sleeve from shut-off valve to final grade for manual access

3.12 POLYVINYL CHLORIDE (PVC) PIPE

A. All PVC pipe specified on the plan shall be virgin, high-impact, polyvinyl chloride (PVC) pipe, having a minimum working pressure rating of Class 200 for pipe sizes specified.

B. All PVC pipe shall be continuously and permanently marked with the manufacturer's name, material, size, and schedule or type. The pipe shall be capable of withstanding a long-term pressure test (1000 hours) of 420 psi and a quick-term burst test of 630 psi.

C. The pipe shall conform to U.S. Department of Commerce Commercial Standard CS 207-60, or latest revi-

sion. Material shall conform to all requirements of Commercial Standard (CS 256-63), or latest revision.

3.13 DUCTILE IRON FITTINGS

A. Mainline fittings, excluding service tees, 2-1/2 inch and larger shall be ductile iron push-on type for IPS size PVC pipe.

B. Fittings shall be manufactured of ductile iron, grade 80-55-0 in accordance with ASTM A-536. Fittings shall have mechanical or push-on joints with gaskets meeting AWWA C-111. Fittings shall have radii of curvatures conforming to AWWA C-110 and shall be cement-lined in accordance with AWWA C-104.

C. Fittings shall be manufactured by HARCO, Lynchburg, Virginia or approved equal.

3.14 PVC FITTINGS

A. Fittings for use with PVC pipe shall be Schedule 40 fittings produced from PVC Type 1, cell classification 12454-B. The fittings shall be listed by the National Sanitation Foundation for potable water services. The fittings shall be listed by IAPMO for water service and gas yard piping in appropriate types and sizes. PVC fittings shall meet the following codes and specifications: ASTM-D1784, ASTM-D2466.

3.15 PVC CEMENT

A. Cement for use on PVC fittings shall be NSF approved, for Type I and Type II PVC pipe, and schedule 40 fittings. Cement is to meet ASTM D2564 and F-493 for potable water, pressure, gas conduit, and drain pipes. Application temperature shall be 35 to 110 degrees Fahrenheit.

3.16 WIRE

A. All controller power wiring shall be two or three conductor cables with a ground wire, type UF, UL listed having polyvinyl chloride (PVC) insulation and a sunlight-resistant PVC overall jacket.

B. All solenoid control and common wiring shall be a single, solid copper conductor, UL listed type PE 600-volt direct burial irrigation wire. Control wires shall be #14 AWG red in color and common wire shall be # 12 AWG white. Each controller shall have separate common wires.

C. All wiring shall be UL listed and OSHA acceptable in accordance with Articles 336 and 339 of the National

Electrical Code (and subject to local code requirements). Type UF cables may be used for underground feeder or branch circuit wiring for installation above or below ground, including direct burial, and in wet or corrosive locations.

D. Cables shall conform to the following standards: UL-83 for Thermoplastic Insulated Wire, UL-493 for Thermoplastic Insulated Underground Feeder and Branch-Circuit Cables, UL-719 for Non-metallic-Sheathed Cables, and Federal Specification J-C30A.

E. Conductors for multiconductor cables, sizes 14, 12 and 10 AWG shall be solid, annealed bare copper. Sizes 8 and 6 AWG shall be concentric, compressed stranded (class B).

F. Conductors for single conductor cables, sizes 14–8 AWG shall be solid, annealed bare copper. Sizes 6–4 AWG shall be concentric, compressed stranded (class B).

3.17 WIRE SPLICES

A. Wire splicing kits for single UF wire connections shall be Direct Burial kits consisting of sealant which shall not set up hard, allowing splices to be reworked without cutting wires.

B. Direct Burial kits shall have an application temperature range of 32 to 120 degrees Fahrenheit and service 600-VAC maximum.

C. D.B.Y. kits shall allow connections of two to five # 18 AWG or two # 12 AWG solid or stranded copper wires.

D. D.B.R. kits shall allow connections of two to five #16 AWG or three #10 AWG solid or stranded copper wires.

E. Splicing kits shall be manufactured by 3M Electrical Products Division, Austin, Texas.

3.18 CAST IRON ISOLATION GATE VALVE

A. The isolation gate valve shall be a non-rising stem valve conforming to specifications of the American Water Works Association.

B. The body and bonnet shall be of high-strength cast iron. They shall be oval shaped with thick metal, well rounded at corners for maximum strength, and thick, well-filleted flanges. Body seat rings shall be of bronze with an extra long thread screwed against shoulders in the body.

C. Discs shall be of cast iron with bronze facings and have deep ribs to resist distortion. Disc rings shall be bronze with accurately formed, forked continuous tongue forced into a dovetailed groove in the cast iron disc.

D. Stems of AWWA valves shall be manganese bronze of 60,000 pounds per square inch tensile strength. The "O" ring seals shall be Buna N Synthetic rubber compound

of 70 Durometer hardness. Bronze glands shall be secured by three or more bolts.

E. Cast iron isolation gate valves shall be manufactured by Kennedy Valve, Elmira, New York or approved equal.

3.19 BRONZE I.P.S. GATE VALVE

A. The bronze gate valve shall be constructed with non-rising stem, solid wedge disc and screwed ends. The body, bonnet and disc shall be made from #85-5-5-5 bronze, ASTM B62. The stem, lock nut, packing nut and gland follower shall be made of brass, ASTM B16, with the gland packing made of asbestos graphite. The hand wheel shall be constructed of cast iron, ASTM A126.

B. The gate valve shall have a working non-shock pressure of 125 psi for saturated steam, and 200 psi for cold water, oil and gas. The body shall have a hydrostatic test pressure of 300 psi and the seat shall be at 200 psi.

C. The bronze I.P.S. gate valve shall be manufactured by Aqua Valve Company, Orinda, California or approved equal.

3.20 VALVE ACCESS BOX

A. Valve boxes shall be constructed of a rigid combination of polyolefin and fibrous components especially compounded for underground enclosures. Superflexon plastic material shall be chemically inert and normally unaffected by moisture, corrosion and the effects of temperature changes. Superflexon shall also have a relatively high tensile strength with light weight because of its solid structural material.

B. The twelve-inch standard valve box shall be 12 inches deep with a measurement of 15 inches long by 10 inches wide. The valve box shall include a locking cover which shall have a slightly mottled, grass-like green color that will blend into turf.

C. The twelve-inch jumbo valve box shall be 12 inches deep with a measurement of 20 inches long by 13 inches wide. The box shall include a locking cover which shall have a slightly mottled, grass-like green cover that will blend into the turf.

D. The ten-inch round valve box shall be 10 inches in diameter by 10 inches deep. The box shall have a lockable cover slightly mottled, grass-like that will blend into the turf.

E. The 5-1/4-inch round valve box shall be 5-1/4 inches in diameter by 24-1/2 inches deep with a green cover which will blend into the turf.

3.21 AIR RELEASE VALVE

A. The air release valves shall be installed at the high points in the system or at points selected by the superintendent. The valves will permit discharging of the surge of air from an empty line when filling and relieve the vacuum when draining the system. The valves shall also release an accumulation of air when the system is under pressure. This shall be accomplished in a single valve body.

B. The valve shall operate through a compound lever system which will seal both the pressure orifice and the air and vacuum orifice simultaneously.

C. The lever system shall permit a 3/16" orifice to release an accumulation of air from the valve body at a pressure of 54.6 psig.

D. The air release valves shall be CRISPIN model IC-10 Universal Air Valve as manufactured by Multiplex Manufacturing Company, Berwick, Pennsylvania.

3.22 PUMPING STATION

A. The pumping station shall be a three (2-40 HP and 1-15 HP) horizontal centrifugal station, 3600 RPM, 230 or 460 Volt/3 Phase/60 Hertz, full voltage, to deliver a nominal flow of 800 GPM at 130 psi, with 10 ft lift.

B. The station shall be fully assembled on a prefabricated steel mounting platform providing support of pumps and all components and accessories.

C. The station shall include a 300-gallon hydropneumatic tank with automatic air ratio control valve and air compressor to control air-water ratio automatically, reduce surges and provide storage capacity.

D. The pump station shall have a 6" automatic station control valve with pressure reducing, pressure sustaining, surge control and flow sensing pilots. The station shall also include a 2" automatic station pressure relief valve.

E. The station shall be constructed with auto-sensory controls which shall include:
  1. Low discharge pressure safety lockout.
  2. High discharge pressure safety shutdown.
  3. Loss of prime safety circuits.
  4. Lightning arrestor.
  5. Phase loss, phase balance and low voltage protection.

F. All electrical components shall conform to Underwriters Laboratory (U.L.), National Electrical Manufacturers Association (N.E.M.A.), and Canadian Standards Association (C.S.A.).

G. The station shall incorporate a 6" wye strainer with automatic purge.

H. A digital flow meter with spool piece and totalizer in a NEMA 4 case shall be installed on the station.

I. Each of the three pumps shall have an individual suction line to the water source. The two 40 HP pumps shall have 6" lines and the 15 HP pump shall have a 4" suction line. All suction lines shall have a foot valve with strainer at intake of line.

SECTION 4

INSTALLATION SPECIFICATIONS

4.01 INSTALLATION REQUIREMENTS

A. The word "piping" in these specifications means pipe, fittings, nipples, and valves, and shall be considered as such in this installation.

B. The arrangements, positions, and connection of piping, drains, valves, and the like indicated on the plan, shall be followed as closely as possible, but the right is reserved by the Owner to change locations and elevations to accommodate conditions which may arise during the progress of the work prior to installation without additional compensation for such changes. The responsibility for accurately laying out the work and coordinating the installation with other trades rests with the Contractor. Should it be found that any work is laid out so that interference will occur, report that to the Owner before commencing work.

4.02 STAKING

A. The Contractor shall stake out all proposed pipe and wire routes, sprinkler, valve and controller locations in accordance with locations shown on the plan. The Contractor shall do all staking with the Owner prior to commencement of work in any area of the installation.

B. The Contractor shall furnish all supplies, equipment and personnel necessary for the staking of the work. A representative of the irrigation products supplier shall be available for assistance upon request.

C. The Contractor shall give a minimum of three (3) days notice to the Owner of the day he wishes to stake a particular section of the work.

4.03 PIPE INSTALLATION

A. The Contractor must provide effective protection at all times to prevent sand, rubbish, or any other debris from entering the piping. When work is stopped at night, or at any other time, the ends of the piping must be closed with plugs properly secured. Sidewalks, cart paths and driveways shall be clear of project debris and equipment at all times and barricades and/or tape shall be installed around any trenches left open.

B. Pipe routing shall be in accordance with the plan, however, the Owner shall have the right to change the route and/or depth of the pipe where rock or other obstacles may interfere with the intended path. In no event shall such changes affect the cost of the work except where those changes greatly alter the quantity of materials and/or labor.

C. The minimum trench width shall provide for a minimum space of 4 inches on each side of the piping. Trench widths shall be held close to these minimums to avoid excess earth loads on piping.

D. Mainline piping 2-1/2 inches and larger in size shall be trenched to a minimum depth of 20 inches and backfilled with rock-free soil completely surrounding the pipe. Any trenches which are in extremely rock-filled soil, or if ledge is present, shall require the trench to be backfilled with a minimum of 4 inches of sand surrounding the pipe. The Owner shall supply all sand required for backfilling.

E. Lateral piping 2-1/2" inches and smaller shall be trenched or pulled to a minimum depth of 15 inches if soil and grade conditions permit. The use of a vibratory plow for pipe installation shall only be allowed as long as minimum cover is maintained.

G. Pipe shall be installed strictly in accordance with recommendations of the manufacturer, including leveling of trench bottoms, bedding of pipe, and securely thrusting any fittings to change direction of gasketed piping.

4.04 ROAD, WALK AND STREAM CROSSINGS

A. Any cutting or breaking of sidewalks, cart paths, and/or roads shall be performed by the Contractor with necessary re-paving as part of the Contract cost. Permission to cut or break sidewalks, cart paths, and/or roads shall be obtained from the Owner.

B. No hydraulic driving or drilling will be permitted under asphaltic concrete paving.

C. All piping and/or wiring through stream crossings shall be installed in galvanized steel or ductile iron piping sleeve. Size of sleeve shall be such as to easily accommodate size of piping and/or wiring through it.

4.05 THRUST BLOCKING

A. Thrust block all gasketed piping including reducers and changes in direction, in accordance with the pipe manufacturer's recommendations. A thrust block shall be installed at the end of all laterals connected to the main line by means of a gasketed fitting.

B. All thrust blocks shall be constructed of concrete as per PVC pipe manufacturer's recommendations. The area of the bearing surface of the thrust block shall meet the pipe manufacturer's specifications based on the fitting and soil involved. All thrust blocks must bear against undisturbed soil.

C. Precast solid concrete blocks are acceptable for 2-inch and smaller pipe provided bearing surface meets above requirements. Approved joint restraints may be accepted in lieu of blocks.

D. All materials required for thrust blocking shall be provided by the Contractor.

4.06 EXCAVATING AND BACKFILLING

A. The Contractor shall do all excavating, vibratory plowing, backfilling and compaction required for the proper installation of the work according to standard acceptable industry practices.

B. When backfilling, all backfill material shall be free from rock, large stone or other unsuitable substances to prevent damage to piping and wiring. Backfilling of trenches containing plastic pipe shall be done when the pipe is cool to avoid excessive contraction in cold weather. All backfill material will be compacted in 6-inch layers as it is brought up to finish grade so as to ensure that no settling results.

C. Excess trench material shall be removed to a readily accessible onsite location as designated by the Owner, at the Contractor's expense.

D. Trenches shall be compacted, left flush with present grade, and raked clean of stone with a fine rake. The Contractor shall then be responsible doing all final trench seed bed preparation and seeding.

4.07 ROCK EXCAVATION

A. If rock is encountered in the alignment and depth shown on the plan, the alignment and/or depth shall be adjusted in order to avoid excavation if at all possible. If alignment and/or depth adjustment cannot be made and it becomes necessary to remove the same, the Contractor shall be paid for all the additional cost incurred in the handling of rock per submitted Rock Clause Bid.

B. Deleterious material shall be removed and hauled to an accessible onsite dumping location, determined by the Owner, at the Bid Price per cubic yard. Determination of deleterious material shall be made jointly by Owner and Contractor. The Owner shall supply replacement backfill material to a central storage area at the Owner's expense. Contractor shall put such replacement backfill material in place at his expense.

C. A separate Bid, "Deleterious Material/Rock Removal Per Cubic Yard," has been included in this proposal.

4.08 BLASTING

A. If considered necessary by the Contractor and Owner as a final alternative, blasting will be performed by licensed personnel. When the use of explosives is necessary, all requirements of local and state laws shall be complied with and all necessary permits shall be obtained by the blasting company which has been hired.

B. The Contractor shall be responsible for all expenses involved with necessary blasting. The material left after blasting will be considered deleterious and paid for by the Owner at a cubic yard rate as quoted on the Bid Form.

C. All parties having structures in proximity to the site of work, and others who may be affected, will be notified by the Owner. Such notice will be given sufficiently in advance to enable others to take proper steps as deemed necessary for protection.

4.09 DEWATERING

A. The Contractor shall provide any necessary pumps for removing water from trenches and other parts of the work to prevent trenches and/or slopes from caving in.

4.10 TRENCH SETTLEMENT

A. During the work period, it shall be the Contractor's responsibility to refill any trenches that may have settled due to incomplete compaction.

B. If within one year from completion date, major settlement due to improper compaction occurs, and an adjustment in pipe, sprinkler heads, topsoil and seed, or paving is necessary to bring the system to the proper level at the permanent grade, the Contractor, as part of the work under this Contract, shall make said adjustments without extra cost to the Owner.

4.11 ELECTRICAL WIRE INSTALLATION

A. The Contract will be responsible to have connections made to the building electrical system as is required for the proper operation of the automatic control system. Subcontracting for an electrician to perform any wiring required by local code shall be the Contractor's responsibility.

B. Wire shall be installed in the same trenches as piping wherever possible and laid on the side of the pipe. All wires shall be bundled or tied together every 8 to 10 feet with electrical tape or tie straps. Wire shall be installed with a minimum slack of 18 inches at all 90-degree bends and at all solenoid connections.

C. All control circuitry passing through the wall of a building or beneath a road or cart path shall be installed in a suitable sleeve, whereas in all other locations they may be installed in the pipe trench. Conduit through construction at building shall be a metal pipe. Other sleeving may be of PVC material.

D. The jointing of all underground wires shall be by the use of wire nuts, covered with Scotchlok or DBY/DBR waterproof connections per installation instructions provided by the manufacturer. Under no circumstances shall wire connections, outside of the satellite controller, be made without the use of a waterproof connector.

E. Any splices on the 120-volt power wiring or communication wiring shall be installed in a 10-inch round valve box. 24-volt common and control wire splices shall be installed in a 4-inch valve cap and sleeve. All splices shall be precisely marked on the as-built drawing.

4.12 VALVE BOXES

A. Valve boxes shall be installed with adequate space for operation and service of equipment in the box. A minimum of 4 inches of gravel shall be placed under each valve box for both drainage and leveling of the box. Gravel shall be furnished by the Owner.

B. All valve boxes for tee, green and fairway valves, main line valves, and splice boxes shall be mounted flush to grade. Extensions shall be used as required for proper installation and setting.

C. Valve boxes shall be installed so that position of box will allow full open and full close of shut-off valves.

4.13 CONTROLLERS

A. The central computer components shall be installed in accordance with the manufacturer's specifications.

B. Field satellite controllers shall be mounted on concrete pads as recommended in manufacturer's controller installation instructions. Controllers shall be mounted securely and flush with to the concrete pad.

C. All satellites and central controller shall be properly grounded by use of two (2) copper grounding plates and #6 bare stranded copper wire at each location. Copper plates shall be installed a minimum of 36 inches below grade. Backfill material shall be free of rocks or large stones and firmly compacted around grounding plate.

D. Satellite controller locations shown on the plan are approximate. Actual locations may be adjusted by the Owner as required.

4.14 SPRINKLER AND VALVE INSTALLATION

A. Sprinklers shall be connected to piping by installation of pre-assembled PVC swing joint assemblies. The swing joint assemblies shall be factory preassembled and installed so that the assembly and sprinkler are not directly over the piping or service tee.

B. If backfill material contains rocks or stones, or is comprised of heavy clay, all swing joint assemblies and sprinklers attached shall be completely backfilled with sand to within 3 inches of final grade. Sand will be supplied by the Owner,

C. Manual shut-off valves shall be installed in the closed position and shall not be opened until main line piping has been pressurized and properly flushed through quick-coupler valves or drains.

D. Quick-coupling valves shall be installed on galvanized swing joint assemblies.

E. All threaded connections into sprinkler heads shall be made with the use of Teflon® tape only. Shut-off valve assemblies and nipples, and galvanized swing joint assemblies shall be made with the use of Teflon® tape or liquid Teflon® sealant. Assembly shall be so that sealant or tape is kept from entering piping or fittings.

4.15 CLEANING THE PREMISES

A. Clean-up shall be performed as each portion of the work progresses. Refuse, rubbish and excess soil shall be removed for the site. Upon completion of the job, the Contractor shall clean up all debris caused by his work and leave the job site in a neat and clean condition. All sidewalks and paving shall be broomed or washed down. All debris removed from the job will be taken away from the premises.

SECTION 5

COMPLETION SPECIFICATIONS

5.01 TESTING THE SYSTEM

A. Work included under this Contract includes all tests required under laws, rules and regulations, and shall be made in accordance therewith.

B. The entire system shall be tested at the normal system working pressure and upon visual inspection of the ground. Should any leaks be found, they shall promptly repaired. The line shall then be retested until satisfactory.

C. The Owner and other required authorities shall be notified at least 48 hours in advance of all tests and all tests shall be conducted to their satisfaction.

D. All irrigation lines shall be tested at a maximum system pressure for a period of at least 24 hours before final approval by the Owner has been given.

5.02 ADJUSTING AND BALANCING THE SYSTEM

A. All areas of the irrigation system shall be inspected to ensure proper coverage and to ensure that proper sprinkler location has been accomplished. If necessary, the Contractor shall adjust or change sprinkler nozzles to correct any mis-located products.

5.03 AS-BUILT DRAWING

A. During the installation, the Contractor shall maintain an as-built drawing of the system updated on a daily basis.

B. After completion of the entire installation, the Contractor shall furnish the as-built drawing showing all sprinkler heads, valves, controllers, drains, etc., to scale with dimensions where required. All splices on 120-volt power, communication, and 24-volt common and control wiring shall be located on as-built plan with precise and detailed measurements indicated.

5.04 INSTRUCTIONS

A. After completion and testing of the system, the Contractor will instruct the Owner in proper operation and maintenance of the system.

B. The Contractor shall supply all manufacturer's Owner's Manuals and Service Manuals to the Owner.

5.05 SERVICE AND GUARANTEE

A. The Contractor shall submit a single guarantee that all portions of the work are in accordance with the Contract requirements and providing for maintenance of the system. The Contractor shall guarantee all work against faulty and improper workmanship for a period of one (1) year from date of final acceptance by the Owner, except where guarantees or warranties for longer terms are specified herein. The Contractor shall correct any deficiencies which occur during the guarantee period at no additional cost to the Owner, all to the satisfaction of the Owner. The Contract shall obtain similar guarantees from Subcontractors.

B. All sprinkler heads shall have a five (5) year, 100 percent manufacturer's warranty against defects in workmanship from date of installation. Central and field satellite controllers shall have a one (1) year, 100 percent manufacturer's warranty from date of installation. Other materials supplied by the Contractor shall be warranted per Distributor and Manufacturer's policies.

C. The Contractor shall provide service necessary to maintain the system for a period of one (1) year from date of final acceptance by the Owner. This service shall include winterizing the system with compressed air and providing all work and equipment necessary to properly start the system in the spring.

SECTION 6

CONCLUSION

6.01 CONCLUSION

A. It has been our purpose in preparing these specifications to supplement the plans and to provide a sprinkler system which is complete in every detail.

B. It has not been our purpose in preparing these plans and specifications to make omissions and/or errors. Such omissions and/or errors, in either the plans or specifications, shall be corrected when called to our attention. Discrepancies of any sort shall not be taken advantage of, as harmony shall be preserved at all times so that construction can be pursued efficiently and rapidly in the letter and spirit of these specifications. The true intent and meaning of the same is that all work of every kind that may be necessary for the complete job be done. This is implied, although the same may not be specifically expressed.

BID FORM

Bidders:

The undersigned, having familiarized themselves with the existing conditions of the project site affecting the cost of the work, and with the Contract Documents, which includes Invitation To Bid, Instructions, Specifications and Drawings, hereby proposes to furnish all materials, equipment, supervision, technical personnel, labor, machinery, tools, installation equipment, and transportation services required to construct and complete the work, all in accordance with the above listed documents, submits herewith in conformity with the specification, the following proposal. All prices include the price of all required bonds and state sales taxes.

Description

Install an 18-hole irrigation system as specified, including sales tax, required bonds, and permit, as indicated on the Plan(s).

Total Price

$_____
(Lump Sum)

Rock/deleterious material removal and replacement per cubic yard.

$_____
(per cubic yard)

Dated this _____ day of _____, 1997.

_____
Company Name of Bidder

By: _____

Title: _____

Index